AF181581

Modern Methods of Plant Analysis

New Series Volume 4

Editors
H. F. Linskens, Erlangen
J. F. Jackson, Adelaide

Volumes Already Published in this Series:

Volume 1: Cell Components
1985, ISBN 3-540-15822-7

Volume 2: Nuclear Magnetic Resonance
1986, ISBN 3-540-15910-X

Volume 3: Gas Chromatography/
Mass Spectrometry
1986, ISBN 3-540-15911-8

Volume 4: Immunology in Plant Sciences
1986, ISBN 3-540-16842-7

Forthcoming:

Volume 5: High Performance Liquid Chromatography
in Plant Sciences
1987, ISBN 3-540-17243-2

Immunology
in Plant Sciences

Edited by
H.F. Linskens and J.F. Jackson

Contributors

J. Brulfert M.-M. Cordonnier D. Ernst H. Grisebach
M.G. Hahn A. Kleinhofs T.M. Kuo D.S. Luthe
D.W. Mc Curdy P. Moesta K.R. Narayanan J.B. Ohlrogge
L.H. Pratt R.J. Robins G. Schmid Y. Shimazaki
D.A. Somers K.C. Vaughn J. Vidal R.L. Warner
E.W. Weiler K.J. Wilson

With 90 Figures

Springer-Verlag
Berlin Heidelberg New York
London Paris Tokyo

Professor Dr. HANS-FERDINAND LINSKENS
Goldberglein 7
D-8520 Erlangen

Professor Dr. JOHN F. JACKSON
Department of Biochemistry
Waite Agricultural Research Institute
University of Adelaide
Glen Osmond, S.A. 5064
Australia

ISBN-13:978-3-642-82855-3 e-ISBN-13:978-3-642-82853-9
DOI: 10.1007/978-3-642-82853-9

Library of Congress Cataloging-in-Publication Data. Immunology in plant sciences. (Modern methods in plant analysis; new ser., v. 4) 1. Plant immunochemistry. 2. Plants–Analysis. I. Linskens, H.F. (Hans F.), 1921– II. Jackson, J.F. (John F.), 1935– . III. Brulfert, J. IV. Series. QK899.I46 1986 581.19'285 86-15532
ISBN-13:978-3-642-82855-3 (U.S.)

This work is subject to copyright. All rights are reserved, whether the whole or part of the material is concerned, specifically those of translation, reprinting, re-use of illustrations, broadcasting, reproduction by photocopying machine or similar means, and storage in data banks.

Under § 54 of the German Copyright Law, where copies are made for other than private use, a fee is payable to "Verwertungsgesellschaft Wort", Munich.

© Springer-Verlag Berlin Heidelberg 1986
Softcover reprint of the hardcover 1st edition 1986

The use of registered names, trademarks, etc. in this publication does not imply, even in the absence of a specific statement, that such names are exempt from the relevant protective laws and regulations and therefore free for general use.

Product Liability: The publisher can give no guarantee for information about drug dosage and application thereof contained in this book. In every individual case the respective user must check its accuracy by consulting other pharmaceutical literature.

2131/3130-543210

Introduction

Modern Methods of Plant Analysis

When the handbook *Modern Methods of Plant Analysis* was first introduced in 1954 the considerations were:

1. the dependence of scientific progress in biology on the improvement of existing and the introduction of new methods;
2. the difficulty in finding many new analytical methods in specialized journals which are normally not accessible to experimental plant biologists;
3. the fact that in the methods sections of papers the description of methods is frequently so compact, or even sometimes so incomplete that it is difficult to reproduce experiments.

These considerations still stand today.

The series was highly successful, seven volumes appearing between 1956 and 1964. Since there is still today a demand for the old series, the publisher has decided to resume publication of *Modern Methods of Plant Analysis*. It is hoped that the New Series will be just as acceptable to those working in plant sciences and related fields as the early volumes undoubtedly were. It is difficult to single out the major reasons for success of any publication, but we believe that the methods published in the first series were up-to-date at the time and presented in a way that made description, as applied to plant material, complete in itself with little need to consult other publications.

Contributing authors have attempted to follow these guidelines in this New Series of volumes.

Editorial

The earlier series *Modern Methods of Plant Analysis* was initiated by Michel V. Tracey, at that time in Rothamsted, later in Sydney, and by the late Karl Paech (1910–1955), at that time at Tübingen. The New Series will be edited by Paech's successor H.F. Linskens (Nijmegen, The Netherlands) and John F. Jackson (Adelaide, South Australia). As were the earlier editors, we are convinced "that there is a real need for a collection of reliable up-to-date methods for plant analysis in large areas of applied biology ranging from agriculture and horticultural experiment stations to pharmaceutical and technical institutes concerned with raw material of plant origin". The recent developments in the fields of plant biotechnology and genetic engineering make it even more important for workers in the plant sciences to become acquainted with the more sophisticated methods, which sometimes come from biochemistry and biophysics, but which also have been developed in commercial firms, space science laboratories, non-university research institutes, and medical establishments.

Concept of the New Series

Many methods described in the biochemical, biophysical, and medical literature cannot be applied directly to plant material because of the special cell structure, surrounded by a tough cell wall, and the general lack of knowledge of the specific behavior of plant raw material during extraction procedures. Therefore all authors of this New Series have been chosen because of their special experience with handling plant material, resulting in the adaptation of methods to problems of plant metabolism. Nevertheless, each particular material from a plant species may require some modification of described methods and usual techniques. The methods are described critically, with hints as to their limitations. In general it will be possible to adapt the methods described to the specific needs of the users of this series, but nevertheless references have been made to the original papers and authors. While the editors have worked to plan in this New Series and made efforts to ensure that the aims and general layout of the contributions are within the general guidelines indicated above, we have tried not to interfere too much with the personal style of each author.

Volume Four – Immunology in Plant Sciences

The New Series in Modern Methods of Plant Analysis was begun in 1985 with a volume on Cell Components, and quickly followed by two further volumes. These dealt with the powerful analytical techniques of Nuclear Magnetic Resonance (Vol. 2) and Gas Chromatography/Mass Spectrometry (Vol. 3). The present volume collects together a series of chapters on the application of the antigen-antibody concept of the animal kingdom to the detection of a large range of plant substances. Those working in the plant sciences a decade or so ago rarely used antibodies experimentally. Today, however, antibodies are used frequently and are becoming indispensable for some purposes, largely due to the specificity of the method and its sensitivity.

Volume 4 begins with a chapter on immunoassay of the plant hormones indole-3-acetic acid, abscisic acid, and the gibberellins, using monoclonal and polyclonal antibodies. Cytokinins and other low molecular weight nonimmunogenic compounds are dealt with in subsequent chapters. The detection of proteins by immunocytochemistry, immunoblotting, and assay by immunoquantification, including radioimmunoassay and Western blot analysis also find a place here. An extremely useful technique dealt with in depth in this volume is that of immunofluorescence, which involves labeling antibodies with fluorescent groups, and utilizes subsequent detection by fluorescence microscopy. Quantitative immunochemical procedures for plant enzymes are also presented, as well as various immunological-cytochemical localization methods applicable to plant cells.

The editors believe that this 4th volume will find immediate use for both undergraduate lectures and laboratory classes, as well as for research laboratories. We hope that this collection of chapters by world experts on immunodetection will further encourage the spread of this technique among scientists, students, and industrial analysts working with plant materials.

Acknowledgements. The editors express their thanks to all contributors for their efforts in keeping to production schedules, and to Dr. Dieter Czeschlik, Ms. K. Gödel and Ms. E. Schuhmacher of Springer publishers for their cooperation with this and other volumes in Modern Methods of Plant Analysis. The constant help of José Broekmans is gratefully acknowledged.

Erlangen and Adelaide, August 1986

H. F. LINSKENS
J. F. JACKSON

Contents

Plant Hormone Immunoassays Based on Monoclonal and Polyclonal Antibodies
E. W. WEILER

Radioimmunoassay and Gas Chromatography/Mass Spectrometry for Cytokinin Determination
D. ERNST (With 14 Figures)

Immunodetection of Phytochrome: Immunocytochemistry, Immunoblotting, and Immunoquantitation
L. H. PRATT, D. W. MCCURDY, Y. SHIMAZAKI, and M.-M. CORDONNIER (With 12 Figures)

Radioimmunoassay for a Soybean Phytoalexin
H. GRISEBACH, P. MOESTA, and M. G. HAHN (With 5 Figures)

The Measurement of Low-Molecular-Weight, Non-Immunogenic Compounds by Immunoassay
R. J. ROBINS (With 11 Figures)

Quantitative Immunochemistry of Plant Phosphoenolpyruvate Carboxylases

J. Brulfert and J. Vidal (With 8 Figures)

Immunochemical Methods for Higher Plant Nitrate Reductase

A. Kleinhofs, K. R. Narayanan, D. A. Somers, T. M. Kuo, and R. L. Warner (With 9 Figures)

Immunological-Cytochemical Localization of Cell Products in Plant Tissue Culture

K. J. Wilson (With 10 Figures)

List of Contributors

BRULFERT, JEANNE, Institut de Physiologie Végétale, CNRS, F-91190 Gif-Sur-Yvette

CORDONNIER, MARIE-MICHÈLE, Biotechnology Research, CIBA-GEIGY Corporation, Research Triangle Park, North Carolina 27709-2257, USA

ERNST, DIETRICH, Max-Planck-Institut für Biochemie, D-8033 Martinsried

GRISEBACH, HANS, Lehrstuhl für Biochemie der Pflanzen, Institut für Biologie II der Universität Freiburg, Schänzlestraße 1, D-7800 Freiburg i. Br.

HAHN, MICHAEL G., Complex Carbohydrate Research Center, University of Georgia, P.O. Box 5677, Athens, GA 30613, USA

KLEINHOFS, ANDRIS, Program in Genetics and Cell Biology and Department of Agronomy and Soils, Washington State University, Pullman, WA 99164-6420, USA

KUO, TSUNG MIN, Northern Regional Research Center, Agricultural Research Service, U.S. Department of Agriculture, Peoria, IL 61604, USA

LUTHE, DAWN S., Department of Biochemistry, P.O. Drawer BB, Mississippi State, MS 39762, USA

MCCURDY, DAVID WILLIAM, Botany Department, University of Georgia, Athens, GA 30602, USA

MOESTA, PETER, BASF, ZH/B A30, D-6700 Ludwigshafen

NARAYANAN, KOMARATCHI R., University of Florida, Tropical Research & Education Center, Homestead, FL 33031, USA

OHLROGGE, JOHN B., Northern Regional Research Center, Agricultural Research Service, U.S. Department of Agriculture, Peoria, IL 61604, USA

PRATT, LEE H., Botany Department, University of Georgia, Athens, GA 30602, USA

ROBINS, RICHARD J., AFRC Institute of Food Research, Norwich Laboratory, Colney Lane, Norwich, NR4 7UA, United Kingdom

SCHMID, GÜNTER, Biochemische Forschung, E. Merck, Frankfurter Straße 250, D-6100 Darmstadt 1

SHIMAZAKI, YUKIO, Botany Department, University of Georgia, Athens, GA 30602, USA

SOMERS, DAVID A., University of Minnesota, Agronomy and Plant Genetics, 411 Borlaug Hall, 1991 Buford Circle, St. Paul, MN 55108, USA

VAUGHN, KEVIN CHRISTOPHER, Southern Weed Science Laboratory, U.S. Department of Agriculture, Agricultural Research Service, P.O. Box 350, Stoneville, MS 38776, USA

VIDAL, JEAN, Université de Paris Sud, Centre d'Orsay, Physiologie Vegetale Moleculaire, Batiment 430, F-91405 Orsay Cedex

WARNER, ROBERT L., Department of Agronomy & Soils, Washington State University, Pullman, WA 99164-6420, USA

WEILER, E. W., Fachbereich Biologie/Chemie, Pflanzenphysiologie, Universität Osnabrück, Postfach 4469, D-4500 Osnabrück

WILSON, KATHRYN J., Purdue University, School of Science at Indianapolis, Department of Biology, 1125 East 38th Street, Indianapolis, IN 46223, USA

Plant Hormone Immunoassays Based on Monoclonal and Polyclonal Antibodies

E. W. WEILER

1 Introduction

This chapter describes the immunoassay of the acidic plant hormones, i.e. the gibberellins (GAs), indole-3-acetic acid (IAA) and abscisic acid (ABA). The other major class of plant hormones for which immunoanalysis is appropriate, the cytokinins is discussed in Chap. 2, this Volume. However, the technical aspects involved in the immunoassay of both basic and acidic plant hormones are very similar.

While the intention of this article is to provide the protocols and necessary technical background information to establish and use immunoassays, more general discussions of the use of this technique have been presented in recent reviews (Weiler 1982a, 1984; Weiler et al. 1985, 1986). General treatments of plant hormone analysis can be found in the books edited by MacMillan (1980) and Crozier and Hillman (1984).

The compounds dealt with in this chapter and the other relevant chapters of this volume exemplify the procedures to handle a wide range of small molecular weight compounds of plant origin, as well. In this respect, no attempt is made here to review the literature on the subject in extenso. Rather, procedures were selected which are based on several years of experience in the author's laboratory. Where appropriate, possible variations or alternative procedures will be referenced.

2 Synthesis of Immunogens

Immunogenicity of low molecular weight compounds is achieved by covalent coupling to a macromolecular support, preferentially a protein (Landsteiner 1945). Hemocyanin or serum albumins are favoured by most workers, the latter being used in the majority of cases. Human (HSA) or bovine (BSA) serum albumins have molecular weights of around 70000, possess a large number of reactive groups such as amino-, carboxyl- or tyrosyl-residues and are sufficiently soluble in mixtures of aqueous and non-aqueous solvents to allow for most coupling reactions in the presence of relatively high concentrations of reactants. The number of hormone residues coupled per molecule of protein carrier can vary substantially without impairing immunogenicity. In most cases, between 5 and 15 mol of hormone per mol of BSA or HSA have been used successfully to immunize rabbits or mice.

2.1 Synthesis of ABA Conjugates

Both the carboxyl group of ABA as well as its carbonyl function have been used to link ABA to proteins. Antisera derived from carboxyl-coupled ABA react preferentially with the protonated hormone (ABAH), its esters and probably other C_1-conjugates in plant extract (e.g. Weiler 1979, 1980) but are less reactive against the dissociated hormone (ABA^-) which is predominantly present under standard immunoassay conditions (pH 7–8). The second type of immunogen (carbonyl-coupled ABA) is to be preferred if ABA^- is to be analyzed with little or no interference from ABA conjugates. A complication results from the optical activity of ABA and the 2-cis-2-trans isomerization of the side chain when exposed to light. It is now widely believed that 2-cis(+)ABA is the physiologically active form (see Addicott 1983) and immunoassay techniques for this form of ABA will be described.

(+/−)ABA should be avoided for immunogen synthesis if antisera are to be produced for use with commercial racemic ^{3}H-ABA, because antibodies against (−)ABA will also be produced. This results in unacceptably high background binding which ist not reversible by (+)ABA from plant material. ^{3}H(+/−)ABA can however be used without problems when monoclonal antibodies against (+)ABA are available. The following procedures can be used to couple ABA to serum albumins:

2.1.1 Procedure 1: Synthesis of ABA-(C_1)-BSA Conjugates (Modified from Weiler 1980)

Dissolve 58 μmol ABA (15.3 mg) in 0.4 ml dimethylformamide (DMF) and add 0.2 ml water. Add this solution to 50 mg BSA in 2 ml water and adjust the pH to 8 with 1 N NaOH. Add – in four aliquots over a period of 2 h – 208 μmol (40 mg) of 1-ethyl-3(3-dimethylaminopropyl)-carbodiimide hydrochloride (EDC) and stir under nitrogen at 4 °C overnight. Dialyze against water (3 × 10 liters per day) until no more free ABA can be detected in the conjugate (typically 4–5 days at 4 °C) and lyophilize. Coupling ratios are best determined by including an internal standard of radiolabeled ABA in the reaction or, if unavailable, by differential UV spectroscopy (reference: EDC reacted protein). This protocol gives coupling ratios typically exceeding 5 ABA per BSA.

2.1.2 Procedure 2: Synthesis of ABA-$C_{4'}$-BSA Conjugates (According to Weiler 1980)

Step 1: Preparation of ABA-$C_{4'}$-Tyrosylhydrazone (ATH). Dissolve 17 μmol (4.5 mg) ABA (and tritium-labeled ABA, if desired) in 1.5 ml methanol and add 78 μmol (15 mg) tyrosine hydrazide, followed by 0.1 ml glacial acetic acid. Incubate at 50 °C for 4 days (darkness, under nitrogen). Check completion of reaction by thin-layer chromatography (TLC) on silica gel (solvent: toluene/ethyl acetate/acetic acid = 75+22+3 (v/v/v), R_f ABA ca. 0.5, R_f ATH = 0). Purify ATH by TLC (silica gel, solvent: methanol/acetic acid = 95+5 (v/v), main product R_f 0.9 with some side product at R_f 0.95). Isolate the main product by TLC (slightly yellow oil) and use as such in step 3.

Step 2: Synthesis of p-Aminohippurate-Substituted BSA. Suspend 200 mg p-aminohippurate in 120 ml water and add 200 mg BSA. Adjust the pH to 8 with 1 N NaOH and warm slightly until a clear solution results. Add 200 mg of EDC and readjust pH to 6.4 with 2 N HCl. Stir 6 h at room temperature (RT), add another 100 mg of EDC and react for 14 h at RT. Dialyze and lyophilize as described in procedure 1 (Sect. 2.1.1.). Coupling efficiency can be monitored by including as tracer p-aminohippurate-(^{14}C) in the reaction mixture. Typical coupling ratios exceed 10 mol of p-aminohippurate per mol of BSA.

Step 3: Diazo-Coupling of ABA-Tyrosylhydrazone to p-Aminohippuric Acid Substituted BSA. Suspend 30 mg of p-aminohippuric acid substituted BSA in 5 ml water and adjust to pH 1.5 with 1 N HCl. Cool the resulting clear solution on ice and add 60 mg $NaNO_2$ in 0.5 ml water (dropwise). After 5 min, add, dropwise, 30 mg ammonium sulfamate dissolved in 0.5 ml water (attention: foaming occurs). Dissolve 8–10 mg of ATH in 0.5 ml methanol, add 10 ml of borate buffer (0.1 M, pH 9) and to this solution, add the diazotized protein dropwise. After the addition is complete, stir another 10 min and dialyze the deep orange solution against water as given in procedure 1, lyophilize. Calculate coupling ratios from the ^{3}H-ABA content of the conjugate. The protocol yields typically 5–10 ATH per BSA.

2.2 Synthesis of Indole-3-Acetic Acid Conjugates

For the selective analysis of IAA, coupling through the side chain should be avoided to minimize potential interference by IAA conjugates (these can be abundant in many plant tissues). This approach was taken by Pengelly and Meins (1977) who used the Mannich reaction to link IAA to BSA via the indole nitrogen. However, possibility due to instability of such conjugates in the immunization process, low titered sera will result and the immune response may be variable. Coupling through the carboxyl function results in more stable conjugates and a somewhat better immune response in rabbits (Weiler 1981) and mice (Mertens et al. 1985). However, IAA has to be converted to the methyl ester to render the compound antigenic and IAA^- will not or very slightly react. Methylation, however, chemically stabilizes the hormone and results in better recoveries. The procedure given below (Weiler 1981) will result in immunogenic IAA-BSA conjugates which can carry from 15 to 30 IAA per BSA:

2.2.1 Procedure 3: Synthesis of IAA-BSA Conjugates Via the Mixed Anhydride

Add, to 300 µmol (52.3 mg) recrystallized, dry IAA, dissolved in 2 ml dry DMF, 75 µl tri-n-butylamine and cool to −15 °C. Add 40 µl isobutylchlorocarbonate and incubate for 8 min in a stoppered vessel. Dissolve 420 mg BSA in 11 ml water and add slowly 11 ml DMF, followed by 0.42 ml 1 N NaOH. Add the IAA-mixed anhydride reaction mixture rapidly with vigorous stirring to the protein solution and continue stirring for 1 h on ice. Add 0.2 ml 1 N NaOH and continue stirring on ice for 5 h. Dialyze, at 4 °C, 1 day against 10% DMF in water (2 × 2 liters), then 4 days against water. Lyophilize the conjugate.

Coupling can be calculated by inclusion of a standard amount of (1-^{14}C)-IAA in the reaction and determination of the radioactivity of the conjugate.

2.3 Synthesis of Gibberellin Conjugates

Some 70 gibberellins (GAs) have been found in nature and against 9 of these (gibberellin A1, A3, A4, A5, A7, A9, A19, A20, A24) specific antisera for use in immunoassays have been produced (Weiler and Wieczorek 1981; Atzorn and Weiler 1983a, b; Kurogochi et al., in preparation; Yamaguchi et al., in preparation; Oden et al. 1985), among them those considered of prime physiological importance (e.g. GA1) as well as their immediate precursors (GA19, GA20). While for certain purposes, special routes to immunogen synthesis may be required, a route employing amide formation between the carboxyl group at C7 and amino groups of the protein is suited for many GAs and yields conjugates highly immunoreactive in rabbits (e.g. Atzorn and Weiler 1983a). The procedure 3 (Sect. 2.2.1) given above can be used with little modifications to link GAs to BSA. These modifications are (1) that slightly more DMF may be required to initially dissolve the GA under study and (2) that coupling ratios – if radiolabeled GAs are unavailable – can be estimated by spectroscopic analysis in concentrated sulphuric acid (see e.g. Atzorn and Weiler 1983a, b). The mixed anhydride procedure yields coupling ratios of typically 4 mol of GA per mol of protein (BSA). This is lower than coupling ratios observed for IAA-BSA conjugates due to the steric inaccessibility of the GA carboxyl group. Higher coupling ratios may be obtained by using a spacer between the GA molecule and the protein (e.g. β-alanine; Yamaguchi et al., in preparation).

3 Radio- and Enzyme-Labeling of Plant Hormones

3.1 Radiolabeled Immunotracers

For competition immunoassay, only high affinity interactions between antibody and antigen should be employed to maximize sensitivity and achieve sufficient selectivity. Tritium-labeled antigens frequently are the first choice for radioimmunoassay (RIA) tracers because tritium allows to achieve the required specific radioactivities while leaving the antigen structure essentially unchanged. It is convenient that a range of suitable immunotracers are in the meantime commercially available at specific activities of over 20 Ci mmol^{-1} (740 GBq mol^{-1}). These include: DL-cis, trans-(G-^{3}H)ABA, (5(n)-^{3}H)IAA, (1,2(n)-^{3}H)GA_1 and (1,2(n)-^{3}H)GA4. Since RIA analysis of GAs requires the methyl esters, an elegant way of introducing the label is via reaction of the GA sodium carboxylates with tritiated methyl iodide (Kurogochi et al., in preparation). Tritium-labeled GA methyl esters can thus be prepared in microgram quantities and with specific activities exceeding 80 Ci mmol^{-1} (3 TBq mol^{-1}).

3.2 Enzyme-Labeled Plant Hormones

Alkaline phosphatase (AP) appears to be the favoured enzyme used in most enzyme immunoassays (EIA) for plant hormones. All EIAs described here make use of this enzyme. IAA (Weiler et al. 1981; Mertens et al. 1985), ABA (Weiler 1982b; Daie and Wyse 1982), some GAs (Atzorn and Weiler 1983b) as well as cytokinins (Hansen et al. 1984; Weiler et al. 1985, see also Chap. 2) have been linked to AP with retention of immunological and enzymatic activity. Two protocols, given below, are used to exemplify coupling of plant hormones to alkaline phosphatase. The procedure given for IAA (procedure 4) can be used to couple other acidic hormones through their carboxyl function.

3.2.1 Procedure 4: Synthesis of Alkaline Phosphatase-Labeled IAA (Modified from Weiler et al. 1981)

Prepare, immediately before use, stock solutions of 100 mM IAA and 140 mM EDC in 50% DMF in water, pH 5.3. Mix equal volumes of both solutions, stir for 15 min at room temperature.

Dilute a 50 μl aliquot (0.5 mg protein) of AP (ELISA grade, Boehringer 567744) as supplied with 100 μl 50% DMF in water. Add 20 μl of the EDC activated IAA solution to the enzyme solution and adjust pH to 6.5. Incubate for 4 h at 4 °C in the dark under nitrogen, preferably with gentle stirring. Dialyze against 2 l 10% aqueous DMF (1 day, 4 °C, darkness) and then 3 days against 5 l day^{-1} Tris-buffered saline (TBS, 50 mM Tris, 1 mM $MgCl_2$, 0.01 M NaCl, pH 7.8). Store the conjugate at −18 °C in 50% glycerol, do not lyophilize. Stability in 50% glycerol in TBS at −18 °C exceeds 1 yr. The conjugate is useful for EIA techniques based on monoclonal antibodies and most antisera raised against carboxyl-coupled IAA conjugates in conjunction with methylated IAA standards and samples.

3.2.2 Procedure 5: Coupling of ABA-$C_{4'}$-Hydrazones to Alkaline Phosphatase (Eberle et al., in preparation)

This conjugate is used for the enzyme immunoassay of ABA using monoclonal antibodies (Mertens et al. 1983; Eberle et al., in preparation) but will also be useful in conjunction with polyclonal antisera:

Dilute 0.5 μmol (0.2 mg) of ABA-4′-(p-aminobenzoyl)hydrazone, dissolved in 20 μl methanol, with 300 μl water and adjust the pH to 1.5 with 1 N HCl. Add 12 mg $NaNO_2$ in 100 μl H_2O and stir 10 min on ice. Add, slowly, 6 mg ammonium sulfamate in 0.1 ml water and stir 10 min.

Add this solution, containing the diazotized ABA-hydrazone dropwise to 1 mg AP (100 μl stabilized solution as supplied commercially, diluted with 1 ml 0.1 M borate buffer, pH 9.6) an stir for 30 min. Purify by dialysis (3 days against 5 l day^{-1} TBS, pH 7.8) and store at −18 °C in 50% glycerol. Stability: over 6 mos.

4 Immunological Procedures

4.1 Immunization of Rabbits

The production of antisera in rabbits is inexpensive and yields sufficient quantities of high-quality antibodies in a relatively short time. The immunization protocol given in Table 1 might serve as an example of how to produce antisera in rabbits. A detectable immune response usually is obtained after the first or second boost already. Per bleed, ca. 40–50 ml of blood may be obtained from an incision of one of the marginal ear veins. From the collected blood, serum is produced by

Table 1. Immunization schedule for rabbits

Phase of immunization	Days after first priming	Action taken[a]
Preimmunization	0	Prime with 0.2 mg conjugate/0.25 ml PBS, emulsified in 0.25 ml CFA, i. d.
	7, 14, 21	Repeat priming
Boost 1	42	Boost with 0.2–0.4 mg conjugate/0.25 ml PBS, emulsified in 0.5 ml CFA, i.m.
Blood collection 1	49 and 56	Collect ca. 35–50 ml blood from an incision of the marginal ear vein
Boost 2	70	See above
Blood collection 2	77 and 84	See above
Boost 3 and so on		

[a] Abbreviations: *PBS*: 0.01 M phosphate, 0.15 M NaCl, pH 7.4; *CFA*: complete Freund's adjuvant; i. d.: intradermal injection; i. m.: intramuscular injection.

Table 2. Immunization schedule for mice

Days after priming	Action taken[a]
0	Prime with 0.2 mg conjugate/0.2 ml PBS, emulsified in 0.2 ml CFA, administer i. p.
7	Boost with 0.2 mg conjugate/0.2 ml PBS, emulsified in 0.2 ml iCFA, administer i. p.
14	Repeat boost
21	Collect blood from tail or retroorbital system, check immune response by RIA
42	Repeat boost
49	Collect blood and check titer
72/79	Repeat boost and blood collection as above; repeat this in monthly intervals until desired immune response is obtained
Days before fusion	
4	Administer 0.2 mg conjugate in 0.4 ml PBS i. p.
0	Fuse

[a] Abbreviations: *iCFA:* incomplete Freund's adjuvant; i. p.: intraperitoneal injection; further abbreviations as in Table 1.

standard procedures and this is kept frozen at −20 °C or lower temperature. Serum fractions from individual bleeds and from different animals must be kept separately until screened for titer and gross selectivity. Only then should appropriate serum fractions be pooled. A single animal usually yields sufficient antiserum from successive bleedings to last for 10^5 to 10^7 immunoassay samples.

4.2 Immunization of Mice

The purpose of immunizing mice is to condition the animals prior to spleen removal for hybridoma production. In the literature, a variety of immunization schedules can be found. In our hands, animals with a matured immune response have given consistently superior yields of positive hybridomas as compared to animals receiving a short-term immunization. As expected, under conditions of long-range immunization, mostly IgG-secreting hybridomas are generated although there may be exceptions (Mertens et al. 1983). An outline of a suitable immunization protocol is given in Table 2.

5 Antiserum Processing

Antisera from rabbits require little or no further treatment before being useful for RIA. There are, however, inevitable variations among serum fractions from different bleeds and also different animals. Serum pools should thus be prepared only after an initial characterization of the individual serum fractions (titers, specificity) has been carried out. It is advisable, on the other hand, to prepare large pools of sera if possible before undergoing the extensive task of detailed characterization of an assay. For EIA techniques, especially in the immobilized-antibody format, immunoglobulin fractions of sera have to be used. Since this is done by standard techniques, this will not be discussed here. Immunoglobulin-enriched serum fractions are best prepared by the rivanol-ammonium sulphate technique detailed by Hurn and Chantler (1980). The procedure yields approximately 80–85% immunoglobulin with major contaminants being serum albumin and transferrin. Further purification, if necessary, should include DEAE-cellulose chromatography (see Hurn and Chantler 1980), but this is frequently not required for EIA tests. Antisera may contain, in addition to those antibodies binding the plant hormone, antibodies recognizing preferentially the hormone-protein link as present in the immunogen. This will inevitably lead to problems, if EIAs are based on such sera and the same, or a similar, link is present in the enzyme-labeled hormone. Removal of the source of interference will be required and this may be achieved by adsorption of "anti-link" antibodies on "immobilized-link" affinity adsorbents or by deliberately changing the link structure of the hormone-enzyme conjugate. General procedures for this cannot be given.

6 Production of Monoclonal Antibodies

For a number of obvious reasons, monoclonal antibodies (MAB) are superior to antisera. Immunoassays based on sera containing mixtures of high- an low-affinity antibodies each exhibiting a different degree of selectivity might show a substantial variation of their properties with changing antigen and contaminant levels which is not easily detected under routine conditions. This ist of course avoided when using MAB. In addition, MAB can be produced in any desired quantity and this is a prerequisite for a widespread distribution of MAB-based immunoassays throughout the scientific community.

6.1 Generation of Hybridomas

A detailed description of the techniques used to generate antibody-secreting hybridomas is beyond the scope of this chapter. Excellent sources of such information are the books by Goding (1984) and Campbell (1984). The importance of applying an appropriate screening procedure to select the desired cell clones cannot be overemphasized. It is strongly recommended to screen by RIA even though other techniques might at first glance seem easier and quicker. RIA allows to challenge each test sample with a low (typically 10 nM) concentration of labeled, yet structurally unchanged hormone. Positive binding immediately indicates that (1) a high-affinity antibody is likely to be present and (2) that it reacts with the free hormone. No other screening procedure yields this important information in a single step. In addition, low affinity antibodies will not be detected. One is left with a very few, but most likely very interesting clones which are being picked up initially and can concentrate one's efforts from the very beginning. Screening assays which initially would pick too many (false) positives, lead to a dramatic but largely superfluous increase in subsequent workload and cost.

6.2 Monoclonal Antibody Production and Characterization

Propagation of hybridomas and the production of MAB can be carried out in vivo or in vitro. Hybridoma cell cultures may yield from 1–100 $\mu g\,l^{-1}$ of specific antibody which is largely secreted during the late stages of the growth cycle. Introducing the hybridoma into the intraperitoneal cavity of mice usually establishes a solid or liquid tumor and the ascites fluid of such mice may yield from 5–15 $mg\,l^{-1}$ of the desired antibody. For a detailed account on MAB production, the reader is again referred to the books of Goding (1984) and Campbell (1984).

For use in RIA, dilute ascites fluid or supernatant medium from hybridoma cultures is frequently perfectly well suited and no clean-up is required. If solid-phase enzyme immunoassays (ELISAs) are to be carried out, dilute ascites fluid can still be used while cell culture media have to be processed as described in Sec. 5. If the protocol given in Sec. 8 is followed, no further purification of the ammonium sulphate precipitated MAB is necessary. Usually, MAB are resistant to storage at sub-zero temperatures in ascites fluid, cell culture media or in lyophilized form. There are reports of unexpected instability of MAB (see Goding 1984).

7 Radioimmunoassay Procedures

It is assumed that the following reagents are available: an appropriate antiserum or MAB checked for titer and cross-reactions and a suitable radiolabeled hormone with a specific activity exceeding 10 Ci $mmol^{-1}$. Given these prerequisites, the RIA protocol can be set up. From extensive experience in the author's lab, the protocol given below was derived. It has proven useful for all hormones so far checked and is also likely to work in the hands of others with little or no need for modification. Due to specific properties of antibodies, variations may be required in assay pH or buffer ionic strength. Other than that, incubation times may be varied (increasing the times is usually unproblematic while decreasing them may be dangerous) and incubation volumes can be adjusted to fit individual needs.

The recommended protocol for the RIA of acidic plant hormones is as follows:

7.1 Procedure 6: Radioimmunoassay of Plant Hormones

Reagents

1. Prepare 0.15 M NaCl in 0.01 M phosphate, pH 7.4 (PBS, phosphate-buffered saline, = assay buffer).
2. Dilute whole bovine serum 6–8-fold with water (final protein concentration approx. 0.7 mg ml^{-1} as assayed with the Biorad kit).
3. Add 1 part water to 10 parts saturated ammonium sulphate solution, pH 7.
4. Prepare a half-saturated ammonium sulphate solution by mixing equal volumes of saturated solution and water.
5. Prepare sufficient RIA incubation mix: 5 parts PBS, 1 part dilute bovine serum (reagent 2) and 1 part aqueous radiotracer dilution to give desired activity (approx. 20 nCi per tube).
6. Dilute antiserum from stock or intermediate dilution appropriately with PBS, keep on ice, use the same day. Store serum stock or tenfold-diluted (PBS) stock frozen in small aliquots. Repeated freezing and thawing might result in loss of activity.
7. Dilute hormone standards from methanolic stock solution (0.1 mg ml^{-1}) with assay buffer.

Assay Protocol

1. Dispense 0.1 ml dilute hormone standard or sample into test tubes (we use 12 × 63 mm glass tubes which fit into standard scintillation vials and into most counters). Run all standards and samples in triplicate. Use buffer instead of

hormone standard for B_o (B_{max}) values. Add 0.7 ml or RIA incubation mix to each tube and vortex gently.

2. Start reaction by adding 0.1 ml dilute antiserum (dilution chosen to bind 30–35% of the radiotracer present under equilibrium conditions; if high sensitivity is not required and precision is a problem, use enough antiserum to bind 50% of the tracer). For determination of unspecific binding (UB values), add 0.1 ml of PBS instead of antiserum dilution. Mix the contents of the tubes by vortexing gently.
3. Incubate for 90 min in the cold (4 °–10 °C), preferably in darkness when assaying for IAA or ABA.
4. Add 1 ml ammonium sulphate solution (reagent 3), incubate in the cold for 30 min. Centrifuge at ca. 2000 × g for 10 min. Decant and discard supernatants.
5. Wash the pellets once with 1 ml half-saturated ammonium sulphate solution and re-centrifuge.
6. Decant thoroughly, leaving for at least 5 min on soft, absorbant tissue to remove as much supernatant as possible.
7. Dissolve pellets in 0.3 ml water.
8. Add 1.5 ml Quickszint 212 and mix well until clear one-phase system results. Note: other scintillation cocktails will require different cocktail to water ratios. Determine radioactivity in a scintillation counter using counts preset if time is not a problem.
9. Calculate dose-response curves (logit/log plot) for standards and logit B/B_0 for the samples using the following equations:

   Relative binding$(B/B_0) = 100\ (cpm_{sample} - cpm_{UB})/(cpm\ B_0 - cpm_{UB})$
   logit $(B/B_0) = \ln\ (B/B_0)/(100 - B/B_0)$

   with: B_0 = maximum binding in the absence of standard or sample
   UB = unspecific binding in the absence of antibody

Logit (B/B_0) vs log antigen plots should be linear in the desired measuring range.

8 Enzyme Immunoassays Based on Monoclonal Antibodies

When antisera are used for EIAs it is usually required to re-optimize available assay protocols extensively. This requires experience and is generally not recommended if no previous experience with the serum in RIA tests is available. Usually, exact timing of all assay steps is of critical importance besides many other factors which cannot be discussed here in detail. It is recommended to consult the original literature for examples. In contrast, the advent of MAB for plant hormones has greatly simplified EIA schemes and a general protocol for MAB-based ELISA has been worked out which should be applicable to most MAB. Most mouse MAB appear to be only weakly adsorbing to polystyrene or similar surfaces. This is the reason why the protocol given here uses as solid-phase immunoreagent first, a polyclonal rabbit anti-mouse Ig antibody to which the specific

MAB is attached. Since this attachment is via a specific antibody-antigen interaction, more efficient use can be made of the MAB reagent. At the same time, the rabbit antibody – being applied in excess – serves as a "buffer" antibody which largely compensates for within-plate and between-plate variations in the adhesive properties of the solid support. We have also observed a consistent drop in unspecific binding to the solid surface when using this "double-immunolayer" assay as compared to "directly-coated" supports. The double-immunolayer ELISA furthermore can be run on almost any polystyrene plate on the market with little or no need to readjust the conditions. Replicate within-plate reproducibility is typically less than 3–5% (coefficients of variation).

The recommended protocol for MAB-based ELISA of plant hormones is as follows:

8.1 Procedure 7: Enzyme Immunoassay of Plant Hormones Based on Mouse Monoclonal Antibodies

Reagents

1. Prepare 50 mM $NaHCO_3$ buffer, pH 9.6.
2. Prepare 150 mM NaCl and 1 mM $MgCl_2$ in 50 mM Tris(hydroxymethyl) – aminomethane – HCl, pH 7.8 (TBS, Tris-buffered saline, = assay buffer).
3. Add 1 g gelatine to 1 liter TBS and autoclave for 30 min at 120 °C (TBS-gelatine).
4. Dilute appropriate amount of lyophilized rabbit anti-mouse immunoglobulin (RAMIG) in reagent 1. (We prepare our RAMIG from rabbits immunized with mouse immunoglobulin. The immunoglobulins are precipitated from the rabbit sera by the rivanol/ammonium sulphate technique (see Hurn and Chantler 1980), dialyzed and lyophilized. This preparation is then dissolved in reagent 1, typically at about 0.25 mg ml^{-1}).
5. Dissolve appropriate amount of MAB fraction (lyophilized ammonium sulphate fraction of cell culture supernatants) in reagent 1, or dilute ascites fluid in reagent 3.
6. Dilute alkaline phosphatase-labeled tracer hormone from stock (kept in TBS/50% glycerol at −18 °C) with reagent 3.
7. Immediately before use, dissolve 1 mg of p-nitrophenylphosphate (e.g. Sigma 104) per ml of reagent 1.

Assay Protocol

1. Coat polystyrene microtitration plates (96-well or similar) overnight at 4 °–10 °C with 0.2 ml RAMIG (reagent 4).
2. Decant thoroughly, add 0.2 ml dilute MAB (reagent 5) and incubate for 24 h at 4 °–10 °C.
3. Decant and rinse twice with tap water.
4. Add, to each cup, 0.05 ml TBS, 0.1 ml hormone standard or sample (in TBS). Add TBS for B_0-values and an excess of hormone (we use 100 pmol) for UB-values. Incubate plates for 60 min at 4 °–10 °C (preferably in the dark for IAA or ABA tests).

5. Add 0.05 ml dilute enzyme tracer (reagent 6), mix and incubate für 3 h at 4 °–10 °C.
6. Decant and rinse with plates twice with tap water.
7. Add 0.2 ml of reagent 7 and incubate for 60 min at 37 °C.
8. Stop reaction with 0.05 ml of 5 N KOH.
9. Read absorbances at 405 nm. Calculate as exemplified in Procedure 6 (Sec. 7.1), step 9. (Note: MAB adsorbed and alkaline phosphatase tracer dilution should be balanced so that the incubation with substrate results in an o.d. of ca. 1 when measured at 405 nm after an incubation with substrate for 1 h at 37 °C).

9 Plant Immunoanalysis

9.1 General Remarks

The analysis of plant hormones by immunoassays consists of the following steps:

1. Extraction of the hormone from the tissue.
2. Suitable clean-up procedure to remove all interference.
3. When appropriate, separation of hormones from related, immunoreactive compounds.
4. Formation of the methyl esters of the GA- and IAA-containing fractions.
5. Immunoassay (see Sect. 8).

Step 1 is common to any technique of plant hormone analysis and will not be considered here. This topic is addressed in recent reviews (e.g. Yokota et al. 1980). Steps 2 to 4 will have to receive some attention because there are specific requirements when applying immunoassays. Furthermore, it is relevant to discuss in this section the various means to validate immunoassays. Checks of assay performance must take notice of the potential sources of interference in immunoassay which are different from sources of interference in bioassays and physicochemical assays.

9.2 Validation of Immunoassays

The most relevant potential sources of interference in immunoassays are the following:

1. The presence of strongly cross-reacting compounds which are structurally similar to the antigen.
2. The presence of excessive amounts of compounds which exhibit only weak cross-reactions.

3. The presence of antibody denaturing or desorbing agents. For example, high levels of phenolic compounds may partially denature antibodies, the presence of surfactants (e.g. saponins) may likewise denature soluble antibodies or may desorb them from solid supports.
4. The presence of factors which prevent the binding of hormone to its binding site (e.g. by complexation).
5. The presence of contaminants which impair the quantitation step.

For example, lipophilic pigments tend to adhere to immunoglobulins when precipitated with ammonium sulphate and introduce quench when liquid scintillation counting is being used to determine radioactivity. Inhibitors of marker enzymes used for ELISA might impair the determination of enzymatic activity and lead to underestimations when present in an extract fraction.

As a general rule, no single test for assay validity is absolutely safe. It is thus strongly recommended to use the maximum number of such controls when dealing with a source of plant material for which this information is not already available.

The following tests are established for quality control in immunoassay:

1. Dilution analysis with internal standardization. In principle, aliquots of the fraction to be analyzed are dosed with increasing amounts of standard hormone. This is repeated with three to four different aliquot sizes (control: standard only). Absence of interference is indicated by the fact that the data points (plot hormone found vs hormone added) fall on straight lines parallel to each other and to the standard line. From this, information about analytical recovery will also be obtained and values should essentially be close to 100% recovery of the added hormone. In addition, proportional variation of hormone level and aliquot size must hold. While this check is easily performed and yields valuable information about gross interference of the types listed above under Nos. 2–5, highly cross-reactive material may be overlooked this way. More precisely, any cross-reactant which gives tracer displacement curves parallel to the standard curve will escape detection.

2. Immunohistograms: for this check, the extract fraction to be analyzed is split. An aliquot is assayed as such (i.e. by dilution analysis, see No. 1, above) while the second aliquot is subjected to further purification by thin-layer chromatography (TLC) or, preferably, by high-pressure liquid chromotography (HPLC). Analysis of the individual fractions yields the immunohistogram. If a single band of immunoreactivity is obtained and the dilution analysis reveals parallelism, accuracy of results is highly probable. Any interference which might still escape detection must consequently exhibit identical immunological and chromatographic properties.

3. Successive approximation (Reeve and Crozier 1980). This is an extension of the approach discussed in No. 2, above, which makes use of a series of different purification steps. It is assumed that if further clean-up of a sample yields consistent values, the sample is clean and the analysis accurate.

4. Comparison with other quantitative techniques. It is quite useful to compare results with those obtained by other techniques. It must be known, however, whether these techniques are accurate by themselves. It is usually assumed that the most conclusive proof of accuracy is by comparison to GC-MS determinations. While a full-scan mass spectrum is apt to prove the identity of the compound under analysis and reveal the absence of other material, care must be taken to prove that the analysis is at the same time quantitative and accurate. In order to get the required sensitivity, single-ion monitoring is frequently being carried out on composite plant fractions. This may carry the danger of interference by other compounds present in the sample.

Available evidence so far suggests that usually, immunoassays yield accurate results when applied to hormone fractions of considerably lower purity than required for other techniques. In some cases, accuracy has been verified in unpurified samples (Wang et al. 1984); while in other cases, HPLC-separated fractions had to be used (Pengelly et al. 1981; Oden et al. 1985; Kurogochi et al., in preparation). Generalizations are not possible, even more so that individual sera or MAB may vary not only in affinity and selectivity, but also in their sensitivity to certain sources of interference such as solutes, detergents, etc.

9.3 Processing of Plant Material

Extensive literature, reviewed recently by Yokota et al. (1980), deals with the extraction and purification of hormones from plant extracts.

Given the diversity of tissues used in plant hormone research, it is impossible to recommend a single work-up protocol which can be applied to every tissue. It is recommended to develop a suitable work-up scheme in the process of validation of assay performance. When no assay interference is detected, pre-analysis processing can be quite simple. An example is the analysis of ABA by MAB-based ELISA (Mertens et al. 1983) in leaves of most species, e.g. *Vicia faba* L. The analysis consists of the following steps:

9.3.1 Procedure 8

1. An 80% methanolic extract containing a suitable amount of ^{3}H-ABA as internal standard is prepared using routine procedures (ca. 30 ml of extract will result per 1 g of fresh weight of tissue).
2. Buffer (KPi, 50 mM, pH 8) is added to reach a final methanol concentration of 70%.
3. The extract is then passed through a SepPak C_{18} (Waters) or similar cartridge to remove lipids and most pigments. The methanol is evaporated.
4. The aqueous phase is either assayed directly or acidified to pH 2.5 and extracted with ethyl acetate. After removal of the organic solvent, the sample is dissolved in RIA buffer and aliquots analyzed by RIA and for recovery of the internal standard. It is especially convenient and strongly recommended to use radiotracers as internal standards in conjunction with ELISA. If RIA is to be used, provisions must be taken to correct for the sample radioactivity. This can best be done by individual tracer dosing (step 1, procedure 6, Sect. 7.1 provided the internal standard and immunotracer are of the same specific radioactivity and structurally identical).

A variation of this protocol employs passage of the extract through a polyvinylpyrrolidone column prior to step 2 to remove phenols. Fractions can then be analyzed directly after step 3. As a further variation, neutral or basic compounds can be eliminated by extracting, prior to step 4, with ethyl acetate from pH 10 to 11, followed by hormone extraction from pH 2.5. The danger of this procedure is that considerable ester hydrolysis might occur which would liberate hormone from base-labile conjugates.

If an analysis of IAA, GAs, and ABA at the same time is attempted, the following general scheme might serve as a guideline in the development of suitable pre-analysis protocols (note: use internal radioactive standards whenever possible):

9.3.2 Procedure 9

1. The methanolic extract from an appropriate amount of tissues is treated as described in procedure 8 (Sect. 9.3.1), steps 2 and 3. The aqueous phase is then acidified to pH 2.5 and passed through another C_{18}-cartridge to retain the hormones (column efficiency should be monitored by checking the radioactivity in the effluent).
2. The hormones are then eluted from the column with neutral or slightly basic methanol.
3. Ca. 20% of the hormone-containing fraction is used for ABA determination directly.
4. Another aliquot (ca. 30%) is treated with ethereal diazomethane (*Warning:* diazomethane and its precursor N-nitrosomethyl urea are highly toxic and dangerous; consult the relevant literature to get familiar with safety precautions required when handling these materials) for 1–5 min at 0 °C. Excess diazomethane is then destroyed by the addition of a drop of 0.2 M acetic acid in methanol. The sample is taken to dryness, redissolved in a small volume of methanol, cleared by filtration through a 0.45 μm filter and subjected to HPLC in the solvent systems indicated by Mertens et al. (1985) or any other suitable system. IAA-containing fractions are pooled and quantitated by ELISA.
5. The remainder of the extract is prepared for HPLC and run e.g. on nucleosil 5N-$(CH_3)_2$ using the solvent system of Yamaguchi et al. (1982) or a similar system. The relevant fractions can then be analyzed by the appropriate RIAs to yield quantitative information about a series of GAs present.

Acknowledgements. The author's work reported here was supported partly by grants of the Deutsche Forschungsgemeinschaft, Bonn, FRG, the Stiftung Volkswagenwerk, Wolfsburg, FRG, the Fonds der Chemischen Industrie, Frankfurt, FRG and the Stifterverband für die Deutsche Wissenschaft, Essen, FRG.

References

Addicott FT (1983) Abscisic acid. Praeger, New York, p 607
Atzorn R, Weiler EW (1983a) The immunoassay of gibberellins I. Radioimmunoassays for the gibberellins A_1, A_3, A_4, A_7, A_9, and A_{20}. Planta 159:1–6
Atzorn R, Weiler EW (1983b) The immunoassay of gibberellins II. Quantitation of GA_3, GA_4, and GA_7 by ultrasensitive solid-phase enzyme immunoassays. Planta 159:7–11
Campbell AM (1984) Monoclonal antibody technology. Elsevier, Amsterdam, pp 265
Crozier A, Hillman JR (1984) The biosynthesis and metabolism of plant hormones. Cambridge University Press, Cambridge, pp 288
Daie J, Wyse R (1982) Adaptation of the enzyme-linked immunosorbent assay (ELISA) to the quantitative analysis of abscisic acid. Anal Biochem 119:365–371
Goding JW (1984) Monoclonal Antibodies: principles and practice. Academic Press, London, p 276
Hansen CE, Wenzler H, Meins FJr (1984) Concentration gradients of trans-zeatinriboside and trans-zeatin in the maize stem. Plant Physiol 75:959–963
Hurn BAL, Chantler SM (1980) Production of reagent antibodies. In: van Vunakis H, Langone JJ (eds) Methods enzymol, vol 70. Academic Press, New York, pp 104–141
Landsteiner K (1945) The specificity of serological reactions. Dover, New York
MacMillan J (1980) Hormonal regulation of development I. Springer, Berlin Heidelberg New York, Encycl Plant Physiol 9:681
Mertens R, Deus-Neumann B, Weiler EW (1983) Monoclonal antibodies for the detection and quantitation of the endogenous plant growth regulator, abscisic acid. FEBS Lett 160:269–272
Mertens R, Eberle J, Arnscheidt A, Ledebur A, Weiler EW (1985) Monoclonal antibodies to plant growth regulators. II. Indole-3-acetic acid. Planta 166:389–393
Oden PC, Weiler EW, Schwenen L, Graebe JE (1985) Comparison of gas chromatography-mass spectrometry, radioimmunoassay and bioassay for the quantification of gibberellin A 9 in Norway spruce (*Picea abies*). Physiol Plant 64:21A
Pengelly WL, Meins F Jr (1977) A specific radioimmunoassay for nanogram quantities of the auxin, indole-3-acetic acid. Planta 136:173–180
Pengelly WL, Bandurski RS, Schulze A (1981) Validation of a radioimmunoassay for indole-3-acetic acid using gas-chromatography-selected ion monitoring-mass spectrometry. Plant Physiol 68:96–98
Reeve DR, Crozier A (1980) Quantitative analysis of plant hormones. In: MacMillan J (ed) Hormonal regulation of development I. Springer, Berlin Heidelberg New York, Encycl Plant Physiol 9:203–280
Wang TL, Futers TS, McGreary F, Cove DJ (1984) Moss mutants and the analysis of cytokinin metabolism. In: Crozier A, Hillman JR (eds) The biosynthesis and metabolism of plant hormones. Cambridge University Press, Cambridge, pp 135–164
Weiler EW (1979) Radioimmunoassay for the determination of free and conjugated abscisic acid. Planta 144:255–263
Weiler EW (1980) Radioimmunoassay for the differential and direct analysis of free and conjugated abscisic acid in plant extracts. Planta 148:262–272
Weiler EW (1981) Radioimmunoassay for picomole quantities of indole-3-acetic acid for use with highly stable ^{125}I- and ^{3}H-IAA derivatives as radiotracers. Planta 153:319–325
Weiler EW (1982a) Plant hormone immunoassay. Physiol Plant 54:230–234
Weiler EW (1982b) An enzyme-immunoassay for cis−(+)−abscisic acid. Physiol Plant 54:510–514
Weiler EW (1984) Immunoassay of plant growth regulators. Annu Rev Plant Physiol 35:85–95
Weiler EW, Wieczorek U (1981) Determination of femtomole quantities of gibberellic acid by radioimmunoassay. Planta 152:159–167
Weiler EW, Jourdan PS, Conrad W (1981) Levels of indole-3-acetic acid in intact and decapitated coleoptiles as determined by a specific and highly sensitive solid-phase enzyme immunoassay. Planta 153:561–571

Weiler EW, Eberle J, Mertens R, Atzorn R, Feyerabend M, Jourdan PS, Arnscheidt A, Wieczorek U (1986) Monoclonal- and polyclonal antibody-based immunoassay of plant hormones. In: Wang TL (ed) Immunology in plant science. Cambridge University Press, Cambridge (in press)
Weiler EW, Eberle J, Mertens R (1986) Immunoassays for the quantitation of plant growth regulators using monoclonal antibodies. Proc IPGSA Conf, Heidelberg (in press)
Yamaguchi I, Fujisawa S, Takahashi N (1982) Quantitative and semiquantitative analysis of gibberellins. Phytochemistry 21:2049–2056
Yokota T, Murofushi N, Takahashi N (1980) Extraction, purification and identification. In: Mac Millan J (ed) Hormonal regulation of development I. Springer, Berlin Heidelberg New York, Encycl Plant Physiol, 9:113–201

Radioimmunoassay and Gas Chromatography/ Mass Spectrometry for Cytokinin Determination

D. ERNST

1 Introduction

During the past few years there has been a rapid increase in the application of modern physicochemical and immunochemical methods to the determination and analysis of plant hormones, especially cytokinins. The classical method for cytokinin determination, the bioassay, can no longer be used for unequivocal cytokinin analysis, due to the limited sensitivity and selectivity (Brenner 1981; Weiler 1982; Ernst et al. 1983a). As the accuracy of the bioassay is always open to question, no exact quantification of cytokinins is possible using this method (Reeve and Crozier 1980). Only in searching for detection of new plant hormones is the bioassay invaluable, but their detailed analysis can be better determined by other methods, e.g. gas chromatography/mass spectrometry (GC/MS) or radioimmunoassay (RIA). These modern methods have, for example, found application in the isolation and identification of 6-(o-hydroxybenzylamino)-9-β-D-ribofuranosylpurine in *Populus robusta* (Horgan et al. 1975), the determination of raphanitin in radish seeds (Summons et al. 1977) and the isolation and identification of 6-(o-hydroxybenzylamino)-2-methylthio-9-β-D-glucofuranosylpurine in *Zantedeschia aethiopica* (Chaves das Neves and Pais 1980). The identification and quantitation of 6-benzylaminopurine riboside in a cytokinin autotrophic cell culture of anise was also performed by gas chromatography/mass spectrometry using single ion monitoring (SIM) (Ernst et al. 1983b).

The advantages of physicochemical and immunochemical methods over bioassays are reviewed by Reeve and Crozier (1980) and were demonstrated for cytokinins by Ernst et al. (1983a). Advantages are: (1) greater sensitivity; (2) greater specificity; (3) better reproducibility of results; (4) shorter analysis times. Several review articles dealing with modern methods of cytokinin analysis have appeared in the literature (Horgan 1978; Reeve and Crozier 1980; Yokota et al. 1980; Brenner 1981; Weiler 1982). In the following article the techniques and practicability of radioimmunoassay and gas chromatography/mass spectrometry for the determination and quantitation of endogenous cytokinins will be discussed, including a description of experiments as a guide for investigators who wish to employ one of these methods.

2 Theoretical Considerations for Analytical Methods

For the determination of cytokinins using radioimmunoassay and/or gas chromatography/mass spectrometry, the theoretical considerations described by

Reeve and Crozier (1980) are worth discussion. The accuracy of an analysis is defined as the concurrence between the estimated and true value, whereas precision results from the reproducibility of the analysis, expressed in terms of variance. Therefore, an adequate number of repetitions of the analysis is required for precision and a control analysis with standards of the compound for accuracy. These two terms have to be kept in mind for a qualitative and quantitative analysis of cytokinins.

The qualitative analysis demonstrates the presence of a cytokinin in a sample. The radioimmunoassay for isopentenyladenosine using a crude plant extract would only demonstrate the presence of an isopentenyladenosine or a 6-benzylaminopurine riboside like compound, as cross-reactions can occur (Milstone et al. 1978; MacDonald et al. 1981; Weiler and Spanier 1981; Ernst et al. 1983a). The identification of the specific cytokinin would only be possible using selective separation procedures, e.g. thin-layer chromatography (Weiler and Ziegler 1981) or high-performance liquid chromatography (HPLC) (MacDonald et al. 1981; Zaerr et al. 1981; Badenoch-Jones et al. 1984; Ernst et al. 1984). However, only mass spectrometry (MS) and gas chromatography/mass spectrometry would permit an exact chemical identification of the corresponding cytokinin, due to the fragmentation pattern of the molecule (Horgan et al. 1973, 1975; Summons et al. 1977, 1983; Dauphin et al. 1979; Regier and Morris 1982; Ernst et al. 1983a, b; Zaerr et al. 1983). The accuracy of quantitative analysis requires internal standards, as losses will occur during the extraction procedure (Hashizume et al. 1979; McCloskey et al. 1979; Summons et al. 1979a, b; Reeve and Crozier 1980; Scott and Horgan 1980, 1982; Brenner 1981; Ernst et al. 1983a; Badenoch-Jones et al. 1984). In GC/MS analysis this is realized by the application of labelled internal standards and in RIA by recovery measurements with radioactive-labelled substances.

3 Radioimmunoassay

3.1 General Principles

The radioimmunoassay is based on the equilibrium of unlabelled antigen and a constant amount of labelled antigen (tracer) with a saturable amount of antibody (Fig. 1). After equilibration, free-labelled antigen is separated from antibody-bound antigen and the radioactivity present in the bound or free form is measured. If an increasing amount of unlabelled antigen is added to known quantities of antibody and tracer, competition of the unlabelled and labelled antigen for the antibody occurs. Increasing amounts of cold antigen results in a decreasing ratio of antibody-bound to unbound-labelled antigen (Table 1). The antibody-dilution

$$\begin{array}{l} \mathrm{Ag} + \mathrm{Ab} \rightleftharpoons \mathrm{AbAg} \\ \qquad\quad + \\ \qquad\quad \mathrm{Ag}^* \rightleftharpoons \mathrm{AbAg}^* \end{array}$$

Fig. 1. Principles of the RIA. *Ab* = antibody; *Ag* = unlabelled antigen; *Ag** = labelled antigen

Table 1. A labelled antigen (Ag*) is bound by an antibody (Ab); increasing amounts of unlabelled antigen (Ag) results in a decreasing ratio of antibody-bound to nonbound-labelled antigen

			Bound/unbound
Ab Ab Ab		AbAg* AbAg* AbAg*	
+		+	
Ag* Ag* Ag*	→		100%
Ag* Ag* Ag*		Ag* Ag* Ag*	
Ab Ab Ab		AbAg* AbAg* AbAg	
+			
Ag* Ag* Ag*	→	+	
Ag* Ag* Ag*		Ag* Ag* Ag* Ag*	
+		+	50%
Ag Ag Ag		Ag Ag	

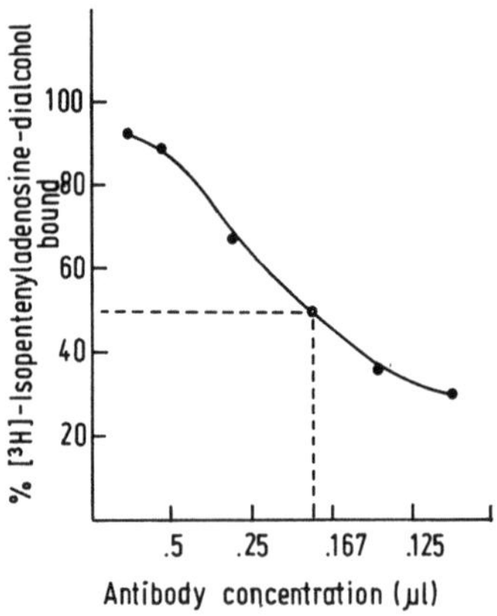

Fig. 2. Antibody dilution curve for isopentenyl-adenosine

curve involves the incubation of a fixed amount of tracer with different concentrations of the antiserum. Usually the amount of antibody yielding approximately 50% of tracer bound is chosen for the standard curve (Fig. 2). This value has shown to be useful, as with higher antibody concentrations the amount of unlabelled antigen required to produce a significant shift in the bound and free fractions will be much greater and the assay will be less sensitive. After construction of an antibody-dilution curve it is necessary to construct a standard curve where fixed amounts of labelled ligand and antibody are incubated with different concentrations of unlabelled ligand. The amount of tracer bound when plotted against the concentrations of the unlabelled antigen results in a sigmoidal curve (Fig. 3). The standard curve can be plotted in various ways, which have been summarized by Chard (1978). The most popular are (1) the ratio of tracer binding in the presence of antigen (B) to tracer binding in the absence of antigen (B_0) (Fig. 3) and (2) the logit transformation (Rodbard 1974), with usually produces a straight line thus simplifying calculation (Fig. 4). For a detail overview of principles of radioimmunoassay, see e.g. Van Vunakis (1980) and references cited therein.

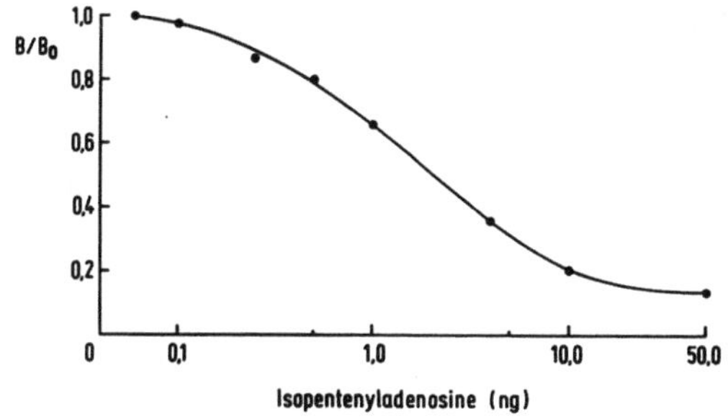

Fig. 3. Standard curve for isopentenyladenosine. B = tracer bound in the presence of antigen; B_0 = tracer bound in the absence of antigen

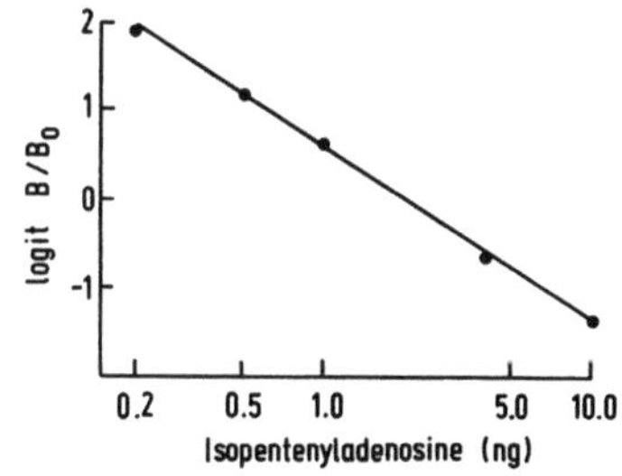

Fig. 4. Logit transformation of the standard curve

$$logit\ B/B_0 = \ln \frac{B/B_0}{1-B/B_0}.$$

B = tracer bound in the presence of an antigen; B_0 = tracer bound in the absence of an antigen

3.2 Cytokinin Antisera

Although the technique of radioimmunoassay has been employed in the determination of human hormones for about 25 yrs (Yalow and Berson 1960), it was another 15 yrs before this technique was used in the determination and quantitation of plant hormones. The first experiments were done by Erlanger and Beiser (1964) indirectly using ribonucleoside antibodies in the study of thermally denaturated DNA. When it was discovered that nucleosides coupled to a protein yielded an immunogenic conjugate, other modified nucleosides were subjected to a corresponding procedure. The first radioimmunoassays with cytokinins were done by Hacker et al. (1972), Humayun and Jacob (1974) and Khan et al. (1977), using antibodies against isopentenyladenosine in the study of t-RNA. These investigators were able to detect isopentenyladenosine in the 10 ng range. Analysis of antiserum specificity directed against isopentenyladenosine, performed by Milstone et al. (1978), yielded cross-reactivities of 100% with the corresponding phosphate and 10% with the free base. Kinetin was 100-fold and adenosine about 3×10^5-fold less effective than isopentenyladenosine.

The most important cytokinins of higher plants, besides those containing the isopentenyladenine group, are those of the zeatin type. Zeatin differs from isopentenyladenine only by the additional presence of a hydroxyl group in the side-chain. The use of an immunoassay in plant cytokinin analysis was first reported by Brandon et al. (1979), however, antibodies against isopentenyladenosine and dihydrozeatin riboside showed a strong cross-reactivity with trans-zeatin riboside, dihydrozeatin riboside and isopentenyladenosine.

The first report on the development of a sensitive and specific radioimmunoassay for the quantification of trans-zeatin, the biological active form of the

Table 2. Cross-reactivities of cytokinin antisera on a molar basis $\left(\% \text{ cross-reactivity} = \frac{\text{pmol standard for } B/B_0=0.5}{\text{pmol competitor for } B/B_0=0.5}\right)^a$

Compound	Zeatin riboside antiserum		Isopentenyl-adenosine antiserum
	^{3}H-assay	^{125}I-assay	
Trans-zeatin riboside	100	100	1.8
Trans-zeatin	44	45	0.9
Isopentenyladenosine	0.10	0.16	100
Isopentenyladenine	ND	ND	56
Cis-zeatin riboside	0.40	ND	1.4
Dihydrozeatin	1.72	ND	0.1
2-Methylthio-trans-zeatin riboside	ND	ND	0
Lupininc acid	ND	ND	0.1
6-Furfurylaminopurine	0.03	0.01	7.4
6-Benzylaminopurine	0.26	0.11	21.5
6-Benzylaminopurine 7-glucoside	0	ND	0.04
6-Benzylaminopurine 3-glucoside	0.24	ND	20.4
6-Benzylaminopurine 9-glucoside	0.29	ND	45.2
6-n-Hexylaminopurine	0.01	0.01	1.1
Adenine	0	0	0
Adenosine	0	0	0
Guanosine	0	0	0
Cytidine	0	0	0
Triacanthine	0.01	0.01	0
4-(2-Ethylhexylamino)-2-methyl-pyrrolo(2,3-d)pyrimidine	0	ND	0
4-Allylamino-2-methylpyrrolo (2,3-d)pyrimidine	0	ND	0
4-(1-Hydroxymethylamino)-2-methyl-pyrrolo(2,3-d)pyrimidine	0	ND	0
4-(2-Hydroxymethylamino)-2-methyl-pyrrolo(2,3-d)pyrimidine	0	ND	0
Diphenylurea	ND	ND	0
3,3-Dimethylacrylic acid	ND	ND	0
Tiglic acid	ND	ND	0

[a] B = tracer bound in the presence of an antigen; B_0 = tracer bound in the absence of an antigen; *ND* = not determined. (From Weiler 1980; Weiler and Spanier 1981).

two possible isomers, was given by Weiler (1980). The detection limit was in the fmol range and the antiserum was highly specific for the trans-isomer (Table 2). Isomerization from trans to cis resulted in almost zero cross-reactivity. Similar results were obtained by MacDonald et al. (1981), Vold and Leonard (1981) and Badenoch-Jones et al. (1984). Antisera raised against isopentenyladenosine in different laboratories (MacDonald et al. 1981; Weiler and Spanier 1981; Ernst et al. 1983a) had a detection limit of 15–100 pg for isopentenyladenosine, depending on the nature of the antiserum and the specificity of the tracer used. The antisera showed only a weak cross-reactivity with the cytokinins from the zeatin group, however, cross-reactivities with cytokinins from the 6-benzylaminopurine group were observed (Tables 2 and 3). The binding of a cytokinin to its antibody is pri-

Table 3. Antisera specificity, as percent cross-reactivity on a molar basis (calculation as in Table 2)

Compound	Anti-isopentenyl-adenosine	Anti-zeatin riboside	Anti-6-benzylaminopurine riboside
Isopentenyladenosine	100	< 1.5	22
Isopentenyladenine	67	< 1.0	6.7
Zeatin riboside	< 1.0	100	< 1.0
Zeatin	< 1.5	41	< 1.0
6-Benzylaminopurine riboside	118	< 1.0	100
Adenosine	0	0.07	0
Adenine	0	0.06	0

marily dependent on the N_6-substituted side-chain, as three antibodies raised against isopentenyladenosine, zeatin riboside and 6-benzylaminopurine riboside showed no cross-reactivity with adenine or adenosine (Tables 2 and 3). The apolar structure of the dimethylallyl chain or the benzyl group does not determine antibody specificity as rigidly as does the side-chain of trans-zeatin. Antisera against trans-zeatin were able to distinguish the cis-zeatin, showing a very weak cross-reactivity (Table 2; Vold and Leonard 1981; Badenoch-Jones et al. 1984). In contrast, antibodies against isopentenyladenosine cross-react very well with 6-benzylaminopurine riboside (over 100%), with 6-benzylaminopurine, and the N_3 as well as the N_9-glucosylated benzylaminopurine (Tables 2 and 3). Additionally, antiserum against 6-benzylaminopurine riboside cross-reacts with isopentenyladenosine (22%), but only weakly with zeatin riboside (Table 3). From Tables 2 and 3 it is clear that besides the side-chain, position N_9 is also important for antibody recognition, as the free bases showed cross-reactivities of only about 50%. Furthermore, 6-benzylaminopurine 9-glucoside showed a better cross-reactivity (45.2%) with an isopentenyladenosine antibody than 6-benzylaminopurine (21.5%), 6-benzylaminopurine 3-glucoside (20.4%) or 6-benzylaminopurine 7-glucoside (0.04%).

The synthetic reactions for the production of cytokinin immunogens are based on a periodate oxidation of the riboside to the corresponding aldehyde, which then is allowed to react with an amino group of the carrier protein. Reduction with borohydride stabilizes the complex (Erlanger and Beiser 1964). A tritiated cytokinin riboside-dialcohol, synthesized as described above for the immunogen, or a ^{125}I-tyramine labelled cytokinin (Weiler 1980), can be used as tracer. In the meantime radioimmunoassays for the quantitation of cytokinins in crude and purified plant extracts have been developed in many laboratories. In 1985 monoclonal antibodies raised against t-zeatin riboside and dihydrozeatin riboside, respectively, were offered commercially (Idetek, Inc.).

3.3 Cytokinin Determination in the Primary Extract

For the evaluation of an RIA it is absolutely necessary to carry out internal checks, since one must assume that there will be interference caused by other substances. In order to avoid adulteration of the results when testing an unknown sample, it is necessary to prepare a dilution graph for this sample. The results of this test should then be compared with the results of a dilution graph prepared for the standard. The graph is then subjected to a logit-transformation (Rodbard 1974), and if it runs parallel to the standard graph any interference on the part of other substances in the primary extract can be excluded. If this is not the case, then it is not possible to carry out cytokinin quantification in the crude extract. The same applies for quantitations carried out during individual purification steps in the course of a cytokinin extraction. This has been demonstrated in detail by Badenoch-Jones et al. (1984) in the application of a ribosylzeatin-antiserum. In the primay extract as well as after various purification steps, such as DEAE cellulose chromatography and HPLC, this parallelism was documented.

The use of RIA with polyclonal antibodies in a crude extract is always restricted to the totality of a cytokinin class (either isopentenyladenine or zeatin), since the presence of substances liable to initiate cross-reactions cannot be eliminated. In addition to the free cytokinin base the corresponding riboside, nucleotides and glucosides might be present, which can produce cross-reactions (Tables 2 and 3). If a primary extract test for isopentenyladenine derivatives is carried out, one must also take the possible presence of cytokinins having an N_6-substituted benzyl ring into consideration (Horgan et al. 1975; Chaves das Neves and Pais 1980; Ernst et al. 1983b), which would mean that a strong cross-reaction with the antiserum would take place (Tables 2 and 3). In the case of RIA of a primary extract, it is thus useful to cite values obtained as "corresponding cytokinin equivalents" of the standard antigen (Ernst et al. 1983a).

3.4 Cytokinin Determination After Preliminary Purification of the Primary Extract

Since an RIA carried out on a primary extract can, at best, only provide the total cytokinin concentration of a given cytokinin class, it is necessary to carry out appropriate preliminary purifications in order to separate individual cross-reacting cytokinins from one another. A partial purification on cellulose phosphate columns is advisable for the determination of cytokinin nucleotides, since although the nucleotides appear in the wash fraction, they are not present in the subsequent elution of the remaining cytokinins with ammonia (Horgan 1978; Badenoch-Jones et al. 1984; Ernst and Oesterhelt 1985). Other conventional purification steps for cytokinins are polyvinylpolypyrrolidone extraction, butanol extraction and Sephadex LH-20 chromatography (Horgan 1978 and articles cited herein). However, since all of these procedures are very time-consuming they are not suitable for an RIA, with the exception of the separation technique for nucleotides. Weiler and Spanier (1981) using thin-layer chromatography on silica gel managed to determine varying quantities of zeatin riboside, zeatin, isopentenyladenosine and isopentenyladenine in crown gall tumors of various species, a separation of polar metabolites was not possible however. Another much better method is the combination of high-performance liquid chromatography (HPLC) with an RIA, first described by MacDonald et al. (1981). If a suitable column material is used it is possible to achieve a complete separation of individual cytokinins, even cis-isomers and trans-isomers, within a very short period of time. The most commonly used material is of the octadecyl silane type, consisting of an octadecyl carbon chain bonded to silica particles. Carnes et al. (1975) separated zeatin riboside, zeatin, isopentenyladenosine and isopentenyladenine with a Bondapak C18/Porasil B column. Holland et al. (1978) were able to separate c-zeatin and t-zeatin from one another using a Lichrosorb RP-8 column. Horgan and Kramers (1979) used a Hypersil ODS column to separate a series of zeatin glucosides. They also separated cis/trans isomers of zeatin and zeatin riboside, dihydrozeatin and dihydrozeatin riboside by changing the run conditions. Andersen and Kemp (1979) used a μBondapak C18 column for the separation of a large number of non-polar cytokinins. Hardin and Stutte (1981) separated on a μBondapak C18 column zeatin derivates and Scott and Horgan (1982) used a Hypersil ODS column for the separation of cytokinin ribonucleoside 5'-monophosphates. One should note that there are great differences between the separating materials which can be purchased (Horgan and Kramers 1979). The use of HPLC in combination with the RIA is an elegant and efficient method for the determination and quantitation of cytokinins. However, since primary extracts cannot always be directly applied to an HPLC column a purification step is necessary. Conventional preliminary purification steps using ion-exchange chromatography, butanol extraction or Sephadex LH-20 chromatography are time-consuming and have the additional disadvantage of cytokinin losses of up to 50% (Horgan 1978; Ernst 1983). Consequently lower values will be generated when subsequent quantification by means of the RIA is carried out and the losses are not corrected for. This problem can be solved if a radioactive-labelled standard with high specificity is used, since one can then directly determine the corresponding losses in each purification step, and the quantification can be corrected accordingly. For a recovery determination investigators usually use only one radioactive-labelled cytokinin compound; the determined loss rates are then applied also to cytokinins with different structures (Ernst 1983; Badenoch-Jones et al. 1984). The fact that individual cytokinins may exhibit different loss rates during an extraction is not taken into consideration here (Scott et al. 1982). The ideal method would be to establish the individual loss rate for each cytokinin by use of corresponding labelled compounds. The use of octadecyl silica columns (Sep-Pak C18), which have a cytokinin recovery of over 90%, is an efficient and time-saving method for carrying out preliminary purification for an HPLC/RIA determination (Morris et al. 1976; MacDonald et al. 1981; Ernst 1983). The methanolic eluate from this column can be injected directly into an HPLC system (Ernst et al. 1984).

The use of immunoaffinity columns is another preliminary purification method which can be applied (Jayabaskaran and Jacob 1982). After the cytokinin antisera have been coupled to a suitable matrix (e.g. CNBr-activated Sepharose) it is possible to extract the cytokinins from the crude extract. The cytokinins can be eluted from such an immunoaffinity column with a recovery of more than 85% and subsequently quantified with the HPLC/RIA method (MacDonald and Morris 1983; Zaerr et al. 1983).

Although the separation of cytokinins in an HPLC system is the optimum method for the subsequent RIA, one should bear in mind that a large number of

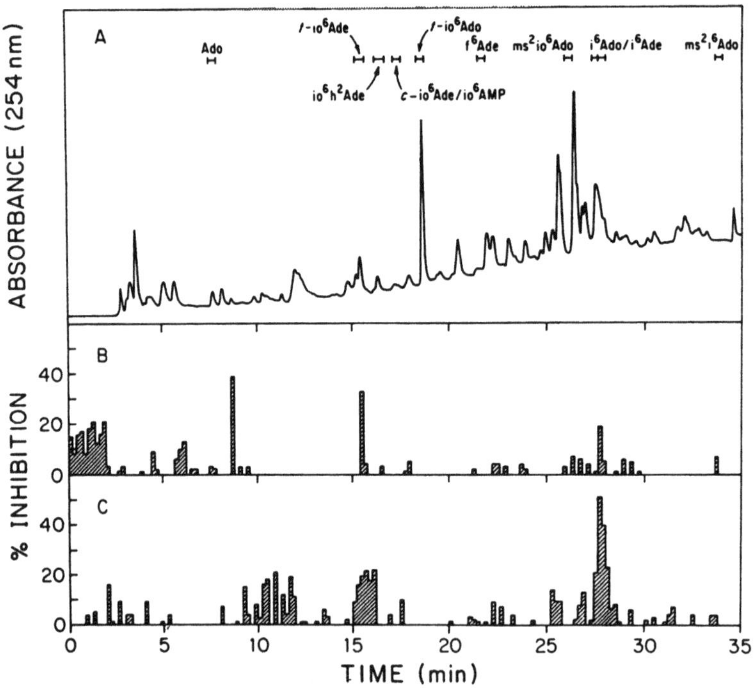

Fig. 5A–C. Cytokinin like substances in *Agrobacterium tumefaciens* culture filtrate after HPLC. **A** UV absorbance; **B** assay with anti-ribosylzeatin; **C** assay with anti-isopentenyl-adenosine (From MacDonald et al. 1981). *Ado* = adenosine; *t/c-io*^{6} *Ade* = t/c-zeatin; *t-io*^{6} *Ado* = zeatin riboside; *io*^{6}*h*^{2} *Ade* = dihydrozeatin; *ms*^{2}*io*^{6} *Ado* = 2-methylthio-zeatin riboside; *io*^{6} *AMP* = zeatin riboside 5′-monophosphate; *i*^{6} *Ado* = isopentenyladenosine; *i*^{6} *Ade* = isopentenyladenine; *ms*^{2}*i*^{6} *Ado* = 2-methylthio-isopentenyladenosine; *f*^{6} *Ade* = kinetin

purines, the chromatographic characteristics of which are unknown, can cross-react with the antiserum. This was demonstrated in the work of MacDonald et al. (1981) and Badenoch-Jones et al. (1984), in which RIA active fractions of the HPLC could not be assigned to any cytokinin structure, e.g. some of RIA active fractions in Fig. 5 have no structural assignement.

In the HPLC/RIA method it is also useful to rechromatograph active HPLC fractions which can be assigned to a cytokinin structure in another HPLC system and to check the identity of the determined cytokinin. This was demonstrated by Badenoch-Jones et al. (1984), who rechromatographed active fractions from a Zorbax C8 column to which zeatin, zeatin riboside and zeatin-9-glucoside could be assigned, on a µBondapak Phenyl column and were able to confirm the determined structures.

If cytokinin structures differing from those of the standard antigen are to be quantified it is necessary to correct for losses due to cross-reaction, e.g. at 50% cross-reaction, the determined value must be multiplied by a factor of two. If this correction is not done the cytokinin values obtained are only equivalents of the cytokinin standard.

The use of monoclonal antibodies which are specific for only one kind of cytokinin could make the separation of cytokinins unnecessary. Previous investigations using monoclonal antibodies (Trione and Morris 1983; Woodsworth et al. 1983) have not been satisfactory, however, since various clones raised against isopentenyladenosine and zeatin riboside, respectively, showed an unacceptably high cross-reaction level with other cytokinins.

3.5 Methods

3.5.1 General Isolation Protocol

The technique has been described in part by Ernst et al. (1983a). The cell material (5 g of an anise cell culture) was homogenized with ice-cold methanol (80%, v/w). Using such an extraction procedure the homogenate has to be checked for the presence of unspecific phosphatases, which could convert cytokinin phosphates into the corresponding cytokinin ribosides and therefore lead to an overestimation of the cytokinin riboside concentration (Ernst et al. 1985). After centrifugation, the organic phase of the supernatant was evaporated and the aqueous phase centrifuged at 10000 × g for 20 min before it was tested by RIA either directly or after HPLC. For HPLC the aqueous phase was loaded on to Sep-Pak-C18 cartridges (Water Ass.), which were washed with water (20 ml) before elution with methanol (10 ml). The eluate was evaporated to dryness and the residue dissolved in methanol (0.5 ml) and filtered through a 0.2 μm filter. The filtrate was subjected to HPLC and single fractions were tested in the RIA.

3.5.2 Synthesis of Cytokinin Protein Conjugates

Cytokinin ribosides were oxidized to the cytokinin riboside-dialdehyde and mixed with $NaBH_4$ (Erlanger and Beiser 1964). The aldehyde group will form a Schiff base with the amino groups of BSA, which then is reduced with $NaBH_4$, yielding a stable secondary amine. For the synthesis of t-zeatin riboside/BSA conjugates, t-zeatin riboside has to be purified by HPLC, as commercial products may contain up to 10% of the cis-isomer (Ernst et al. 1984).

The cytokinin riboside (30 μmol) was dissolved in 2 ml methanol, to which 5 ml of a 10 mM $NaIO_4$ solution were added dropwise under stirring, and then incubated for 20 min. Excess $NaIO_4$ was destroyed by the addition of 30 μmol ethylene glycol (0.3 ml) over 5 min. BSA (115 mg) was dissolved in water and the pH adjusted to 9.3 with 5% K_2CO_3 (total volume 5 ml). The cytokinin riboside-dialdehyde solution was added dropwise to the BSA solution, maintaining the pH at 9.3 with K_2CO_3. After incubation for 1 h at room temperature, the aldehyde groups were reduced by the addition of 10.8 mg $NaBH_4$ for 2 h. To destroy excess of $NaBH_4$ the pH was adjusted to 6.3 with 1 M acetic acid for 1 h. The conjugate was dialyzed exhaustively against water at 4 °C. The protein content was then adjusted to 4 mg ml^{-1} and the molar binding ratios cytokinin/BSA were determined by difference spectra recorded at 268 nm (for isopentenyladenosine: 4.9, zeatin riboside: 2.8, 6-benzylaminopurine riboside: 5.7). The conjugates were stored at −25 °C.

3.5.3 Immunization

One ml of the cytokinin/BSA conjugate (= 4 mg protein) was mixed by ultrasonication with 1 ml complete Freund's adjuvant. Rabbits were immunized by subcutaneous injections of 0.5 ml (= 1 mg protein) antigen emulsion, followed by a weekly injection of 1 mg protein in incomplete Freund's adjuvant over 2 weeks. After an interval of 3 weeks, a booster injection of 1 mg protein in incomplete adjuvant was given. At the end of the seventh week, 30 ml blood were collected from the ear vein, incubated for 2 h at room temperature and stored overnight at 4 °C. The clotted material was removed, the serum centrifuged at 10000 × g for 1 h and the supernatant stored at −25 °C. The crude serum can be used without further purification if the titer is high enough. The serum titer was determined by an antibody dilution curve (Fig. 2) according to the RIA protocol. (Serum titer, 50% of tracer bound; for isopentenyladenosine: 1:5800, zeatin riboside: 1:1000, 6-benzylaminopurine riboside: 1:1470.)

3.5.4 Synthesis of ^{3}H-Cytokinin Riboside-Dialcohols

The method used was as described by Weiler and Spanier (1981) with some modifications. The cytokinin (10 μmol) in 0.5 ml methanol was oxidized to the aldehyde by the dropwise addition of 25 μmol $NaIO_4$ in water over 15 min. To destroy excess $NaIO_4$, 0.1 ml of 0.1 M ethylene glycol was added and the mixture stirred for an additional 5 min. The cytokinin riboside-dialdehydes were purified by thin-layer chromatography (Merck silica gel 60; chloroform/methanol 9:2; v/v). The corresponding R_f-regions were eluted with methanol, the eluate evaporated to dryness and the residue dissolved in 1 ml ethanol. To 2.6 μmol ^{3}H-$NaBH_4$ (6.1×10^{9} Bq; $1.48–2.22 \times 10^{12}$ Bq $mmol^{-1}$) in ethanol 0.6 μmol cytokinin riboside-dialdehyde in ethanol was added. After an incubation time of 10 min excess ^{3}H-$NaBH_4$ was destroyed by the addition of 10 μmol retinal in ethanol.[1] (The use of acetic acid is not recommended, as tritiated hydrogen would by formed.) After a further 30 min, the reaction mixture was brought to dryness with nitrogen and the residue dissolved in 50 μl water and 500 μl petroleum ether. The cytokinin riboside-dialcohols are present in the water phase, which was dried under nitrogen and the residue redissolved in ethanol before purification by thin-layer chromatography as described above. Zones containing the cytokinin riboside-dialcohol (isopentenyladenosine: $R_f = 0.21$, zeatin riboside: $R_f = 0.13$, 6-benzylaminopurine riboside: $R_f = 0.26$) were eluted with methanol, the eluate was concentrated under sterile conditions and stored at −25 °C under nitrogen. (Specific radioactivity; isopentenyladenosine: 13.69×10^{10} Bq $mmol^{-1}$, zeatin riboside: 5.18×10^{10} Bq $mmol^{-1}$, 6-benzylaminopurine riboside: 10.36×10^{10} Bq $mmol^{-1}$). The synthesis of the labelled cytokinin riboside-dialcohols should be done with caution in a well-functioning fume hood and under nitrogen.

---

[1] For another purpose retinal was used, yielding the corresponding retinol (Lanyi and Oesterhelt 1982).

3.5.5 RIA Protocol

The reactions were carried out in polystyrol tubes, as glass might show an unspecific adsorption for cytokinins (MacDonald et al. 1981). The incubation mixture contained the following: 500 µl 10 mM phosphate buffer pH 7.4 containing 150 mM NaCl, 100 µl cytokinin standard or sample solution, 100 µl tracer (5.55×10^3 Bq ml^{-1}) and 100 µl BSA solution (20 mg ml^{-1}, RIA grade). After mixing, 100 µl of diluted antiserum, enough to bind about 50% tracer in the absence of a competitor, was added and the mixture incubated for 1 h at room temperature. Immunoglobulins were precipitated by the addition of 1.2 ml 91% $(NH_4)_2SO_4$ solution and pelleted by centrifugation at $5200 \times g$ for 30 min. A 1.6 ml aliquot of the supernatant was mixed with 3 ml of scintillation cocktail and then counted. Bound cpm are defined as the difference of the cpm-value in the sample, containing no antibody minus the cpm-value in the sample containing the antibody.

Antibody Dilution Curve. Antiserum was diluted serially in 10 mM phosphate buffer pH 7.4 containing 150 mM NaCl. 100 µl was incubated without a standard according to the RIA protocol. The amount of serum, binding about 50% of the tracer (Fig. 2) was used for the construction of the standard curve.

Standard Curve. Cytokinin standards were dissolved in methanol and diluted serially. After evaporation of the methanol, the incubation mixture was added according to the RIA protocol. Figure 3 shows the standard curve for isopentenyladenosine antiserum with a detection limit of about 100 pg. The measuring range extends from 0.2–10 ng. Within range a strictly linear relationship exists between the logit-transformation of the binding parameters and the isopentenyladenosine concentration (Fig. 4).

RIA in Combination with HPLC. A 5 µm Lichrosorb RP8 column (250×4.6 mm) was washed for 30 min with water/acetonitrile (70:30, v/v) + 0.5% acetic acid. The sample was injected in a volume of 50 µl methanol. The solvent was water/acetonitrile (70:30) + 0.5% acetic acid. Having a flow rate of 2 ml min^{-1}, 4 ml fractions were pooled, evaporated and the residue tested in the RIA directly or after an appropriate dilution. Figure 6 shows the result obtained after the injection of a pre-cleaned anise cell extract corresponding to 0.75 g fresh weight, yield-

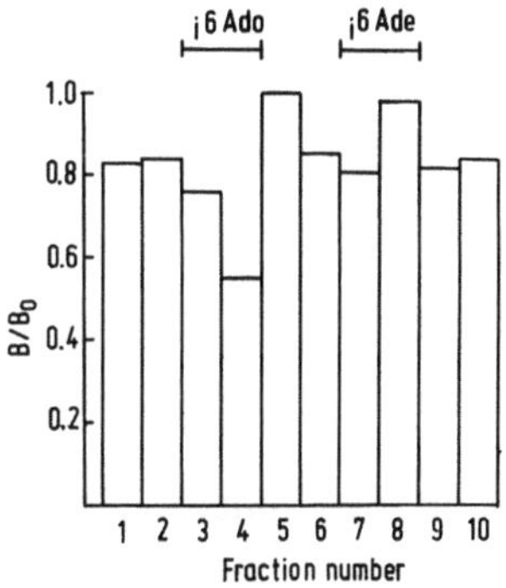

Fig. 6. RIA activity in HPLC fractions of a purified anise cell extract, eluting from the Lichrosorb RP8 column. Assay with anti-isopentenyladenosine. An aliquot corresponding to 0.75 g fresh weight was injected. i^{6Ade} = isopentenyladenine; i^{6Ado} = isopentenyladenosine

ing 3.2 ng isopentenyladenosine g^{-1} fresh weight after recovery correction of 66% on the basis of (^{14}C)6-benzylaminopurine. After a run of 30 min the column was equilibrated with the solvent for 30 min before a new sample was injected. In serial analysis the retention times of the cytokinins were controlled by standard cytokinins, detected at 254 nm. 6-benzylaminopurine or kinetin are recommended as standards, as these plant growth regulators have not yet been found in plants. Furthermore, only small amounts of the standard (ng range) should be used, to exclude any memory effect of the column.

4 Gas Chromatography/Mass Spectrometry

4.1 General Principles

Gas chromatography/mass spectrometry (GC/MS) is a combination of two analytical systems. Gas chromatography is an excellent tool for the separation and detection of cytokinins in a complex mixture (Most et al. 1968; Kemp and Andersen 1981; Kemp et al. 1982, 1983; Stafford and Corse 1982). A gas chromatographic capillary column should be used, because the retention time is more finely diagnostic of identity in this case than it is with a packed column. However, for a good qualitative analysis, gas chromatography should be combined with mass spectrometric analysis, as the accuracy of retention measurements is not sufficient to eliminate other compounds that might elute with the observed time period. After separation on a GC-column most of the carrier gas is removed and the sample introduced into the ionization chamber of a mass spectrometer. The mass spectrometer produces ions and then separates them according to their mass-to-charge ratio (m/z or m/e), yielding molecular ions and characteristic fragment ions. In most mass spectrometers ionization is performed by a high energy electron beam (EI) or by chemical ionization (CI). In an electron impact source the sample is bombarded with a beam of electrons in the vapour phase, using a beam energy of about 70 eV. Most of the ions formed are singly charged, but only a small proportion of these are negatively charged. The most important process is the formation of positive molecular ions:

$$M + e \rightarrow (M)^{+\cdot} + 2e.$$

Molecular ions having enough energy are able to decay into fragment ions by homolytical or heterolytical cleavage:

$$(M)^{+\cdot} \rightarrow A^{+\cdot} + B$$
$$(M)^{+\cdot} \rightarrow C^{+} + D^{\cdot}.$$

Dependent on the fragmentation pattern, it is possible to identify a compound independent of the relative intensities of single ions. However, for a quantitative analysis it is necessary to operate under constant conditions so that the relative intensities of peaks in the spectrum do not change from one determination to the other.

In chemical ionization mass spectrometry, the sample is ionized by the aid of a reactant gas, such as methane, isobutane or ammonia. The reactant ions are produced by a combination of electron impact ionization and ion molecule collisions. If methane is the reactant gas, the most important ions formed are $(CH_5)^+$ and $(C_2H_5)^+$. The reactant ions react with the sample mainly by proton transfer:

$$(CH_5)^+ + MH \rightarrow (MH_2)^+ + CH_4.$$

To produce a mass spectrum it is necessary to separate the ions according to their ratio mass-to-charge. For that two types of mass spectrometers, a magnetic instrument or a quadrupole instrument are used in GC/MS. The latter is performed in multiple ion selection (MIS) as well as in single ion monitoring (SIM). After leaving the analyzer, ions are detected by an appropriate photorecorder. Initially, data were coded manually and then entered into a computer system for processing. Further improvements in computer systems led to an on-line connection with the GC/MS system. This opened possibilities of (1) a short output time after data acquisition; (2) the storage of data; (3) the manipulation and selection of required data; (4) the computer control of the GC/MS system; and (5) easier library research. Practical aspects of GC/MS applications were published by Rose and Johnstone (1982) and Message (1984).

4.2 Identification of Cytokinins

GC/MS of cytokinins was first demonstrated by Horgan et al. (1973). These investigators were able to demonstrate the presence of trans-zeatin riboside in a partially purified extract of sycamore sap by comparing the retention time, and the principle ion peaks of a biological active cytokinin fraction with a synthetic TMS-trans-zeatin riboside. The most important peaks were observed at m/z 639, 624, 551, 550, 536, 520, 276, 259, 245, 230, 202, 201, 200, 188, 156, and 103. Young (1977) reported the identification of zeatin riboside in extracts of chinese gooseberry using permethylated derivatives. From the major ions, he selected two ions for multiple ion detection (MID). In this method the intensity of only a few ions is recorded and selection is performed by changing the ion-accelerating voltage at a fixed magnetic field. This technique is very sensitive, as during the whole elution of a GC-peak only a few ions of the compound and not the whole mass spectrum are focused at the collector. In single ion monitoring (SIM) only one ion is monitored and in combination with the retention time it is possible to determine the compound of interest. However, in this case investigators have to keep in mind that only known cytokinins can be identified in the plant extract, being analyzed for the presence of cytokinins.

In general, trimethylsilyl (TMS) derivatives are used for GC of cytokinins. However, at levels below 100 ng, zeatin was completly lost as the TMS-derivative (Young 1977). Similar results were obtained by Ernst (1983) in a quantitative analysis of isopentenyladenine below 50 ng. A permethylation procedure was successful in overcoming this problem. Ludewig et al. (1982) reported the successful application of trifluoroacetyl derivatives of the most common cytokinins to GC/MS with a detection limit of 1 pg.

The first successful application of GC/MS techniques to cytokinin identification was reported by Horgan et al. (1973), as mentioned above, and by Shindy and Smith (1975) who identified isopentenyladenine, isopentenyladenosine, zeatin, dihydrozeatin and zeatin riboside in extracts of cotton ovules. In the meantime many metabolites of the isopentenyladenine as well as the zeatin group were identified. Summons et al. (1977) reported the occurrence of raphanitin in radish seeds, this is a zeatin derivative in which the glucose is linked to the 7-position of the purine (7-β-D-glucopyranosyl zeatin). Morris (1977) published the identification of glucosyl zeatin in *Vinca rosea* crown gall tissue using permethyl derivatives. However, Morris was not able to determine the position of the glucose moiety. This was carried out by Scott et al. (1980), who demonstrated the occurrence of zeatin 9-glucoside as the major endogenous cytokinin of *Vinca rosea* crown gall tissue. In *Zea mays* a number of glucosyl derivatives of the zeatin family were detected by electron impact as well as chemical ionization mass spectrometry (Summons et al. 1980). Cytokinins in t-RNA of *Phaseolus vulgaris* were identified as cis-ribosylzeatin, 2-methylthio-ribosylzeatin and isopentenyladenosine (Edwards et al. 1981), and in t-RNA of *Agrobacterium tumefaciens* as isopentenyladenosine, 2-methylthio-isopentenyladenosine, cis/trans-zeatin riboside and cis-methylthio-zeatin riboside (Morris et al. 1981). Cis-zeatin riboside was identified in t-RNA from both normal and crown gall tissue of *Vinca rosea* L., whereas the trans-isomer was found only in crown gall t-RNA (Palni and Horgan 1983). Tsoupras et al. (1983) reported the presence of isopentenyladenosine-mononucleotide linked to ecdysone in eggs of *Locusta migratoria* using TMS-derivatives and chemical ionization mass spectrometry. Besides the identification of naturally occurring cytokinins, GC/MS analysis is also useful in metabolic pathway studies of cytokinins (Summons et al. 1980; McGaw et al. 1984; Palni et al. 1984).

As it is not possible to inject a crude extract into the GC/MS system, a purification procedure is necessary. Besides conventional procedures such as ion-exchange chromatography, solvent partitioning, Sephadex LH-20 chromatography and thin-layer chromatography (Horgan 1978; Yokota et al. 1980; Ernst et al. 1983b), short octadecyl silica-columns may be used with HPLC systems (Morris et al. 1976; Yokota et al. 1980; MacDonald et al. 1981; Ernst et al. 1983a). The last is the preferred method (see also the purification procedures described in Sect. 3.5).

The mass fragmentation of nucleosides, as well as of cytokinins have been thoroughly examined (Minden and McCloskey 1973; McCloskey 1974; Hashizume and McCloskey 1976). As an example, the fragmentation pattern of TMS-isopentenyladenine is discussed below. Isopentenyladenine was isolated from an anise cell culture using a methanolic extraction procedure, cellulose phosphate chromatography, butanol extraction and Sephadex LH-20 chromatography. After trimethylsilylation the sample was injected into a gas chromatograph. The GC-peak according to a synthetic TMS-isopentenyladenine exhibited the mass spectrum as shown in Fig. 7. The fragmentation pattern is given in Fig. 8. The ion m/z 332 resulted from the molecular ion m/z 347 by the loss of a methyl group.

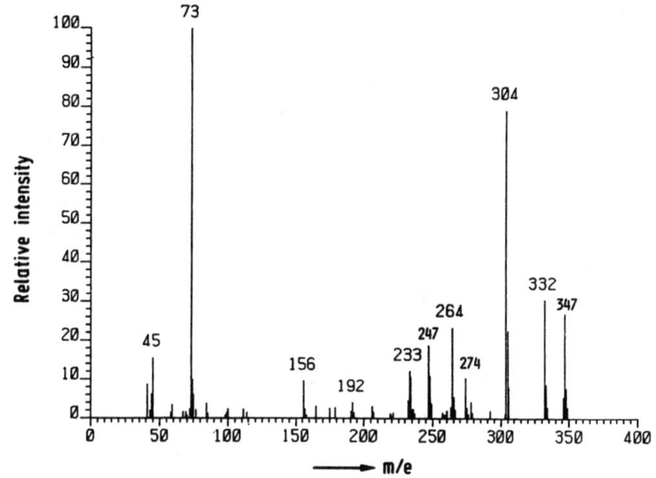

Fig. 7. Mass spectrum of naturally isolated isopentenyladenine from anise, as the trimethylsilyl derivative

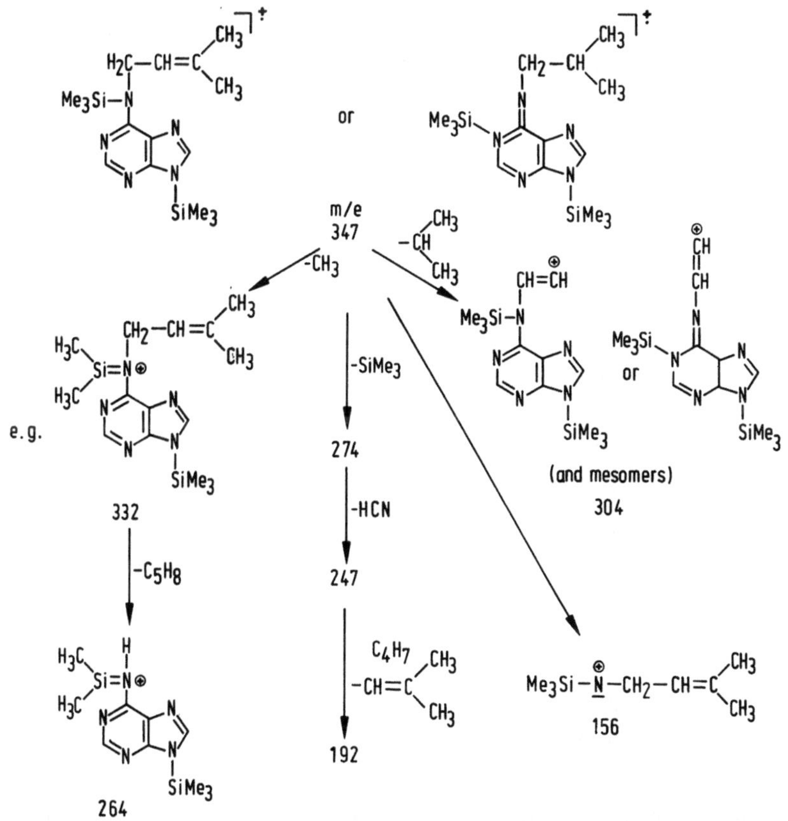

Fig. 8. Fragmentation pattern of isopentenyladenine as TMS derivative

Subsequent release by cleavage of the side-chain yielded the ion m/z 264 (m/z 347 → m/z 332 → m/z 264). The fragment ion m/z 274 resulted from the separation of a silylgroup (m/z 347 → m/z 274) and the ion m/z 192 generated by the loss of the side-chain, including the nitrogen in 6-position of the purine (m/z 347 → m/z 274 → m/z 247 → m/z 192). The ion m/z 304, a mesomeric from, resulted from the loss of $-CH-(CH_3)_2$ from the molecular ion (m/z 347 → m/z 304). The appearance of the ion m/z 156 proved the existence of the trimethylsilylated side-chain. From these data and the corresponding retention time the identity of isopentenyladenine was proved.

The identification of new, naturally occurring cytokinins is only possible by GC/MS techniques. Horgan et al. (1975) reported the existence of 6-(o-hydroxybenzylamino)-9β-D-ribofuranosylpurine in leaves of *Populus robusta*. This was the first report of a naturally occurring purinyl cytokinin having an aromatic side-chain. 6-(o-hydroxybenzylamino)-2-m-ethylthio-9β-D-glucofuranosylpurine was identified in fruits of *Zantedeschia aethiopica* by Chaves das Neves and Pais (1980). 6-Benzylaminopurine riboside was isolated from an anise cell culture and the structure determined from the retention time and the mass fragmentation pattern (Fig. 9). It has been observed that the molecular ion (m/z 573) may lose a methyl radical (m/z 573 → m/z 558) and two molecules of trimethylsilanol (m/z 573 → m/z 483 m/z 393). The detection of ions, m/z 349, 259, 243, 230, 217, and 147 proved the presence of a trimethylsilylated riboside (Ernst et al. 1983b). In 1984, 9-(hexosyl)-β-D-ribofuranosyl-1-(4-hydroxy-3-methyl-but-2-enyl-amino)-purine, a glucoside of zeatin riboside in which the glucosyl moiety is attached directly to the ribosyl moiety was identified in *Pinus radiata* buds by Taylor et al. (1984).

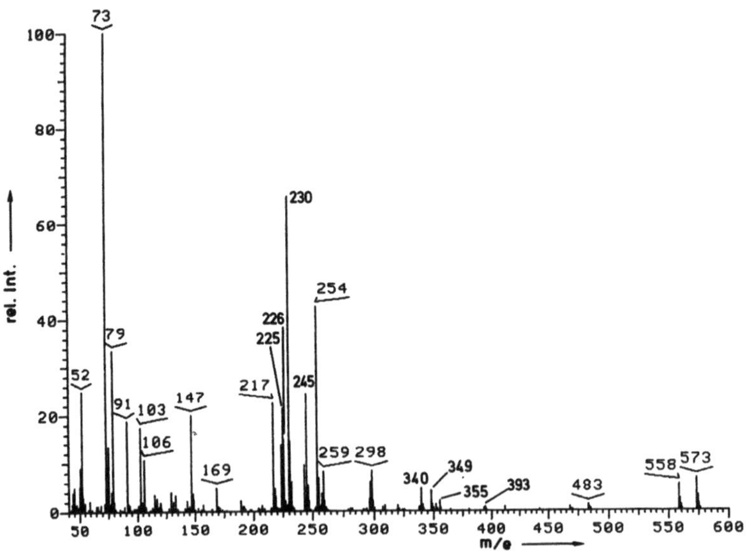

Fig. 9. Mass spectrum of naturally isolated 6-benzylaminopurine riboside from *Pimpinella anisum*, as the trimethylsilyl derivative

4.3 Quantitation of Cytokinins

As mentioned in the previous section multiple ion detection (MID) or single ion monitoring (SIM) are used in the identification of known cytokinins. These methods are also used for the quantitation of cytokinins in the nanogram range. The relative amount of cytokinins can be estimated with an external standard cytokinin by MID or SIM: a comparison of the intensity of a characteristic ion in the mass spectrum of a standard and the intensity of the same ion of the extraction product is computer analyzed. Such a computer analysis was demonstrated by Dauphin et al. (1979, 1980) for the quantitation of different cytokinins in male as well as in female apices of *Mercurialis*.

A preferred method is the application of internal standards to cytokinin quantitation, which allows an absolute measurement of the concentration. This technique is based on stable isotope dilution. For cytokinin quantitation, D_2-labelled or ^{15}N-labelled internal standards are used (Summons et al. 1977, 1979a, b; McCloskey et al. 1979, 1981; Hashizume et al. 1979, 1982; Scott and Horgan 1980; Ernst et al. 1983a, b). It is important to note that the labelled and unlabelled compounds have an analogous mass fragmentation and the same GC-retention time. Impurities coeluting with the cytokinin of interest will not affect the assay as long as they do not have ions at the m/z values selected for monitoring. Hashizume et al. (1979), Summons et al. (1979a) and Sugiyma et al. (1983) reported that deuterium-labelled standards eluted approximately 5 s prior to the endogenous cytokinin. However, this change is insufficient to interfere with the cytokinin characterization and quantitation in a GC/MS run processed by a computer, as sufficient full scans can be taken and stored in a data system during peak elution for an area calculation of the appropriate ion. The labelled and unlabelled cytokinin can be distinguished by analysis, because the m/z values of their mass spectral fragments will differ, depending on the number of labelled atoms. From Fig. 10 it is obvious that the molecular ion of isopentenyladenosine m/z 551 is shifted to m/z 553 and the characteristic fragment ion m/z 232 to m/z 234 for D_2-isopentenyladenosine. During a GC/MS run the m/z values of the molecular ions of the characteristic fragment ions of the unlabelled and labelled cytokinin are continuously recorded and the peak areas are compared. A computer coupled to the mass filter recorded mass spectra as the standard mixture eluted from the chromatograph and printed out SIM chromatograms of the molecular ion (m/z 551 and m/z 553) or the selected fragment ions (Fig. 11). The ratio of areas of the peaks was plotted against known concentrations of isopentenyladenosine/D_2-isopentenyladenosine to give the calibration plot. A typical calibration curve is illustrated in Fig. 12 for isopentenyladenosine and D_2-isopentenyladenosine using the peak area ratios m/z 551 to m/z 553 or m/z 232 to m/z 234. A known quantity of the labelled standard can then be added to a tissue extract, and after purification the ratio of the natural cytokinin to labelled internal standard is determined. For precise measurements it is necessary to analyze at least three aliquots of the extract.

Unlike the RIA determination it is not necessary to account for losses in endogenous cytokinins during purification, because losses will affect the labelled standard as well as the endogenous cytokinin. Quantitation is performed by the

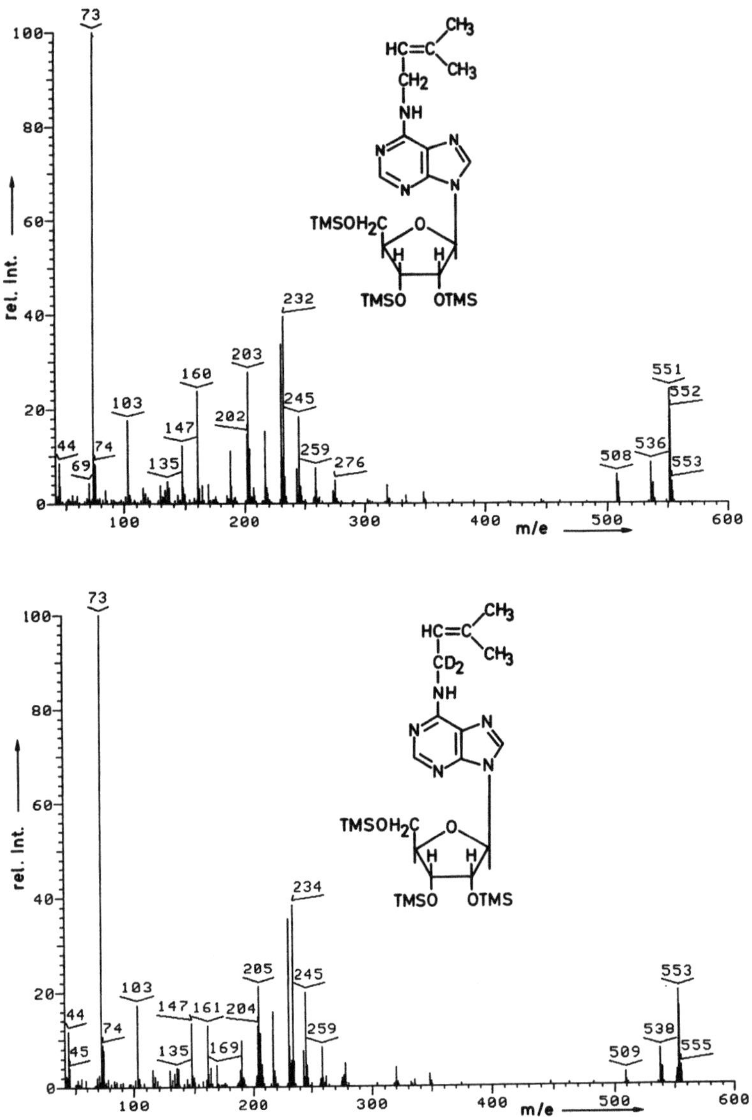

Fig. 10. Mass spectrum of isopentenyladenosine (*above*) and D_2-isopentenyladenosine (*below*)

determination of the ratio of unlabelled to labelled cytokinin and therefore losses will not affect the accuracy of the GC/MS method. Although it is not strictly necessary to determine the concentration of the labelled isotope present in the internal standard before the latter can be used, knowledge of the isotopic purity is important in deciding whether or not it will be a useful standard, as a low isotopic incorporation can generate confusing data (Millard 1978).

The synthesis of labelled cytokinin standards is well described in the literature. In deuterium labelling, the 6-chloropurine skeleton is condensed with the respec-

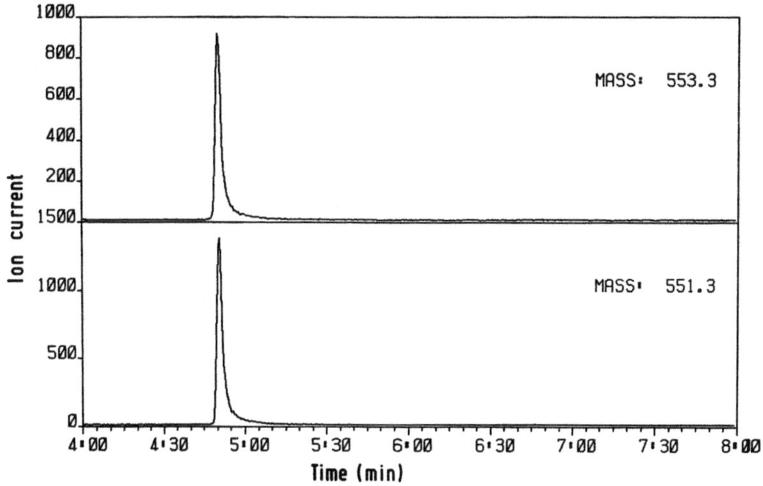

Fig. 11. Mass spectrometric quantification of isopentenyladenosine. For SIM analysis the molecular ion of unlabelled m/z551 and labelled m/z553 isopentenyladenosine was used. (Isopentenyladenosine: 0.89 ng, D_2-isopentenyladenosine: 0.74 ng)

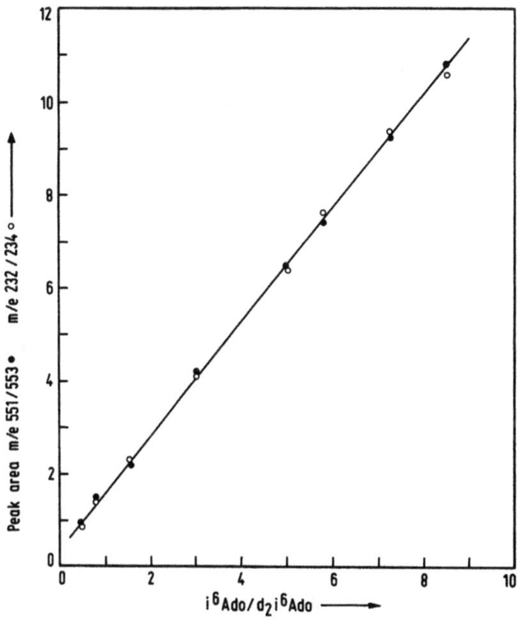

Fig. 12. Calibration plot for isopentenyladenosine, as the TMS derivatives

tive labelled amine. For the isopentenyladenine derivatives, D_2- or D_6-labelled amines are used, while for the zeatin derivatives D_2- or D_5-labelled amines are used (Summons et al. 1979a, b; Hashizume et al. 1979). The (D_6)- and the (D_5)-derivatives fit better than the (D_2)-derivatives, as the latter show an overlap with ^{28}Si, ^{29}Si, ^{30}Si or ^{13}C peaks (Summons et al. 1979a; Ernst et al. 1983a). Figure 10 shows such an overlap using the molecular ion of isopentenyladenosine (m/z 551) and D_2-isopentenyladenosine (m/z 553). Therefore the ratio m/z 551 / m/z 553 had to be corrected for calibration purposes. The use of lower fragment ions is not recommended, because interferences from low-mass ions of by-products not separable by GC lead to difficulties (Ernst et al. 1983a). Ions suitable for mass spectrometric quantitations of cytokinin metabolites are summarized by Summons et al. (1979a). After synthesis of the labelled cytokinin, it must be identified by comparison with an authentic unlabelled sample: there should be no difference in thin-layer chromatography and HPLC, whereas nuclear magnetic resonance analysis (NMR) and MS should show the incorporation of the D-atoms.

Another possibility for quantitation of cytokinins is the use of (^{15}N)-cytokinins as internal standards. The synthesis of (^{15}N)-zeatin was described in detail by Scott and Horgan (1980). The detection limit according to their calibration curve was in the range of 5 ng. The content of different zeatin glucosides in *Vinca rosea* crown gall tissue was determined using ($^{15}N_4$)-labelled standards by Scott et al. (1982).

The first report of the application of GC/MS to the identification and quantitation of endogenous cytokinins was by Summons et al. (1977). These workers quantitated raphanitin in radish seeds using (D_2)-raphanitin as an internal standard. The detection limit of the GC/MS technique so far reported is in the range of 1–5 ng (Young 1977; Hashizume et al. 1979; Ernst et al. 1983a). Usually 5–10 g of a tissue are enough for the detection of about 2–3 ng cytokinin g^{-1} fresh weight. These low quantities of plant material permit a very short extraction procedure using a Sep-Pak-C18 pre-cleaning step followed by HPLC as described by Ernst et al. (1983a). However for plants having only traces of cytokinins a larger amount of tissue has to be extracted using extensive purification procedures. Yamane et al. (1983) extracted kilograms of fronds from *Equisetum arvense* and obtained 25 ng isopentenyladenine kg^{-1} fresh weight. For a precise quantitation it is necessary to add the internal standard at the beginning of an extraction procedure and not to individual fractions during the purification, otherwise the determined values would represent only minimal levels (McCloskey et al. 1980, 1981; Palni et al. 1983).

The determination and quantification of cytokinin nucleotides was not carried out for a long time. Nucleotide analysis usually requires a dephosphorylation step with alkaline phosphatase after separation of the cytokinin nucleotides from other cytokinins by cellulose phosphate chromatography. However this approach cannot distinguish the position of the phosphate group, nor can it distinguish between mono-, di-, and tri-phosphates. To overcome this problem Summons et al. (1983) analyzed *Datura innoxia* crown gall tissue with (D_5)-ribofuranosylzeatin 5′-monophosphate as the internal standard. To confirm that the nucleotide was a 5′-phosphate a chemical degradation of the nucleotide fraction was carried out, yielding the expected zeatin. From their calibration curve and the mass spectra, which are shown in Fig. 13, they calculated a zeatin riboside monophosphate content of 64 ng g^{-1} fresh weight. Unfortunately the internal standard was not added at the beginning of their extraction, so losses might have occurred. Another elegant method was reported by Scott and Horgan (1984) using (D_2)-zeatin riboside 5′-monophosphate and ($^{15}N_4$)-zeatin riboside as internal standards. After enzymatic hydrolysis of the nucleotide fraction, zeatin riboside 5′-monophosphate was analyzed as zeatin riboside (m/z 390:m/z 392). The riboside fraction

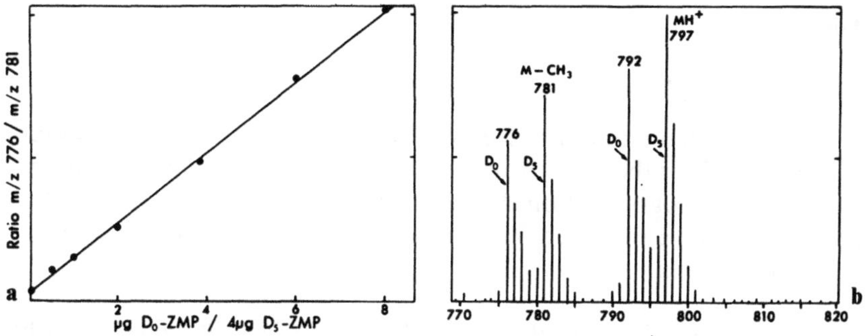

Fig. 13. a Calibration line for the M-CH_3 ion of D_0+D_5 TMS-ZMP. **b** Molecular ion region in the CI mass spectrum of TMS-ZMP from the tissue extract. (From Summons et al. 1983). *ZMP* = zeatin riboside 5′-monophosphate

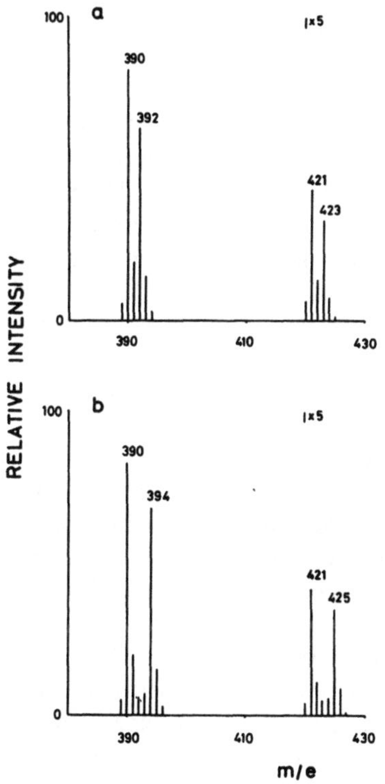

Fig. 14. Peak clusters in mass spectra obtained by GC/MS of permethyl zeatin riboside samples. Internal standards of (^{2}H)ZMP and ($^{15}N_4$)ZR were added to the same tobacco crown gall tissue extract. **a** $M^{+\cdot}$ and ($M^{+\cdot}$-$\cdot OCH_3$) peak clusters in the mass spectrum obtained for the hydrolyzed, permethylated ZMP sample. **b** $M^{+\cdot}$ and ($M^{+\cdot}$-$\cdot OCH_3$) peak clusters in the mass spectrum obtained for the permethylated ZR sample. The peaks at m/e 421 and 390 are derived from the natural compound, while those occurring 2 or 4 m/e units higher are derived from the $^2H_2$ or $^{15}N_4$ internal standards, respectively (From Scott and Horgan 1984). *ZR* = zeatin riboside; *ZMP* = zeatin riboside 5′-monophosphate

yielded only the ($^{15}N_4$)-zeatin riboside (m/z 390:m/z 394) and showed no contribution of (D_2)-zeatin riboside, indicating that no hydrolysis of zeatin riboside 5′-monophosphate occurred during the purification (Fig. 14). The correct position of the phosphate group also was demonstrated by chemical degradation of zeatin riboside 5′-monophosphate to zeatin.

In conclusion, as it is in principle possible to synthesize labelled standards of all known cytokinins, quantitative analysis by the GC/MS method of any of these plant hormones is possible.

4.4 Methods

4.4.1 General Isolation Procedure and GC/MS

After addition of the deuterium-labelled cytokinin standard the anise cells were extracted as described for the RIA. However, HPLC fractions containing the cytokinin were rechromatographed on the same column. This is necessary as it is not possible to redissolve the dried fractions from the first run for the separate derivatization procedures.

GC/MS for Isopentenyladenosine and 6-Benzylaminopurine Riboside. A Carlo Erba gas chromatograph model 2110 equipped with an injector for split-splitless (280 °C) injections and a flame-ionization detector (280 °C) was used. Cytokinin containing fractions were evaporated and the dried residues dissolved in ethanol/water (35%, v/v), before transferring to 150 µl Teflon-capped vials and drying using a speed-vac concentrator. The residues were dissolved in 2–5 µl acetonitrile (10–400 ng cytokinin μl^{-1}) to which was added 2–5 µl of hexamethyldisilazane/trimethylchlorosilane/pyridine (2:1:10, by vol). The vials were mixed and kept at 90 °C for 30 min. After cooling, the mixture was centrifuged and portions (0.5–2 µl) of the supernatant injected into the gas chromatograph. Separation was achieved on a 25-m fused silica capillary (i.d. 0.32 mm) coated with CP^{tm}Sil5. The injection was splitless for 1 min at 150 °C oven temperature. The temperature was increased to 280 °C at a rate of 5 °C min^{-1}. Helium was used as a carrier gas at 1.8 ml min^{-1}. The retention time for isopentenyladenosine was 10–12 min and for 6-benzylaminopurine riboside 11 min. The gas chromatograph was connected to a Varian CH7A mass spectrometer by an all glass open split and mass spectra were recorded at 70 eV. The GC/MS system was directly coupled to a Varian SS200MS data system.

For SIM measurements the ions for isopentenyladenosine (m/z 551, m/z 553) and for 6-benzylaminopurine riboside (m/z 254, m/z 256) were used. The calibration curve (Fig. 12) was prepared by mixing standard solutions of isopentenyladenosine (11.6 ng μl^{-1}) and D_2-isopentenyladenosine (48 ng μl^{-1}) in acetonitrile, at varying ratios: isopentenyladenosine/D_2-isopentenyladenosine from 0.1 to 8.0 and the total cytokinin content adjusted to 5–50 ng μl^{-1}. Portions of this mixture (2–5 µl) were silylated as described and injected into the gas chromatograph. Measurements of the peak area ratios (Fig. 11) were made, using the SIM software from the Varian MAT for the SS200 data system. For 6-benzylaminopurine riboside/D_2-6-benzylaminopurine riboside the same procedure was used.

It is important to note that the time between derivatization and measurements was always constant to avoid different decompositions of the derivatives. The glass inserts of the injector were cleaned routinely and are deactivated with silyl-8 (Pierce).

GC/MS for Isopentenyladenine. As mentioned in Sect. 4.2 trimethylsilylation was not possible. For a quantitative measurement a permethyl derivative of isopentenyladenine was used. Standards or dried HPLC fractions were incubated with 10 µl dimethylsulfinyl anion in dimethylsulfoxide and 10 µl methyliodide. After 30 min 10 µl water and 20 µl chloroform were added and then mixed. The tubes were centrifuged (Minifuge, Beckman) and the upper phase was removed. The lower phase was washed with 100µl water and dried, first in a speed-vac concentrator and then over P_2O_5. The residue was dissolved in 10 µl chloroform and injected into the gas chromatograph. Separation was achieved on a 10-m fused silica column (i.d. 0.32 mm) coated with CP^{tm}Sil5. The injection was splitless for 1 min at 80 °C. The temperature was increased to 190 °C at a rate of 10 °C min^{-1} and the retention time was 3.5–4 min. All other conditions were as described for isopentenyl-adenosine. For SIM measurements the ions m/z 231 and m/z 233 were used.

4.4.2 Synthesis of Deuterium-Labelled Cytokinins

a) D_2-3-Methyl-2-Butenylamine. To $LiAlD_4$ (90 mmol) in ether (200 ml), 3,3-dimethylacrylonitrile (70 mmol) in ether (30 ml) was slowly added with stirring. After an incubation time of 30 min at room temperature the reaction was stopped by the addition of 2.5 ml water and 50 ml 10% NaOH. The ether was decanted from the precipitated aluminium hydroxide and the residue washed with ether (4 × 20 ml). The combined ether phases were dried over Na_2SO_4, the ether was evaporated and the residue distilled at 100 °–150 °C (Hall and Robins 1968; 110 °C) yielding D_2-3-methyl-2-butenylamine. The expected deuterium incorporation was determined by NMR measurements.

*b) D_2-*Benzylamine. To $LiAlD_4$ (100 mmol) in ether (200 ml), benzonitrile (100 mmol) in ether (20 ml) was added slowly with stirring and cooling. The reaction was stopped by the addition of 4 ml water, 3 ml 20% NaOH and then an additional 14 ml of water. The ether was decanted, the residue washed, the combined ether phases were dried and distilled as described above. The D_2-benzylamine formed boiled at 184 °C (Windholz 1976; 185 °C) and showed the expected D_2-incorporation by NMR analysis.

c) D_2-Isopentenyladenosine. D_2-3-methyl-2-butenylamine (0.6 mmol) and 6-chloropurine riboside (0.3 mmol) were refluxed in 6 ml n-butanol for 2 h. The butanol was evaporated at 40 °C and the residue crystallized from ethanol. The product was identical in thin-layer chromatography (Merck silica gel 60, n-butanol/acetic acid water 12:3:5, by vol; R_f=0.59), NMR and GC/MS measurements (see Fig. 10) except isotopic composition to an authentic unlabelled sample.

d) D_2-Isopentenyladenine. D_2-3-methyl-2-butenylamine (2 mmol) and 6-chloropurine (1 mmol) were refluxed in 19 ml n-butanol, then evaporated and crystallized as described above. Identity and purity of the product was confirmed by thin-layer chromatography (R_f=0.63), NMR and GC/MS measurements (important ions of the permethyl derivative: m/z 233, 218, 189, 162, 134, 107, 100).

e) D_2-6-Benzylaminopurine Riboside. D_2-benzylamine (1.2 mmol) and 6-chloropurine riboside (0.6 mmol) were refluxed in 12 ml n-butanol for 1 h, then evaporated and crystallized as described above. Thin-layer chromatography (R_f=

0.59), NMR and GC/MS analysis (trimethylsilyl derivate, m/z 575, 560, 485, 349, 348, 259, 256, 243, 230, 226, 217, 108, 93) showed the identity and expected D_2 incorporation of the product.

4.4.3 Practical Approach

Ten g anise cells were purified as outlined for the RIA. As an internal standard 100 ng or 20 ng D_2-isopentenyladenosine was added to the tissue prior to the extraction. The measured and corrected ratio m/z 551 / m/z 553 was 0.76 and 1.71, respectively. This yielded a mean value of 2.9 ng isopentenyladenosine g^{-1} fresh weight. As mentioned in Sec. 4.3, isopentenyladenosine showed a signal at m/z 553, due to isotopic impurities. A correction was necessary to avoid an overestimation of D_2-isopentenyladenosine in the mixture of both compounds. From 15 SIM measurements of isopentenyladenosine it was calculated that the area of m/z 553 was, on average, 16% of the area of m/z 551. Hence the ratio (R) m/z 551 / m/z 553 was calculated as follows:

$$R = \frac{\text{peak area m/z 551}}{\text{peak area m/z 553} - 16/100 \times \text{peak area m/z 551}}$$

The calibration plot of the corrected values was identical to that for the ions m/z 232 and m/z 234 (Fig. 12). In another experiment, 17 g anise cells of the log-growth phase were extracted in the presence of 208 ng D_2-isopentenyladenosine and 156 ng D_2-isopentenyladenine, showing a ratio of 1.4 for m/z 551/m/z 553 (isopentenyladenosine) and of 0.41 for m/z 231/m/z 233 (isopentenyladenine). This yielded 9.6 ng isopentenyladenosine and 2.2 ng isopentenyladenine g^{-1} fresh weight.

Quantitations with D_2-6-benzylaminopurine riboside are described by Ernst et al. (1983b). Examples for the zeatin group are given by Hashizume et al. (1979), Horgan et al. (1981), and Summons et al. (1981).

5 Concluding Remarks

For the identification and quantitation of cytokinins the methods of choice are RIA in combination with HPLC or GC/MS using MID. The detection limit of the RIA depends on the nature of antibody, as well as on the specificity of the labelled cytokinin. Usually the detection limit is in the range of 10–100 pg. The RIA is a very rapid and sensitive method, easy to carry out for an investigator. As antisera are elicited against a cytokinin riboside-dialcohol, cross-reactions with other members of each cytokinin class may occur. Therefore a purification procedure, e.g. HPLC, is absolutely necessary for the determination of the cytokinin. Such a purification is as time-consuming as the GC/MS method (see Sects. 3.5.5 and 4.4.1). Losses of cytokinins will occur during extraction, for which correction must be made. In RIA this may be done using a radioactive-labelled analogue, (or preferably a homologous), compound. In GC/MS measurements losses during purification do not affect the accuracy of the method because quantitation is based on stable isotopic dilution. The detection limit of GC/MS is not as good as for the RIA, ranging from 0.5–5 ng. An exact identification of the chemical nature of the respective cytokinin is possible depending on the retention time, the molecular ion and characteristic fragment ions. Identification of a new, naturally occurring cytokinin is possible only by GC/MS, according to the interpretation of the mass spectrum. However, the sophisticated and expensive equipment required for GC/MS makes such analysis impractical for many investigators.

A comparison of cytokinin concentrations determined by RIA as well as by GC/MS is given in Table 4. It is quite clear that these two methods give comparable results. This is especially true for authentic tissue, analyzed by the two methods. In the author's laboratory analysis of anise tissue by GC/MS and RIA yielded isopentenyladenosine levels of 2.9 ng and 3.2 ng g^{-1} fresh weight, respectively. Cytokinin concentrations in *Vinca rosea* crown gall tissue, which was developed by infection with *Agrobacterium tumefaciens*, were also comparable when determined by these two methods. The slightly higher values obtained by the RIA might be due to the use of different tissues. Weiler and Spanier (1981) used 21-day-old plants infected with *Agrobacterium tumefaciens* B6, whereas Scott and Horgan (1980) used a culture of *Vinca rosea* crown gall tissue. Furthermore, Weiler and Spanier (1981) analyzed the crude extract, and it was shown by Scott et al. (1982) that an extract of *Vinca rosea* crown gall tissue also contained zeatin 9-glucoside and zeatin, compounds cross-reacting with a zeatin riboside antiserum (Badenoch-Jones et al. 1984). Horgan et al. (1981) found by GC/MS analysis a considerable variation in the level of zeatin riboside in crown gall tissue depending on the age of the tissue. The average level in young tissue was 50 ng g^{-1} fresh weight, while in mature tissue it was 400 ng g^{-1} fresh weight. These data correspond to those reported by Weiler and Spanier (1981) for a RIA after thin-layer chromatography of a crude methanolic tissue extract of infected plants (79 ng g^{-1} fresh weight). The t-zeatin content in the culture medium of *Agrobacterium tumefaciens* C58, analyzed by Regier and Morris (1982), also showed a good similarity between the RIA and the GC/MS. Surprisingly, Badenoch-Jones et al. (1984) found no t-zeatin in the medium. However, Badenoch-Jones et al. (1984) detected 720 ng t-zeatin riboside l^{-1} medium during the mid-log phase in the medium of *Agrobacterium tumefaciens* C58. In contrast, Regier and Morris (1982) found only traces of t-zeatin riboside irrespective of the growth phase. McCloskey et al. (1980) reported a content of 310 ng l^{-1} of t-zeatin riboside at the stationary phase using GC/MS techniques. The reasons for these discrepancies in the data are not clear. Possible explanations might be (1) the investigators analyzed different strains of *Agrobacterium tumefaciens;* (2) the culture conditions used were different; or (3) the initial cell density was different. The sets of data for *Datura innoxia* crown gall tissue are very similar as the analysis were done with the same tissue. The slightly lower values reported by Palni et al. (1983) using GC/MS when compared to the results obtained by Badenoch-Jones et al. (1984) using a RIA are due to the addition of the labelled compounds to partially purified fractions and not at the beginning of the extraction. The values for zeatin

Table 4. A comparison of the cytokinin concentrations as determined by GC/MS in different tissues, with the concentrations determined by RIA[a]

Cytokinin	*Pimpinella anisum* L. Cell culture; Present study		*Catharanthus roseus* L. Normal tissue	*Catharanthus roseus* L. Crown gall tissue	*Catharanthus roseus* L. Normal tissue	*Catharanthus roseus* L. Crown gall tissue
	GC/MS	RIA	GC/MS; Scott and Horgan (1980) (ng g^{-1} FW)		RIA; Weiler and Spanier (1981)	
i^{6}Ado	2.9	3.2	ND	ND	18	65
t-io^{6}Ado	ND	ND	2.5	402	8	667

	Agrobacterium tumefaciens C58 Culture medium GC/MS;		*Agrobacterium tumefaciens* C58 Culture medium RIA;	
	McCloskey et al. (1980)	Regier and Morris (1982) (ng l^{-1})	Regier and Morris (1982)	Badenoch-Jones et al. (1984)
t-io^{6}Ade	ND	800	200–800	n. d.
t-io^{6}Ado	310	ND	Traces	720
c-io^{6}Ado	1100	ND	ND	ND

	Datura innoxia Crown gall tissue			*Lupinus luteus* L. seeds		*Zea mays* kernels	
	GC/MS; Palni et al. (1983)	Summons et al. (1983)	RIA; Badenoch-Jones et al. (1984) (ng g^{-1} FW)	GC/MS; Summons et al. (1981)	RIA; Badenoch-Jones et al. (1984)	GC/MS; Summons et al. (1979)	RIA; Badenoch-Jones et al. (1984)
t-io^{6}Ado	78	ND	81	393	576	530	152
t-io^{6}Ade	27	ND	46	18.5	10	220	30
io^{6}h^{2}Ado	7	ND	NA	666	1385	ND	NA
t-io^{6}Ade-9-glu	250	ND	282	n. d.	n. d.	15	78
t-io^{6}Ado-5′-P	47	64	82	ND	120	ND	212

[a] *i^{6}Ado* = isopentenyladenosine; *t-io^6 Ade* = t-zeatin; *c/t-io^{6}Ado* = c/t-zeatin riboside; *io^{6}h^{2}Ado* = dihydrozeatin riboside; *t-io^{6}Ade-9-glu* = t-zeatin 9-glucoside; *io^{6}Ado-5′-P* = t-zeatin riboside 5′-monophosphate; *NA* = not possible to quantitate accurately; *ND* = not determined; *n. d.* = not detected; *FW* = fresh weight. (Part of the data was adapted from Badenoch-Jones et al. 1984).

riboside 5′-monophosphate are in good agreement with those of Summons et al. (1983), who found 64 ng g^{-1} fresh weight using the labelled compound. The differences for GC/MS and RIA in samples of *Lupinus luteus* seeds are not great and may depend on a different seed age. Badenoch-Jones et al. (1984) used the term immature seeds, whereas Summons et al. (1981) used the term developing seeds. Furthermore, Summons et al. (1981) demonstrated a change in the cytokinin content during seed development. For the *Zea mays* study the consistency of the two methods is not as good. The great difference, up to a factor of seven for zeatin, may be based on (1) different seed age or (2) most probably, a difference in the plant variety used. Badenoch-Jones et al. (1984) analyzed *Zea mays*, FI hybrid Iochief, while Summons et al. (1979a) used *Zea mays*. However, the possibility that different varieties of one species will show a different cytokinin content cannot be excluded.

A valid comparison between RIA and GC/MS data can only be made using identical plant material. Doing so the two methods agree very well, as was shown by Ernst et al. (1984) for the determination of endogenous levels of isopentenyladenosine in an anise cell culture during development with a maximum level during logarithmic growth (RIA: 58 ng, GC/MS: 47 ng g^{-1} fresh weight). Similar results are shown in the present study for *Pimpinella anisum* L. and for *Datura innoxia* crown gall tissue (Palni et al. 1983, Badenoch-Jones et al. 1984) (Table 4).

From the data presented here it is evident that RIA and GC/MS are very useful techniques in the determination and quantitation of cytokinins. Which of the two methods is preferred depends on the investigator.

Acknowledgements. I am grateful to Prof. D. Oesterhelt and Prof. W. Schäfer, Martinsried, for their helpful discussions and constructive comments on the manuscript. I also wish to thank Dr. T. Mock and Dr. K. Woolley for improving the English. This work was supported by a grant of the Max-Planck-Gesellschaft.

References

Andersen RA, Kemp TR (1979) Reversed-phase high-performance liquid chromatography of several plant cell division factors (cytokinins) and their cis and trans isomers. J Chromatogr 172:509–512

Badenoch-Jones J, Letham DS, Parker CW, Rolfe BG (1984) Quantitation of cytokinins in biological samples using antibodies against zeatin riboside. Plant Physiol (Bethesda) 75:1117–1125

Brandon DL, Corse JW, Layton LL(1979) Reagents for immunoassay of cytokinins. Plant Physiol (Bethesda) 63(S):82

Brenner LM (1981) Modern methods for plant growth substances analyses. In: Briggs WR, Green PG, Jones RL (eds) Annu Rev Plant Physiol, vol 32. Annual Reviews, Palo Alto California, USA, pp 511–538

Carnes MG, Brenner ML, Andersen CR (1975) Comparison of reversed-phase high-pressure liquid chromatography with Sephadex LH-20 for cytokinin analyses of tomato root pressure exudate. J Chromatogr 108:95–106

Chard T (1978) An introduction to radioimmunoassay and related techniques. In: Work TS, Work E (eds) Laboratory techniques in biochemistry and molecular biology. North-Holland, Amsterdam

Chaves das Neves HJ, Pais MSS (1980) A new cytokinin from the fruits of *Zantedeschia aethiopica*. Tetrahedron Lett 21:4387–4390
Dauphin B, Teller G, Durand B (1979) Identification and quantitative analyses of cytokinins from shoot apices of *Mercurialis ambigua* by gas chromatography-mass spectrometry computer system. Planta (Berl) 144:113–119
Dauphin-Guerin B, Teller G, Durand B (1980) Different endogenous cytokinins between male and female *Mercurialis annua* L. Planta (Berl) 148:124–129
Edwards CA, Armstrong DJ, Kaiss-Chapmann RW, Morris RO (1981) Cytokinin-active ribonucleosides in *Phaseolus* RNA. Plan Physiol (Bethesda) 67:1181–1184
Erlanger BF, Beiser SM (1964) Antibodies specific for ribonucleosides and ribonucleotides and their reaction with DNA. Proc Natl Acad Sci USA 52:68–74
Ernst D (1983) Endogene Cytokinine in einer Zellsuspensionskultur von Anis (*Pimpinella anisum* L.) Thesis, Ludwig-Maximilians-Universität, München
Ernst D, Oesterhelt D (1985) Changes of cytokinin nucleotides in an anise cell culture (*Pimpinella anisum* L.) during growth and embryogenesis. Plant Cell Rep 4:140–143
Ernst D, Schäfer W, Oesterhelt D (1983a) Isolation and quantitation of isopentenyladenosine in an anise cell culture by single-ion monitoring, radioimmunoassay and bioassay. Planta (Berl) 159:216–221
Ernst D, Schäfer W, Oesterhelt D (1983b) Isolation and identification of a new, naturally occuring cytokinin (6-benzylaminopurineriboside) from an anise cell culture (*Pimpinella anisum* L.).Planta (Berl) 159:222–225
Ernst D, Oesterhelt D, Schäfer W (1984) Endogenous cytokinins during embryogenesis in an anise cell culture (*Pimpinella anisum* L.). Planta (Berl) 161:240–245
Hacker B, Van Vunakis H, Levine L (1972) Formation of an antibody with serologic specificity for $N^6(\Delta^2$-isopentenyl) adenosine. J Immunol 6:1726–1728
Hall RH, Robins MJ (1968) N-(3-methyl-2-butenyl)adenine. In: Zorbach WW, Tibson RS (eds) Synthetic procedures in nucleic acid chemistry, vol 1. Interscience, New York, pp 11–12
Hardin JM, Stutte CA (1981) Analyses of plant hormones using high-performance liquid chromatography. J Chromatogr 208:124–128
Hashizume T, McCloskey JA (1976) Electron impact-induced reactions of N^6-(3-methyl-2-butenyl)adenosine and related cytokinins. Biomed Mass Spectrom 3:177–183
Hashizume T, Sugiyama T, Imura M, Cory HT, Scott MF, McCloskey JA (1979) Determination of cytokinins by mass spectrometry based on stable isotope dilution. Anal Biochem 92:111–122
Hashizume T, Suye S, Sugiyama T (1982) Isolation and identification of cis-zeatin riboside from tubers of sweet potato (*Ipomoea batatas* L.). Agric Biol Chem 46:663–665
Holland JA, McKerrell EH, Fuell KJ, Burrows WJ (1978) Separation of cytokinins by reversed-phase high-performance liquid chromatography. J Chromatogr 166:545–553
Horgan R (1978) Analytical procedures for cytokinins. In: Hillman JR (ed) Isolation of plant growth substances. Cambridge University Press, pp 97–114 (Soc Exp Biol, seminar ser 4)
Horgan R, Kramers MR (1979) High-performance liquid chromatography of cytokinins. J Chromatogr 173:263–270
Horgan R, Hewett EW, Purse JG, Horgan JM, Wareing PF (1973) Identification of a cytokinin in sycamore sap by gas chromatography-mass spectrometry. Plant Sci Lett 1:321–324
Horgan R, Hewett EW, Horgan JM, Purse J, Wareing PF (1975) A new cytokinin from *Populus robusta*. Phytochemistry 14:1005–1008
Horgan R, Palni LMS, Scott I, McGaw B (1981) Cytokinin biosynthesis and metabolism in *Vinca rosea* crown gall tissue. In: Guern J, Peaud-Lenoel C (eds) Metabolism and molecular activities of cytokinins. Springer, Berlin Heidelberg New York, pp 56–65
Humayun MZ, Jacob TM (1974) Specificity of anti-nucleoside antibodies. Biochem J 141:313–315
Jayabaskaran C, Jacob TM (1982) Isolation and estimation of cytokinins and cytokinin-containing transfer RNAs from *Cucumis sativus* L. var. Guntur seedlings. Plant Physiol (Bethesda) 70:1396–1400

Kemp TR, Andersen RA (1981) Separation of modified bases and ribonucleosides with cytokinin activity using fused silica capillary gas chromatography. J Chromatogr 209:467–471

Kemp TR, Andersen RA, Oh J, Vaughn TH (1982) High-resolution gas chromatography of methylated ribonucleosides and hypermodified adenosines. Evalution of trimethylsilyl derivatization and split and splitless operation modes. J Chromatogr 241:325–332

Kemp TR, Andersen RA, Oh J (1983) Cytokinin determination in tRNA by fused-silica capillary gas chromatography and nitrogen-selective detection. J Chromatogr 259:347–349

Khan SA, Humayun MZ, Jacob TM (1977) A sensitive radioimmunoassay for isopentenyladenosine. Anal Biochem 83:632–635

Lanyi JK, Oesterhelt D (1982) Identification of the retinal-binding protein in halorhodopsin. J Biol Chem 257:2674–2677

Ludewig M, Dörffling K, König WA (1982) Electron-capture capillary gas chromatography and mass spectrometry of trifluoroacetylated cytokinins. J Chromatogr 243:93–98

MacDonald EMS, Moris RO (1983) Rapid, single-pass isolation of analytically pure plant hormones by immunoaffinity chromatography. Plant Physiol (Bethesda) 72(S):26

MacDonald EMS, Akiyoshi DE, Morris RO (1981) Combined high-performance liquid chromatography-radioimmunoassay for cytokinins. J Chromatogr 214:101–109

McCloskey JA (1974) Mass spectrometry. In: Ts'o POP (ed) Basic principles in nucleic acid chemistry, vol 1. Academic Press, New York, pp 209–309

McCloskey JA, Hashizume T, Basile B, Sugiyama T, Sekiguchi S (1979) Determination of cytokinins in bamboo shoots by mass spectrometry using selected ion monitoring. Proc Jpn Acad (Ser B) 55:445–450

McCloskey JA, Hashizume T, Basile B, Ohno J, Sonoki S (1980) Occurrence and levels of cis- and trans-zeatin ribosides in the culture medium of a virulent strain of *Agrobacterium tumefaciens*. FEBS Lett 111:181–183

McCloskey JA, Basile B, Kimura K, Hashizume T (1981) Presence of levels of cis- and trans-methylthiozeatin riboside and cis- and trans-zeatin riboside in tobacco plant determined by mass spectrometry. Proc Jpn Acad (Ser B) 57:276–281

McGaw BA, Heald JK, Horgan R (1984) Dihydrozeatin metabolism in radish seedlings. Phytochemistry (Oxf) 23:1373–1377

Message GM (1984) Practical aspects of gas chromatography/mass spectrometry. Wiley, New York

Millard BJ (1978) Quantitative mass spectrometry. Heyden, London

Milstone DS, Vold BS, Glitz DG, Shutt N (1978) Antibodies to N^6-(Δ^2-isopentenyl)-adenosine and its nucleotide: interaction with purified tRNAs and with bases, nucleosides and nucleotides of the isopentenyladenosine family. Nucleic Acids Res 5:3439–3455

Minden DL, McCloskey JA (1973) Mass spectrometry of nucleic acid components. N,O-permethyl derivatives of nucleosides. J Am Chem Soc 95:7480–7489

Morris RO (1977) Mass spectroscopic identification of cytokinins. Plant Physiol (Bethesda) 59:1029–1033

Morris RO, Zaerr JB, Chapman RW (1976) Trace enrichment of cytokinins from Douglas-fir xylem exudate. Planta (Berl) 131:271–274

Morris RO, Regier DA, Olson RM, JR, Struxness LA, Armstrong DJ (1981) Distribution of cytokinin-active nucleosides in isoaccepting transfer ribonucleic acids from *Agrobacterium tumefaciens*. Biochemistry 20:6012–6017

Most BH, Williams JC, Parker KJ (1968) Gas chromatography of cytokinins. J Chromatogr 38:136–138

Palni LMS, Horgan R (1983) Cytokinins in transfer RNA of normal and crown-gall tissue of *Vinca rosea*. Planta (Berl) 159:178–181

Palni LMS, Summons RE, Letham DS (1983) Mass spectrometric analyses of plant tissues. Plant Physiol (Bethesda) 72:858–863

Palni LMS, Palmer MV, Letham DS (1984) The stability and biological activity of cytokinin metabolites in soybean callus tissue. Planta (Berl) 160:242–249
Reeve DR, Crozier A (1980) Quantitative analyses of plant hormones. In:MacMillan J (ed) Hormonal regulation of development I. Molecular aspects of plant hormones. Springer, Berlin Heidelberg New York, pp 203–280 (Encycl Plant Physiol, New Ser Vol 9)
Regier DA, Morris RO (1982) Secretion of trans-zeatin by *Agrobacterium tumefaciens:* A function determined by the nopaline Ti plasmid. Biochem Biophys Res Commun 104:1560–1566
Rodbard D (1974) Statistical quality control and routine data processing for radioimmunoassay and immunoradiometric assays. Clin Chem 20/10:1255–1270
Rose ME, Johnstone RAW (1982) Mass spectrometry for chemists and biochemists. Cambridge University Press, Cambridge
Scott IM, Horgan R (1980) Quantification of cytokinins by selected ion monitoring using ^{15}N labelled internal standards. Biomed Mass Spectrom 7:446–449
Scott IM, Horgan R (1982) High-performance liquid chromatography of cytokinin ribonucleoside 5′-monophosphates. J Chromatogr 237:311–315
Scott IM, Horgan R (1984) Mass-spectrometric quantification of cytokinin nucleotides and glycosides in tobacco crown-gall tissue. Planta (Berl) 161:345–354
Scott IM, Horgan R, McGaw BA (1980) Zeatin 9-glucoside, a major endogenous cytokinin of *Vinca rosea* L. crown gall tissue. Planta (Berl) 149:472–475
Scott IM, Martin GC, Horgan R, Heald JK (1982) Mass spectrometric measurements of zeatin glycoside levels in *Vinca rosea* L. crown gall tissue. Planta (Berl) 154:273–276
Shindy WW, Smith OE (1975) Identification of plant hormones from cotton ovules. Plant Physiol (Bethesda) 55:550–554
Stafford AE, Corse J (1982) Fused-silica capillary gas chromatography of permethylated cytokinins with flame-ionization and nitrogen-phosphorus detection. J Chromatogr 247:176–179
Sugiyama T, Suye S, Hashizume T (1983) Mass spectrometric determination of cytokinins in young sweet-potato plants using deuterium-labeled standards. Agric Biol Chem 47:315–318
Summons RE, MacLeod JK, Parker CW, Letham DS (1977) The occurence of raphanatin as an endogenous cytokinin in radish seed. FEBS Lett 82:211–214
Summons RE, Colin CC, Eichholzer JV, Entsch B, Letham DS, MacLeod JK, Parker CW (1979a) Mass spectrometric analysis of cytokinins in plant tissues. Biomed Mass Spectrom 6:407–413
Summons RE, Entsch B, Parker CW, Letham DS (1979b) Mass spectrometric analyses of cytokinins in plant tissues. FEBS Lett 107:21–25
Summons RE, Entsch B, Letham DS, Gollnow BI, MacLeod JK (1980) Regulators of cell division in plant tissues. Planta (Berl) 147:422–434
Summons RE, Letham DS, Gollnow BI, Parker CW, Entsch B, Johnson LP, MacLeod JK, Rolfe BG (1981) Cytokinin translocation and metabolism in species of the leguminoseae: studies in the relation of shoot and nodule development. In: Guern J, Peaud-Lenoel C (eds) Metabolism and molecular activities of cytokinins. Springer, Berlin Heidelberg New York, pp 69–79
Summons RE, Palni LMS, Letham DS (1983) Determination of intact zeatin nucleotide by direct chemical ionisation mass spectrometry. FEBS Lett 151:122–126
Taylor JS, Koshioka M, Pharis RP, Sweet GB (1984) Changes in cytokinins and gibberellin-like substances in *Pinus radiata* buds during lateral shoot initiation and the characterization of ribosyl zeatin and a novel ribosyl zeatin glycoside. Plant Physiol (Bethesda) 74:626–631
Trione EJ, Morris RO (1983) Monoclonal antibodies against cytokinins. Plant Physiol (Bethesda) 72(S):114
Tsoupras G, Luu B, Hoffmann JA (1983) A cytokinin (isopentenyl-adenosylmononucleotide) linked to ecdysone in newly laid eggs of *Locusta migratoria*. Science 220:507–509
Van Vunakis H (1980) Radioimmunoassay: An overview. In: Van Vunakis H, Langone JL (eds) Immunochemical techniques part A. Academic Press, New York, pp 201–209 (Methods Enzymol, Vol. 70)

Vold SB, Leonard NJ (1981) Production and characterization of antibodies and establishment of a radioimmunoassay for ribosylzeatin. Plant Physiol (Bethesda) 67:401–403
Weiler EW (1980) Radioimmunoassays for trans-zeatin and related cytokinins. Planta (Berl) 149:155–162
Weiler EW (1982) Plant hormone immunoassay. Physiol Plant 54:230–234
Weiler EW, Spanier K (1981) Phytohormones in the formation of crown gall tumors. Planta (Berl) 153:326–337
Weiler EW, Ziegler H (1981) Determination of phytohormones in phloem exudate from tree species by radioimmunoassay. Planta (Berl) 152:168–170
Windholz M (1976) The Merck Index, 9th edition. Merck, Rahay, NY, USA
Woodsworth ML, Latimer LJP, Janzer JJ, McLennan BD, Lee JS (1983) Characterization of monoclonal antibodies specific for isopentenyladenosine derivatives occuring in transfer RNA. Biochem Biophys Res Commun 114:791–796
Yalow RS, Berson SA (1960) Immunoassay of endogenous plasma insulin in man. J Clin Invest 39:1157–1175
Yamane H, Watanabe M, Satoh Y, Takahashi N, Iwatsuki K (1983) Identification of cytokinins in two species of pteridophyte sporophytes. Plant Cell Physiol 24:1027–1031
Yokota T, Murofushi N, Takahashi N (1980) Extraction, purification and identification. In: MacMillan (ed) Hormonal regulation of development I. Molecular aspects of plant hormones. Springer, Berlin Heidelberg New York, pp 113–201 (Encycl Plant Physiol, New Series, Vol 9)
Young H (1977) Identification of cytokinins from natural sources by gas liquid chromatography/mass spectrometry. Anal Biochem 79:226–233
Zaerr JB, Akiyoshi D, MacDonald EMS, Morris RO (1981) Combined high performance liquid chromatography-radioimmune assay for cytokinins: a rapid and sensitive substitute for bioassay. Plant Physiol (Bethesda) 67(S):101
Zaerr JB, Durley RC, Morris RO (1983) Application of automated immunoaffinity chromatography to the analyses of cytokinins. Plant Physiol (Bethesda) 72(S):26

Immunodetection of Phytochrome: Immunocytochemistry, Immunoblotting, and Immunoquantitation [1]

L. H. PRATT, D. W. MCCURDY, Y. SHIMAZAKI, and M.-M. CORDONNIER

1 Introduction

Although as recently as 10 to 15 years ago plant scientists seldom made use of antibodies as research tools, that is no longer the case. Antibodies are quickly becoming indispensible for many purposes, including one-step purification of antigens, their visualization in situ by immunocytochemistry and on nitrocellulose blots of polyacrylamide gels, and their quantitation by radioimmunoassay or enzyme immunoassay. The relatively recent development of monoclonal antibodies produced by hybridoma technology (Köhler and Milstein 1975; see Goding 1983 for thorough treatment) expands the utility of antibodies as research tools by at least an order of magnitude. Whereas a typical antiserum, which is conveniently referred to as a polyclonal antibody preparation, can be specific for a given antigen, a monoclonal antibody is specific for a given domain, known as an antigenic determinant or epitope, on that antigen. Consequently, while polyclonal antibodies provide information about the location or quantity of an antigen, a monoclonal antibody provides information about the location or quantity of a small portion of that antigen. Furthermore, a panel of monoclonal antibodies directed to a single, complex antigen, such as a protein, can be used to dissect that antigen and to identify its structure-function relationships.

1.1 Antibody Specificity

To take maximum advantage of antibodies as research tools it is essential that each antibody preparation be appropriately tested for specificity. The controls that are often performed for this purpose are inadequate in many instances. In particular, while Ouchterlony double immunodiffusion and immunoelectrophoresis are often convenient preliminary tests of specificity, they are so insensitive compared to applications such as those described here that they are inadequate by themselves. Moreover, polyclonal and monoclonal antibodies pose such different problems regarding specificity that they need to be considered independently.

[1] *Abbreviations:* ELISA, enzyme-linked immunosorbent assay; Ig, immunoglobulin; kD, kilodalton; PAGE, polyacrylamide gel electrophoresis; PBS, 10 mM sodium phosphate, 140 mM NaCl, pH 7.4; Pr and Pfr, phytochrome in the red- and far-red-absorbing forms, respectively; SDS, sodium dodecyl sulfate.

Polyclonal Antibodies. It can be exceedingly difficult to establish the specificity of a polyclonal antibody preparation to a degree that is appropriate for the applications described here. Specific controls relevant to each assay will be mentioned where appropriate. It is important at the outset, however, to recognize that while contaminating antibodies might be present at very low concentration relative to the antibodies of interest, they may nevertheless become a problem, especially when the antigen of interest is of low abundance, while the antigen recognized by the contaminating antibodies is of high abundance. The most useful preliminary step, therefore, can be to purify potentially monospecific antibodies from an antiserum with an affinity column made from highly purified antigen (e.g., Pratt 1984a). Even in this instance, however, the monospecificity of the antibodies is inherently limited by the purity of the antigen with which the affinity column was made. Contaminants in an antigen preparation that are too low to detect by common methods, such as sodium dodecyl sulfate polyacrylamide gel electrophoresis (SDS-PAGE), may still be sufficiently abundant to purify enough contaminating antibodies to produce unwanted activity in the final assay.

Ultimately, ad hoc controls, designed around unique characteristics of the antigen of interest, are needed to establish specificity. For example, in the case of phytochrome its photoreversible redistribution within the cell (Mackenzie et al. 1975) and its rapid depletion from etiolated tissue following initial photoconversion from Pr to Pfr (i.e., phytochrome in the red- and far-red-absorbing forms, respectively; Shimazaki et al. 1983) make it possible to devise good internal controls for the specificity of an antibody preparation, since it is highly unlikely that an unrelated antigen would respond to red and far-red light in the same way.

Monoclonal Antibodies. Controls for monoclonal antibodies are for a different purpose than are those for polyclonal antibodies. If a monoclonal antibody is shown to be directed to a given antigen, then one can be absolutely certain that it is specific for its epitope on that antigen (Goding 1983). Nevertheless, if that same epitope happens to exist on another unrelated antigen, whether because of functional homology or by chance, then the antibody is not specific for the antigen to which it is directed. Even though antibodies similar to the hypothetical case just mentioned are likely to be present in polyclonal antibody preparations, they would normally be present in such small proportion that they would go undetected. In pure form, as in the case of a monoclonal antibody, however, the unwanted activity would be detected readily.

The often-used control with a polyclonal antibody preparation, which is to adsorb the preparation with demonstratably pure antigen and to determine whether it still exhibits activity in the assay in question, is of no value with a monoclonal antibody. Rather, one must again rely on ad hoc controls that derive from some unique property of the antigen and/or on other controls. For example, if one documents that three different monoclonal antibodies bind to different epitopes on the same antigen, and that all three yield the same outcome in a given assay, then one can have confidence that only the antigen of interest is being detected. It would be highly unlikely that all three antibodies would independently exhibit the same antigenic nonspecificity.

1.2 Immunodetection Methods

The three methods described here, while perhaps appearing superficially unrelated, are in principle quite similar. Each of these methods, immunocytochemistry, immunoblotting, and immunoquantitation, relies on the high affinity and specificity of antibodies to visualize a polypeptide antigen in the presence of a high background of unrelated protein. While immunocytochemistry and immunoblotting are at best only semiquantitative, immunoquantitation can be highly precise.

Immunocytochemistry was the first of the three methods to be introduced. Immunoblotting effectively applies the principles of immunocytochemistry to selective staining of an antigen on nitrocellulose rather than in a tissue section. Immunoquantitation is also in some respects a logical extension of immunocytochemistry. In the protocol described here, the antigen to be "stained" is artificially applied to a solid support (the surface of an enzyme-linked immunosorbent assay (ELISA) well in the application described below) and antibodies are then used for its visualization by a protocol that produces label in proportion to the amount of antigen that is bound to the solid surface. Thus, while each of these three methods provides different information about an antigen, they are conceptually similar. Presentation of the three together is intended in part to emphasize the common aspects of the methods.

2 Immunocytochemistry

2.1 Rationale

Immunocytochemistry provides the unique capability of visualizing in situ virtually any antigen at both tissue and subcellular levels. Such information can often provide important insights into the function of the antigen under investigaion. For example, the immunocytochemical detection of phytochrome in several grass species (Pratt and Coleman 1971, 1974; Coleman and Pratt 1974) has yielded valuable information concerning the cellular distribution of the pigment with a degree of resolution not attainable by more conventional phytochrome detection methods (e.g., Briggs and Siegelman 1965).

The most important criteria for successful immunocytochemistry are: (1) a fixation protocol that not only provides good structural preservation of the tissue, but also preserves and immobilizes antigenic sites within that tissue, and (2) a method of tissue preparation and sectioning that allows adequate penetration of antibodies to these antigenic sites. Within these two constraints, many different methodological approaches have proven useful. These useful experimental protocols vary widely in terms of fixation conditions and the type of tissue embedding used for sectioning, as well as in immunolabeling techniques, whether pre- or postembedding. The success of any one method depends entirely on the properties of the antigen under investigation, properties such as the relative abundance of the antigen, and its retention of antigenic activity following fixation (Sternberger 1979; Knox et al. 1980; Bullock and Petrusz 1982).

We are currently employing freeze-sectioning of formaldehyde-fixed tissue coupled with fluorescence microscopy for the immunolocalization of phytochrome. The requirements presented above for successful immunocytochemistry are satisfied by this approach. The fixation protocol retains sufficient antigenic activity to enable subsequent identification by monoclonal antibodies to oat phytochrome. In addition, the cryosectioning procedure, which uses only sucrose as both a support medium and cryoprotectant during sectioning, provides good access of antibodies to antigenic sites within the tissue. This approach eliminates potential problems imposed by a lack of penetration of antibodies into sections of tissue embedded in methacrylate or epoxy resins. The thin sectioning also exposes antigenic sites in the tissue that might otherwise have been enclosed by cellular membranes. Moreover, fluorescence microscopy offers several advantages over the commonly employed enzyme-linked immunocytochemical procedures (for reviews see Sternberger 1979; Bullock and Petrusz 1982), which use conventional bright field microscopy. The former is not limited by the resolution capabilities of transmitted light microscopy, and thus facilitates light microscopic observation of sequestered phytochrome (see below).

Phytochrome is a relatively low abundance protein, even in etiolated tissue. To visualize this protein successfully it is therefore essential to establish an immunostaining protocol that produces a high phytochrome-specific signal against a low background of nonspecific staining. To achieve this goal we use a three-step immunostaining protocol (Fig. 1) that comprises first a mixture of three mouse monoclonal antibodies against phytochrome (MAP), followed by a bridge of rabbit antibodies to mouse IgG (RAM), and finally rhodamine-conjugated goat an-

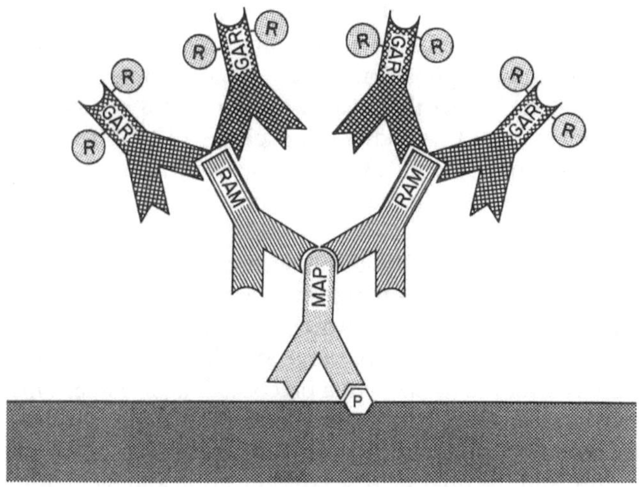

Fig. 1. Protocol for immunolabeling phytochrome in frozen thin sections. *P* = phytochrome in the tissue section; *MAP* = mouse-derived monoclonal antibody to phytochrome: *RAM* = rabbit antibodies to mouse IgG; *GAR* = goat antibodies to rabbit IgG; *R* = rhodamine. Use of RAM allows the attachment of more rhodamine per antigenic site

tibodies to rabbit IgG (GAR-R). The bridge antibodies (RAM) amplify the phytochrome-specific signal by enabling the attachment of more rhodamine per antigenic site. Also, the three monoclonal antibodies in the mixture each recognize a different antigenic site on the phytochrome polypeptide (Cordonnier et al. 1985), thereby enabling more than one monoclonal antibody to bind to each phytochrome molecule.

Reducing the level of nonspecific background stain also effectively increases immunocytochemical sensitivity. We routinely incubate sections with undiluted nonimmune lamb serum prior to the application of each antibody, and include nonimmune lamb serum in the diluent for all antibodies. These precautions effectively block sites that have nonantigenic affinity for immunoglobulins (Pratt and Coleman 1974), which solves a common problem when using plant tissue for immunocytochemistry. Also, both RAM and GAR-R are used at the highest concentrations that give no nonspecific staining. These concentrations are determined by first titrating with a series of GAR-R dilutions and choosing the highest concentration that can be used without giving nonspecific activity. Once the appropriate dilution of GAR-R is obtained, the optimal concentration of RAM is determined similarly.

2.2 Method

Freeze-Sectioning. Following appropriate irradiation of dark-grown tissue, split 3- to 4-day-old oat coleoptiles longitudinally and excise 1- to 2-mm apical segments with a sharp razor blade. Fix the tissue overnight on ice in 4% (w/v) formaldehyde in 0.1 M sodium phosphate, pH 7.4. Dilute the formaldehyde from a 10% stock, which is freshly prepared by dissolving paraformaldehyde in hot distilled water containing a few drops of 0.5 N NaOH. Cool the paraformaldehyde on ice, neutralize it with an equal volume of 0.5 N HCl, and filter it before using. Use a stock solution of 0.2 M sodium phosphate, pH 7.4, to accommodate the dilution of the 10% formaldehyde. Wash the fixed tissue for 6 to 8 h at 4 °C in 1.0 M sucrose, 0.4% formaldehyde, 0.1 M sodium phosphate, pH 7.4, followed by a further 3 to 4 h at room temperature in 1.3 M sucrose, 0.1 M sodium phosphate. The sucrose-infiltrated tissue is then ready for immediate cryosectioning or may be stored for 5 to 6 days at 4 °C.

Cryosection tissue with a Sorvall FTS Cryoultramicrotome according to the methods developed by Tokuyasu (1980). Mount fixed tissue pieces in a small drop of 1.3 M sucrose on the face of a copper block, which is then rapidly immersed in liquid nitrogen. Transfer the frozen block with tissue attached to the cryobowl of the microtome and roughly trim the tissue with a precooled razor blade. Final trimming of the tissue face is performed with a glass knife. Cut 1-µm thick sections with a new edge of the knife. We routinely use temperatures of −50 ° to −55 °C, but this temperature and the final sucrose concentration and incubation times used for tissue infiltration will vary depending on the type of tissue and the thickness of sections required.

Remove sections from the cryostage individually using a small droplet of sucrose (e.g., 1.7 M sucrose in 0.1 M sodium phosphate, pH 7.4) suspended by a

Table 1. Protocol for immunofluorescent localization of phytochrome in frozen thin sections using monoclonal antibodies[a]

Reagent	Incubation time (min)
2% Gelatin in PBS	10
PBS	10
Lamb serum	15
PBS-10% lamb serum	Rinse
Monoclonal antibodies to phytochrome, 1 μg ml^{-1} each	60
PBS-10% lamb serum	2×10
Lamb serum	10
PBS-10% lamb serum	Rinse
Rabbit antibody to mouse IgG, 1 μg ml^{-1}	30
PBS-10% lamb serum	2×10
Lamb serum	10
PBS-10% lamb serum	Rinse
Rhodamine-conjugated goat antibodies to rabbit IgG, diluted appropriately	30
PBS	3×15
Mount sections	

[a] All solutions contain 0.02% sodium azide

wire loop. Precool the sucrose drop in the cryobowl so that it is firm, but not frozen solid. Take care to ensure that the droplet does not touch the glass knife before the sucrose begins to freeze. If the sucrose drop is positioned properly, the section should "jump" from the knife and adhere to the sucrose. Bring the frozen section to room temperature before transferring it to a clean silicone-coated glass slide. Prepare slides for this purpose by spraying them with Slipicone (Dow Corning, Cat. No. 316), followed by extensive buffing with tissue paper to remove excess surface silicone.

Immunostaining of Frozen Sections. After transfer to glass slides, wash the sections gently with PBS (10 mM sodium phosphate, 140 mM NaCl, pH 7.4) to remove excess sucrose, and treat them with 2% (w/v) gelatin in PBS for 10 min to make them hydrophilic (Table 1). After a second 10-min wash with PBS, cover the sections with lamb serum (e.g., GIBCO Laboratories, 200-6070) and incubate for 15 min. Finally, rinse the sections with PBS containing 10% (v/v) lamb serum. Dilute all of the antibodies used for immunostaining with PBS-10% lamb serum. Add a mixture of three mouse monoclonal antibodies to oat phytochrome, each at 1 μg ml^{-1}. Treat control sections identically, except that nonimmune mouse IgG (e.g., Sigma, I-5381), used at 3 μg ml^{-1}, is substituted for the monoclonal antibody mixture. After a 1-h incubation, wash sections twice for 10 min each with PBS-10% lamb serum, 10 min with undiluted lamb serum and then rinse them with PBS-10% lamb serum. Add immunopurified rabbit antibodies to mouse IgG (Cordonnier et al. 1983) at 1 μg ml^{-1}. After a 30-min incubation at room temperature, wash the sections as before. Finally, apply rhodamine-labeled goat antibodies to rabbit IgG (e.g., Miles-Yeda, 61–266), at a 1:400 dilution.

After another 30-min incubation at room temperature, wash sections several times with PBS, mount them in 90% (v/v) glycerol, 10% PBS containing 5% (w/v) *n*-propyl gallate, and observe them with a fluorescence microscope (e.g., Zeiss IM) fitted with filter combinations for rhodamine visualization (e.g., Zeiss filter pack 487715). The propyl gallate retards photobleaching of rhodamine (Giloh and Sedat 1982).

Photomicrographs are recorded on Tri-X film (400 ASA) developed in HC-110 (dilution B), or Plus-X film (125 ASA) push developed in HC-110 (dilution B) to 1000 ASA with Factor 8 (Saunders et al. 1983). Exposure times for the fluorescence photomicrographs are determined empirically. Differential interference contrast images are taken using the automatic exposure capability of an Olympus OM-4 camera.

2.3 Controls

When using mouse monoclonal antibodies as the primary antibody, substitution with an equal concentration of nonimmune mouse Igs of the same isotype(s) serves as an appropriate control. As discussed in the Introduction, however, such a control cannot eliminate the possibility that a particular monoclonal antibody is by chance recognizing an epitope on an unrelated molecule. It becomes necessary, therefore, to develop controls that are specific to the properties of the antigen under investigation. In the case of phytochrome, each antibody is used individually to immunostain the protein in both its diffuse and red-light-induced sequestered condition (see next Section). If the same results are obtained with each of the antibodies, then it can reasonably be assumed that they are recognizing phytochrome specifically in the sections.

2.4 Evaluation of Method

Phytochrome-specific immunostaining in dark-grown *Avena* coleoptiles reveals a diffuse localization for this chromoprotein throughout the cytoplasm (Fig. 2). No immunodetectable phytochrome is present in plastids or nuclei, although in the latter case the outlines of nuclei are often made visible by the staining of phytochrome present in the thin layer of cytoplasm that surrounds these organelles (Fig. 2, arrow). In contrast to this diffuse distribution, exposure of the etiolated tissue to as little as 1 s of saturating red light (McCurdy and Pratt 1986) results in redistribution (termed "sequestering" by Mackenzie et al. 1975) of the chromoprotein into discrete areas within the cell (Fig. 3). These regions of sequestered phytochrome average about 1 μm in size, although some are larger (Fig. 3, arrows). They do not correspond with any organelles that can be identified at the light microscopic level. The resolution of immunofluorescence microscopy is demonstrated here by its ability to detect readily the small point sources of fluorescence that represent sequestered phytochrome. Control sections incubated with nonimmune mouse IgG show no comparable staining (Fig. 4), indicating the specificity that can be achieved by this protocol.

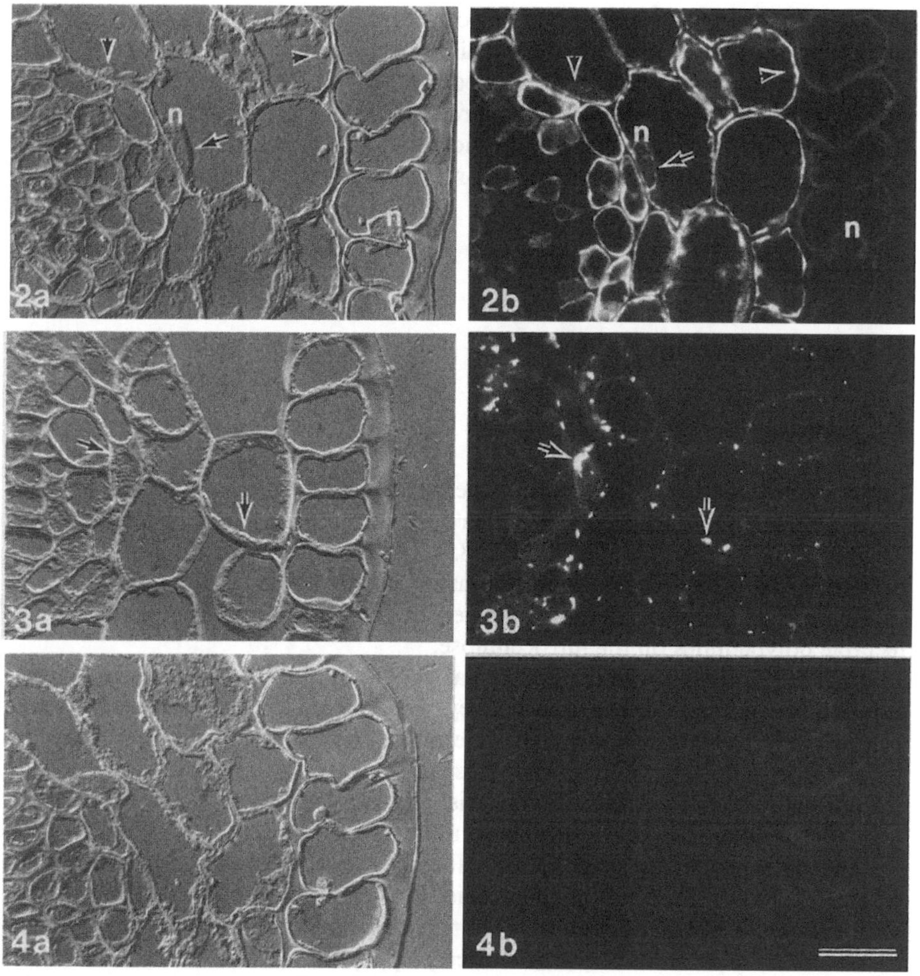

Figs. 2–4 (a, b). Paired differential interference contrast (**a**) and fluorescence (**b**) micrographs of 1 μm-thick sections of dark-grown *Avena* coleoptiles. *Bar* = 20 μm; × 550 (Adapted from McCurdy and Pratt 1986)

Fig. 2. Nonirradiated tissue immunolabeled with a mixture of monoclonal antibodies to phytochrome, Oat-25, Oat-22, and Oat-13 (1 μg ml^{-1} each). Nuclei (*n*) and plastids (*arrowheads*) do not stain for phytochrome. *Arrow* indicates staining of phytochrome in cytoplasm around a nucleus

Fig. 3. Red-irradiated (1 s red light, 10 s dark, 4 s far-red light) tissue immunolabeled with a mixture of Oat-25, Oat-22, and Oat-13 (1 μg ml^{-1} each). The size of the loci of sequestered phytochrome vary from less than 1 μm to greater than 3–4 μm (*arrows*)

Fig. 4. Control section incubated with nonimmune mouse IgG (3 μg ml^{-1})

The methods described here should be generally useful for immunocytochemical applications with plant tissue. The limiting feature is the requirement, both in terms of equipment and technical expertise, to prepare cryosections of the tissue of interest. This limitation can represent a considerable problem when attempting to examine specific cells in morphologically complex tissues, but the unique advantages offered by cryosectioning make this technique worthy of consideration.

3 Immunoblotting

3.1 Rationale

Immunoblotting is an adaptation of the principles of immunocytochemistry to the visualization of antigens on an adsorptive surface (Lin and Kasamatsu 1983; Towbin and Gordon 1984; Gershoni 1985). The most common application of immunoblotting is the electrophoretic transfer of polypeptides from an SDS polyacrylamide gel to nitrocellulose, followed by immunostaining antigenically detectable polypeptides on the nitrocellulose.

Immunoblotting is often referred to as "Western blotting", a term that is best avoided because it is jargon that conveys little meaning (Gershoni and Palade 1983). Descriptive terminology is preferable. Thus, the transfer step may be termed electroblotting, at least when it is effected electrophoretically. When electroblotting is combined with immunovisualization of polypeptides on the blot, the entire process may be termed immunoblotting.

The immunostaining protocol described here has been refined to provide an optimal signal-to-noise ratio, where noise is the sum of (1) nonspecific staining of nonantigenic polypeptides and (2) nonspecific background color development unrelated to the presence of electroblotted polypeptides (Pratt 1984b; Cordonnier et al. 1986). The goal, of course, is to detect the smallest quantity of antigen in the presence of the maximal amount of antigenically unrelated polypeptides.

The protocol is in principle identical to that which we use for immunocytochemistry (Fig. 1). Phytochrome, or polypeptides derived from phytochrome, are electrotransferred to nitrocellulose (Towbin et al. 1979). Subsequent nonspecific adherence of antibodies to the nitrocellulose is minimized by a blocking or quenching step, after which mouse-derived, monoclonal antibody to phytochrome (MAP) is added, followed by unlabeled rabbit antibody to mouse IgG (RAM). Quantitative immunoblotting experiments indicate that a threefold increase in sensitivity is achieved by inclusion of this "bridge" antibody. Goat antibody to rabbit IgG (GAR), which has conjugated to it alkaline phosphatase in place of rhodamine (R), is used as the label. Alkaline phosphatase is visualized with a substrate mixture that yields an insoluble, brightly colored reaction product.

The protocol includes a variety of precautions to minimize noise. Of special importance are the inclusion in diluent and wash buffer of the detergent Tween

20 (polyoxyethylene sorbitan monolaurate), which reduces nonspecific adsorption of antibodies to nonantigenic polypeptides, and inclusion in all solutions of both bovine serum albumin and sheep serum, which minimizes nonspecific adherence of antibodies to the nitrocellulose.

3.2 Method

The initial electrophoretic step can be performed by a variety of methods. Methods for the SDS-PAGE that we employ are standard (Laemmli 1970; Studier 1973; Pratt 1984a; Cordonnier et al. 1985). Consequently, no attempt will be made to deal with this step here. The protocol described below begins with the electroblotting step, which we have found to work well for polypeptides from 3 to more than 200 kilodaltons (kD) in size.

Sandwich Construction. A variety of commercial electrotransfer apparatuses, which include the sandwich holders and electrophoretic chamber, are available. Begin by preparing a clean, shallow tray with a bottom surface slightly larger than the holder for the sandwich. Add enough freshly prepared electrotransfer buffer [25 mM Tris(hydroxymethyl)amino methane, 192 mM glycine, 20% methanol, pH adjusted to 8.3 with HCl] to keep the sandwich wet as it is being constructed. Place one-half of the holder in the container and place over it a piece of Scotch-Brite or sponge pad. Over the Scotch-Brite place a sheet of filter paper, such as Whatman No. 1, that is slightly larger than the gel. Cut a piece of nitrocellulose (e.g., Bio-Rad Nitrocellulose Membrane, 0.45 μm) so that it is slightly larger than the gel. Throughout, handle the nitrocellulose either with forceps or gloves. Forceps marketed for this purpose (e.g., Nalgene, 399-0001) are most convenient. Wet the nitrocellulose by laying it on the surface of the transfer buffer. Allow it to wet slowly by capillary action. Once fully wetted, it may then be transferred to the sandwich, placing it over the filter paper.

Remove the stacking gel, if any, from the polyacrylamide gel. Cut away a small corner from the lower right-hand corner of the gel (i.e., on the opposite end from lane 1) so that its orientation will be unambiguous. Place the prepared polyacrylamide gel on the nitrocellulose so that the marked corner is to the lower right. In this way, lane 1 will be to the left on the nitrocellulose after transfer. Removal of all air bubbles, especially from between the nitrocellulose and the gel, is critical at this step. If any are trapped, they may be removed by gently pushing them out with a gloved finger. Complete the sandwich by placing a second sheet of prewetted filter paper over the gel, and a second piece of prewetted Scotch-Brite or sponge pad over the filter paper. Close the sandwich holder and place it in a transfer chamber such that the nitrocellulose will be nearer the positive electrode than is the polyacrylamide gel, ensuring that the negatively charged polypeptides will transfer to the nitrocellulose. Cover with freshly prepared electrotransfer buffer. We have had difficulties on occasion when transfer buffer was used more than once.

Electroblotting. Electrotransfer is effected at 40 V (ca. 150 mA) for 3 h using a Bio-Rad apparatus (equivalent to a gradient of 5 V cm^{-1}). As we normally perform the electrophoretic separation on the same day as the electrotransfer, we operate the electrotransfer power supply from a timer. Thus, upon completion of electrotransfer, the apparatus remains assembled overnight, during which time additional transfer by diffussion can occur.

Preparation for Immunostaining. Remove the sandwich from the transfer chamber and open it carefully, such that the nitrocellulose and polyacrylamide gel do not move relative to one another. With a ball-point pen that will write on wet nitrocellulose and that contains water-insoluble ink, mark the nitrocellulose to identify the position of the top and bottom of the separating gel. Remove the nitrocellulose from the sandwich and transfer it to a shallow dish (e.g., Nalgene, 5700-0500) containing 0.2% Ponceau S stain in 3% trichloroacetic acid. After about 5 min, transfer the nitrocellulose to another shallow dish containing water. Rinse briefly until background color is removed.

Place the stained nitrocellulose with adsorbed protein up on a piece of clean glass previously wetted with water. Draw lines to mark permanently positions of both the top and bottom of the separating gel and the position of the electrophoretic front. Mark also the position of lane 1 (Fig. 5).

Transfer the marked nitrocellulose to a clean light-box with wetted surface and photograph immediately. We use color print film that is processed commercially by a local printer, who adjusts the printing process to give the proper color balance. Ponceau S stains total protein, but is relatively insensitive. Consequently, it does not visibly stain polypeptides that are present in an amount that yields a satisfactory immunostain.

Mark enough lanes on the nitrocellulose so that after cutting it into strips, the original position of each strip in the blot will be clearly identified (Fig. 5). Remove

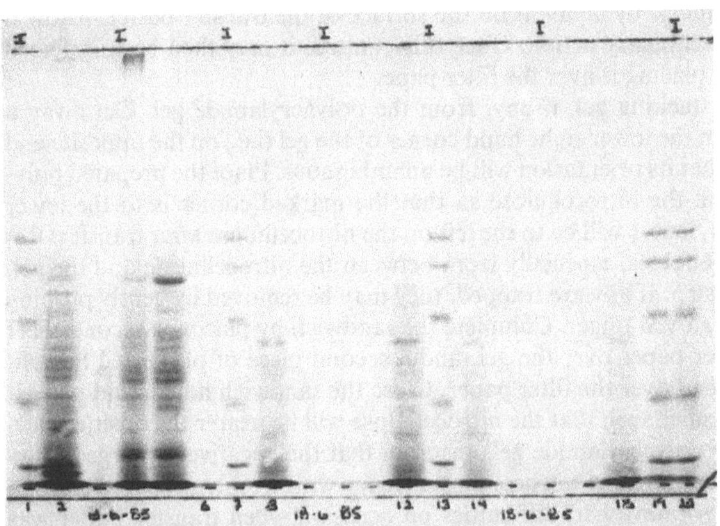

Fig. 5. Nitrocellulose stained with Ponceau S after electrotransfer of polypeptides to it from a 5% to 10% gradient SDS polyacrylamide gel. It has been marked with ink in preparation for immunostaining as described in the text. Molecular weight standards are in *Lanes 1, 7, 13, 19,* and *20*. (Previously unpublished data)

excess nitrocellulose from around the margins and cut the nitrocellulose into strips as needed. A surgical scalpel with disposable, curved cutting edges works well. Transfer any strips containing molecular weight standards back into the Ponceau S for restaining. After a few moments, they can be destained as before, blotted dry, and then allowed to air dry completely. Nitrocellulose strips to be immunostained are transferred to blocking solution (PBS, to which is added bovine serum albumin to 1%, sheep serum to 1%, and NaN_3 to 0.02%). The nitrocellulose should be left in the blocking solution for a minimum of 1 h and, in some cases, may be stored in it at 4 °C for at least several months. Since ink will transfer from one piece of nitrocellulose to another during storage, the pieces should be separated.

While transient staining of the nitrocellulose with Ponceau S is not necessary, it does offer some advantages. (1) It provides a permanent record of total protein distribution. Since the Ponceau S washes out of the nitrocellulose into the blocking solution, it does not interfere with subsequent evaluation of immunostain. (2) It provides permanently stained molecular weight standards that can be compared directly to immunostained polypeptides. Since no organic solvent is involved, the strip with molecular weight standards does not shrink appreciably with respect to immunostained strips. (3) It simplifies cutting the nitrocellulose into strips. While it is at least theoretically possible that staining with Ponceau might reduce subsequent antigenicity of a polypeptide relative to a given antibody, this has so far not been found to be the case.

Immunostaining. A sheet of clean, level glass provides a good working surface. Place on the glass a piece of Parafilm slightly larger than the piece of nitrocellulose to be immunostained. By rubbing it firmly against the dry glass, it will adhere and provide an immobile, hydrophobic working surface.

With forceps, remove the nitrocellulose from the blocking solution and rinse briefly with a stream of wash solution [10 mM Tris(hydroxymethyl)amino methane-Cl, pH 8.0 at 25 °C, 0.05% Tween 20, 0.02% NaN_3, 1% sheep serum, and 0.1% bovine serum albumin] from a squeeze bottle. Drain excess liquid from the nitrocellulose and place it on the Parafilm, ensuring that the surface with adsorbed protein is up. Add sufficient primary antibody diluted in diluent (PBS with 0.05% Tween 20, 0.02% NaN_3, 10% sheep serum, and 1% bovine serum albumin) to provide a layer at least 1 mm thick. The Parafilm is sufficiently hydrophobic and the nitrocellulose sufficiently hydrophilic that the liquid will form a bubble on the nitrocellulose. It is not necessary to agitate the nitrocellulose during this and subsequent incubations (Table 2), nor is it usually necessary to enclose it. A monoclonal antibody concentration of 1 $\mu g\ ml^{-1}$ is usually in excess. When affinity is low, however, a concentration of up to 5 $\mu g\ ml^{-1}$ may be used. Higher concentrations begin to increase noise as well as signal and thus may not be satisfactory. Incubate for 1 h at room temperature. For most antibodies, longer incubations result in an increase in background, with no discernible improvement in immunospecific staining.

After the 1-h incubation with primary antibody, pick up the nitrocellulose with forceps, rinse it with a stream of wash solution from a squeeze bottle, place it in a tray, and cover it with wash solution. Agitate gently for 15 min, replace the

Table 2. Protocol for immunostaining nitrocellulose blots with monoclonal antibodies

Reagent	Incubation time (min)
Blocking solution	60
Monoclonal antibody to phytochrome, 1 μg ml^{-1}	60
Wash solution	2×15
Rabbit antibody to mouse IgG, 1 μg ml^{-1}	60
Wash solution	2×15
Alkaline phosphatase-conjugated goat antibody to rabbit IgG, diluted 1000-fold	60
Wash solution	2×15
Substrate buffer	15
Substrate solution	Indefinite

wash solution, and agitate for another 15 min. If different primary antibodies are used, segregate the nitrocellulose strips according to primary antibody during this and all subsequent washes. Otherwise, primary antibody from one strip may enter the wash solution and bind to its antigen in another strip.

Briefly rinse the nitrocellulose as before, place on clean Parafilm, add antigen-specific rabbit antibody to mouse IgG diluted to 1 μg ml^{-1} with diluent, and incubate for 1 h at room temperature. Although we prepare our own RAM (Cordonnier et al. 1983), these antibodies are commercially available. They function equally well whether the primary antibody is an IgM or an IgG since both have light chains in common. Suitable commercial antibodies would be goat antibodies to mouse IgG (e.g., Sigma, M-8642). If these antibodies are used in place of RAM, then the GAR-AP conjugate (see below) should be replaced by alkaline phosphatase-conjugated rabbit antibodies to goat immunoglobulins (e.g., Sigma, A-7650).

Wash two times for 15 min each as before.

Again, briefly rinse the nitrocellulose, place it on clean Parafilm, and then add alkaline phosphatase-conjugated goat antibody to rabbit IgG (e.g., Sigma, A-8025). These conjugates in principle need to be titrated to determine suitable working dilutions. Without exception, however, we have found that a 1000-fold dilution in diluent works well. Incubate as before for 1 h.

Wash again two times for 15 min each. Replace the wash solution with substrate buffer (0.15 M sodium bicarbonate, 4 mM $MgCl_2$, pH 9.6) and agitate for a further 15 min. Prepare fresh substrate solution (Blake et al. 1984) as follows. To 45 ml of substrate buffer add 0.5 ml bromochloroindolyl phosphate stock solution (5 mg ml^{-1} in dimethylsulfoxide, store at 4 °C) and 5 ml nitroblue tetrazolium stock solution (1 mg ml^{-1} in substrate buffer, store at 4 °C in darkness, replace each 2 weeks). The substrate solution is layered over the nitrocellulose on clean Parafilm. Color development begins within minutes and is near saturation within 1 h. Development may be prolonged, however, if bands are weak, in which case the blot should be incubated with substrate solution in a tightly sealed box (e.g., Nalgene, 5700-0500) to reduce evaporation.

Rinse the developed blot with water to remove excess substrate, blot dry with paper towels, and dry thoroughly on a clean, nonadherent surface. Cover the nitrocellulose with a piece of glass or plastic when almost dry to prevent it from curling. Once dry, the nitrocellulose blot may be reconstituted, together with the Ponceau S-stained standards. The blots may then be stored in a notebook under a thin sheet of plastic.

Alternatives. Polyclonal antibodies may be used as primary antibody in place of the mouse monoclonal antibody used in the above protocol. With rabbit antibodies, a concentration of 1 $\mu g\ ml^{-1}$ is satisfactory if they are antigen specific (i.e., affinity purified with a column of immobilized antigen; Pratt 1984a). If they are in a purified IgG fraction, a concentration of about 10 $\mu g\ ml^{-1}$ should serve as a good starting point. If in a crude serum, then a dilution of 100- to 1000-fold should prove satisfactory. In the latter two cases, the precise concentration to use must be determined by testing several dilutions. For maximum sensitivity, a goat antibody to rabbit IgG (e.g., Sigma, R-2004) is a satisfactory "bridge" antibody, which may then be followed by alkaline phosphatase-conjugated rabbit antibody to goat IgG (e.g., Sigma, A-7650). Alternatively, the primary antibody may be followed immediately by alkaline phosphatase-conjugated goat antibody to rabbit IgG. Finally, enzyme labels other than phosphatase are available, such as peroxidase, as well as radiolabels, most notably ^{125}I. With imagination, a wide variety of other protocols can be devised given the range of commercially available unlabeled and labeled antibodies.

3.3 Controls

Controls are needed for two purposes: to detect (1) possible antigenic nonspecificity and (2) possible nonspecific adherence of antibodies to antigenically unrelated polypeptides or to the nitrocellulose itself.

If the primary antibody is a mouse-derived monoclonal antibody, substitution of nonimmune mouse IgG (or IgM if the monoclonal antibody is an IgM) can serve as a routine control. Since each monoclonal antibody is unique, however, there is no wholly satisfactory, generally applicable control. Ultimately, ad hoc controls designed around some unique property of the antigen must be devised.

If the primary antibody is a polyclonal preparation, substitution of an equivalent, but nonimmune, preparation (i.e., IgG fraction or serum) from the same animal species is routinely satisfactory for detection of possible nonspecific adherence of antibodies. It is much more difficult, however, to determine whether a polyclonal antibody preparation is monospecific for a given antigen, given the sensitivity of immunoblotting. As mentioned already, assays such as Ouchterlony double immunodiffusion and immunoelectrophoresis are too insensitive to be useful by themselves. A more appropriate control would be to adsorb the antibody preparation with demonstrably pure antigen and determine whether it still exhibits activity by immunoblotting. Alternatively, immunostaining can be done with monospecific antibodies that have been immunopurified with demonstrably pure antigen. It must still be stressed, however, that since immunoblotting is so sensitive, and since a potential contaminant in the blot might easily be present at a concentration several orders of magnitude greater than the antigen of interest, results must always be interpreted with caution.

3.4 Evaluation of Method

The high signal-to-noise ratio that can be obtained with this protocol is documented by immunoblotting a dilution series of whole shoot extracts that were prepared by extracting lyophilized, etiolated oat shoots with SDS sample buffer (Fig. 6; see Cordonnier et al. 1985 for details). The Ponceau S stain reveals the total protein profile (Fig. 6b). The same piece of nitrocellulose was then cut into two pieces, one of which was immunostained with Oat-22 and the other with Pea-25 (Fig. 6a). Oat-22 is a monoclonal antibody directed to oat phytochrome (Cordonnier et al. 1983); Pea-25 is a monoclonal antibody directed to pea phytochrome (Cordonnier et al. 1986). Note that the antibodies detect phytochrome, even though it is present at such a low abundance that it cannot be seen with Ponceau S, even with the highest sample load. The faintest bands represent an original sample load containing 10 ng of phytochrome. With longer development of the blot, sample loads as low as 1 ng are readily visible (Cordonnier et al. 1986). Under optimal conditions, we estimate that we can detect as little as 1 part of phytochrome in the presence of 100000 parts of other protein.

The epitope specificity of monoclonal antibodies is apparent when proteolytically digested samples of phytochrome are examined (Cordonnier et al. 1985).

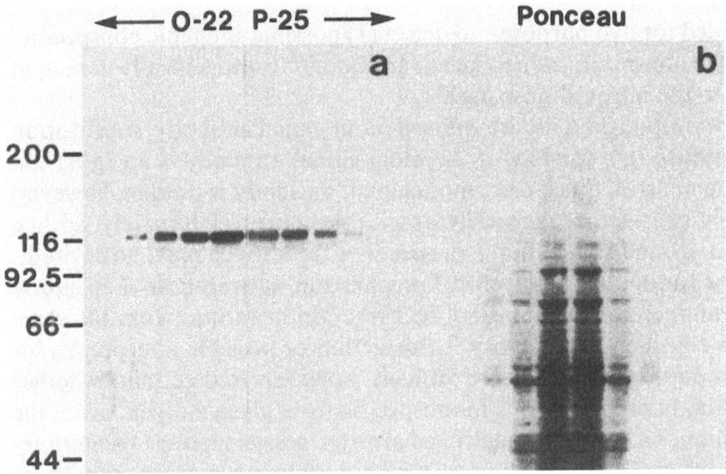

Fig. 6. Nitrocellulose first stained with Ponceau S (**b**) and then immunostained as indicated with two monoclonal antibodies to phytochrome, Oat-22 and Pea-25 (**a**). SDS sample buffer extracts of lyophilized etiolated oats were applied to a 5% to 10% gradient gel, subjected to electrophoresis, and transferred to nitrocellulose. The *center two lanes* contain 300 ng phytochrome. Phytochrome quantity decreases outwards in both directions, to 100, 30, 10, and 3 ng. (Adapted from Cordonnier et al. 1986)

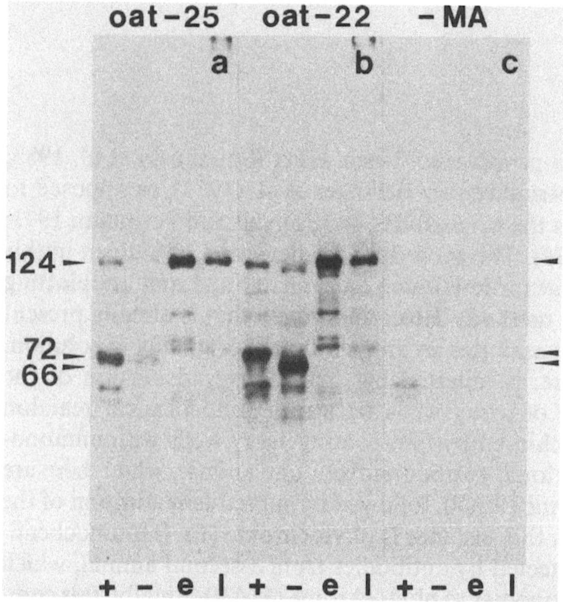

Fig. 7. Immunoblot of a 5% to 10% gradient SDS polyacrylamide gel stained with Oat-25 (**a**), Oat-22 (**b**), or nonimmune mouse IgG (**c**). *Lanes l*: 1 µl of lyophilized, etiolated oat shoots extracted directly into SDS sample buffer, containing an estimated 20 ng of phytochrome. *Lanes e*: 5 µl of an aqueous crude extract of etiolated oat shoots mixed with SDS sample buffer immediately after extraction, containing about 50 ng of phytochrome. *Lanes* −: 5 µl of the same sample shown in *lanes e*, but after 7-h incubation at 22 °C with phytochrome as Pr. *Lanes* +: as *lane* −, but with 3.3 mol Oat-25 added per mole of phytochrome in the extract prior to beginning the 7-h incubation. (Adapted from Cordonnier et al. 1985)

Phytochrome was digested in crude extracts of etiolated oats in the presence or absence or a monoclonal antibody (Oat-25) that apparently binds to a site about 6 kD from its amino terminus. For this reason, Oat-25 immunostains undegraded, 124-kD phytochrome (Fig. 7a, lanes e, l), but not 118-kD phytochrome (Fig. 7a, lane −), which is missing an amino terminal, 6-kD fragment (Daniels and Quail 1984). That 118-kD phytochrome was present in this lane is evident when a replicate lane is immunostained with Oat-22 (Fig. 7b, lane −). Moreover, when Oat-25 was present during the digestion, its presence protected a cleavage site about 6 kD from the amino terminus. This protection results in the production of a 72-kD peptide that can be immunostained by both Oat-25 and Oat-22 (Fig. 7a and b, lanes +), rather than the 66-kD peptide that is obtained in its absence (Fig. 7b, lane −). As would be predicted, this smaller degradation product is missing the epitope recognized by Oat-25 (Fig. 7a, lane −). The control strip (Fig. 7c), which received as a control nonimmune mouse IgG in place of monoclonal antibody, shows negligible background stain. The only bands visible are in the lanes containing added Oat-25 (lane +). These bands are of the correct size to be residual IgG derived from the added Oat-25.

4 Immunoquantitation

4.1 Rationale

The ELISA described here is a positive sandwich assay (Shimazaki et al. 1983; Pratt 1984b), similar to that introduced by Belanger et al. (1973), as opposed to a competitive assay, which was the original design (Engvall and Perlmann 1971; Van Weemen and Schuurs 1971). The procedure for detection of antigen in this positive ELISA is similar to that for immunocytochemistry and immunoblotting (Fig. 1). While these latter two methods detect an antigen that is already present on a solid phase, the ELISA quantitates an antigen that is originally in solution. This positive ELISA, therefore, is initiated by quantitative adsorption of the antigen to the plastic surface of assay wells by an immunochemical reaction (Fig. 8a). This adsorption is achieved by first coating assay wells with immunopurified, antigen-specific polyclonal antibodies from one animal, which here are rabbit antibodies to phytochrome (RAP), followed by subsequent addition of the antigen to be assayed, which in this instance is phytochrome (P). Immunochemically bound antigen is then detected by antibodies from a second animal, which here are mouse monoclonal antibodies to phytochrome (MAP). Finally, this com-

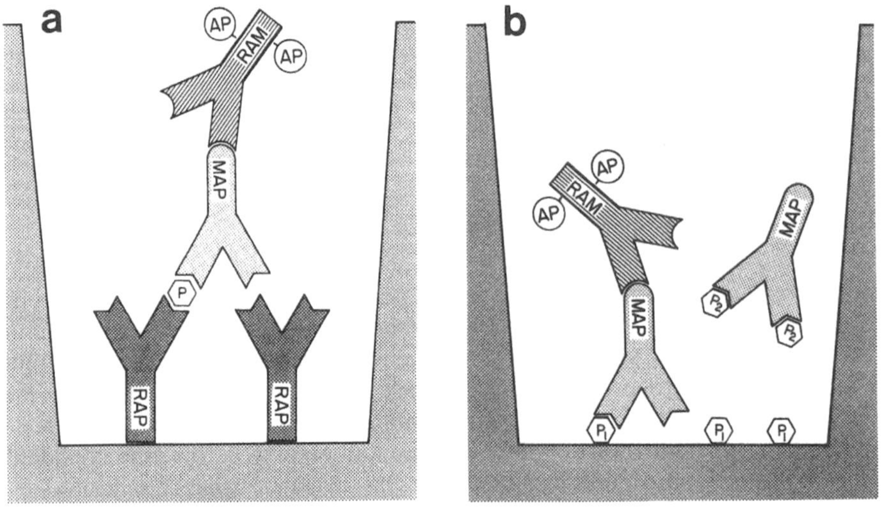

Fig. 8. Positive (**a**) and competitive (**b**) ELISA protocols for phytochrome quantitation. *RAP* = rabbit antibodies to phytochrome; *P* = phytochrome; *MAP* = mouse-derived monoclonal antibodies to phytochrome; *RAM* = rabbit antibodies to mouse IgG; *AP* = alkaline phosphatase. In the positive assay, reaction product is directly related to the amount of phytochrome assayed, while in the competitive assay, the correlation is inverse (see Fig. 9). The positive assay is performed as described in the text. The competitive assay is initiated by coating ELISA plate wells with purified antigen (P_1). Mouse-derived monoclonal antibody to phytochrome is independently preincubated with phytochrome of known amount (to prepare a standard curve), or with a sample containing an unknown amount of phytochrome (P_2). The extent of competition that results from P_2 is determined with alkaline phosphatase-conjugated rabbit antibodies to mouse IgG

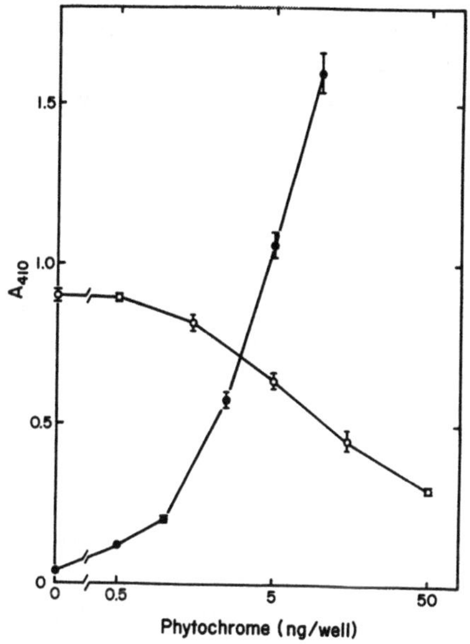

Fig. 9. Comparison of standard curves for the positive (●) and the competitive (o) ELISA. Both were done with the same monoclonal antibody to oat phytochrome (Oat-22), and with immunopurified oat phytochrome as antigen. For the competitive ELISA, appropriate concentrations of monoclonal antibody and of immunopurified oat phytochrome, with which assay wells were coated, were predetermined to yield maximal sensitivity to competition. The positive ELISA was performed by the standard protocol described in the text, except that one monoclonal antibody rather than a mixture was used. The concentration of alkaline phosphatase-conjugated rabbit antibodies to mouse IgG was the same for both ELISAs. *Bars* are standard errors of four replicates from two independent experiments. (Previously unpublished data)

plex is quantitated by alkaline phosphatase-labeled rabbit antibodies to the Igs from the second animal (AP-RAM).

The most important advantage of this ELISA is that of being a positive assay, which has the potential to be more sensitive. In a positive assay, antigen is detected by an increase in reaction product over a low background, rather than by a decrease in reaction product over a high background, as is the case with more conventional competitive assays (Voller et al. 1980). A direct comparison of a positive with a competitive ELISA illustrates this point (both protocols are shown in Fig. 8). To obtain a significant change in ELISA activity, the positive assay requires four fold less antigen (0.5 ng vs 2 ng; see Fig. 9). Moreover, this change in activity is detected, as noted, over a low background in the positive assay ($A=0.04$), but over a high background in the competitive assay ($A=0.90$).

To optimize the protocol presented here, the following points were considered (Schuurs and Van Weemen 1977; Saunders 1979; Shimazaki et al. 1983). (1) A mixture of monoclonal antibodies was used to increase assay sensitivity, as well as to reduce the possibility that the results might reflect solely a property unique to only one epitope on the antigen. (2) Antibodies for coating the assay wells and labeled second antibody were from the same animal species in order to minimize cross-reaction between them. Any such cross-reactivity would increase background activity, with a concomitant decrease in sensitivity. (3) Optimal antibody concentrations were selected as compromises among: (a) conserving relatively scarce reagents, (b) yielding minimum background activity so that maximal sensitivity can be achieved, and (c) providing high activity. (4) Appropriate times for incubation were chosen as a compromise between providing maximum signal, while at the same time minimizing (a) experimental error resulting from deviations in incubation times, (b) nonspecific adherence of unrelated substances, (c) the possibility of modification of antigen by other substances, for example phenolics and proteases, during incubation of crude plant extracts, and (d) total assay time. When antigen in a crude plant extract is assayed, it is also important to incubate the assay sample at 0 °–4 °C, rather than at room temperature or 37 °C, to further minimize antigen modification during assay.

4.2 Method

Sample Preparation. Grind 1.5 g of frozen oat shoots in liquid nitrogen with a mortar and pestle. Disperse the resultant powder by homogenization with an Ultra Turrax in 3 ml of ice-cold 50 mM Tris(hydroxymethyl)amino methane, 0.2 M 2-mercaptoethanol, pH adjusted to 8.5 with HCl at room temperature. Clarify the crude extract by centrifugation at 40 000 × g for 15 min. Use the clarified extract immediately, without further preparation.

ELISA. Wash vinyl assay plates (Costar, 2596) by filling wells with 95–100% ethanol, soaking at least 2 h, rinsing with a stream of ethanol, and drying. Coat the wells of an ethanol-washed ELISA plate with immunopurified rabbit antibodies to oat phytochrome (RAP in Fig. 8a) by adding to each well 50 μl of a 5 μg ml^{-1} preparation (Table 3), which is made by diluting a 30-μg stock aliquot (stored at −20 °C prior to use) into 6 ml of 0.2 M sodium borate, 75 mM NaCl, pH 8.5. Incubate overnight at 4 °C.

The next morning, empty the wells by inverting the plate and shaking it forcefully. Wash the plate three times. Each wash consists of filling all wells with wash solution [10 mM Tris(hydroxymethyl)amino methane-Cl, pH 8.0 at room temperature, 0.05% Tween 20 and 0.02% NaN_3], and emptying the wells as above. Block remaining nonspecific protein binding sites by filling the wells with block-

Table 3. Protocol for phytochrome quantitation with an ELISA that uses antigen-specific antibodies from two different animals

Reagent	Incubation time and temperature
Rabbit antibody to phytochrome, 5 μg ml^{-1} Wash 3 times	Overnight, 4° C
Blocking solution Wash 1 time	30 min, ambient
Phytochrome to be assayed Wash 3 times	3 h, 4° C
Monoclonal antibodies to phytochrome, mixture, 10 μg ml^{-1} each Wash 3 times	2 h, 37° C
Alkaline phosphatase-conjugated rabbit antibody to mouse Igs, diluted 500-fold Wash 3 times	2 h, 37° C
Substrate solution	30 min, ambient

ing solution (PBS supplemented with 1% bovine serum albumin) and incubating the plate for 30 min at room temperature. Wash the plate once with wash solution. Add 50 µl of test sample to each well (P in Fig. 8a). When required, dilute test samples with diluent (PBS, containing 0.05% Tween 20, 0.02% NaN_3 and 1% bovine serum albumin). Incubate the plate for 3 h at 4 °C. Wash three times with wash solution and add to each well 50 µl of a mixture of four mouse monoclonal antibodies to oat phytochrome (MAP in Fig. 8a). The monoclonal antibodies are diluted in diluent and each is present at 10 µg ml^{-1}. After an additional 2-h incubation at 37 °C, wash the plate three times as before. Add to each well 50 µl of alkaline phosphatase-conjugated rabbit antibody to mouse IgG (AP-RAM in Fig. 8a). This second antibody preparation is diluted with diluent by typically 500-fold, either from a commercially available preparation (e.g., Sigma, A-1902) or from the concentration obtained following enzyme conjugation as described by Voller et al. (1980). After incubation at 37 °C for 2 h, wash the plate three times as before.

Add to each well 50 µl of substrate solution, which is prepared by dissolving *p*-nitrophenyl phosphate at a concentration of 0.6 mg ml^{-1} in 9.6% (v/v) diethanolamine, 0.5 mM $MgCl_2$, adjusted to pH 9.6 with HCl. Incubate for 30 min at room temperature. Stop the enzyme reaction by adding 50 µl of 3 N NaOH to each well. The extent of enzyme activity is monitored by measuring either absorbance at about 400 nm, or the absorbance difference between 400 and 500 nm.

Even with up to 192 samples in two assay plates, the entire protocol (excluding the preparatory overnight coating step) can be completed within 10 h (Table 3). A repeating dispenser (e.g., Tridak Stepper, model 4003-050) facilitates multiple additions of 50-µl aliquots.

Correction for Nonspecific Interference. Crude oat extracts exhibit nonspecific interference in the ELISA (Shimazaki et al. 1983). The amount of interference varies as a function both of the tissue being extracted, and of the extent of its dilution. Neither ammonium sulfate fractionation, filtration through Sephadex G-25, nor addition of the protease inhibitor, phenylmethylsulfonyl fluoride, eliminates the interference. To keep the procedure as simple as possible, so that a maximum number of extracts can be assayed in a single experiment, and to minimize the possibility that phytochrome might be lost during sample preparation, a method to correct for this interference has been developed in place of making further efforts to eliminate it.

This correction makes use of an ELISA for quantitation of pea phytochrome and is based on the following observations. Oat phytochrome does not itself exhibit any activity in the ELISA for pea phytochrome. Crude oat extracts, however, do interfere with the assay for pea phytochrome, presumably to the same extent as with the assay for oat phytochrome.

The correction is made as follows. Prepare a standard curve with known quantities of purified oat phytochrome by following the standard ELISA protocol (Fig. 10, curve a). Prepare two standard curves with known quantities of purified pea phytochrome. One is prepared by diluting pea phytochrome with diluent (Fig. 10, curve b). The other is prepared by diluting the same pea phytochrome with crude oat extract (Fig. 10, curve c), which is used for this purpose at the same

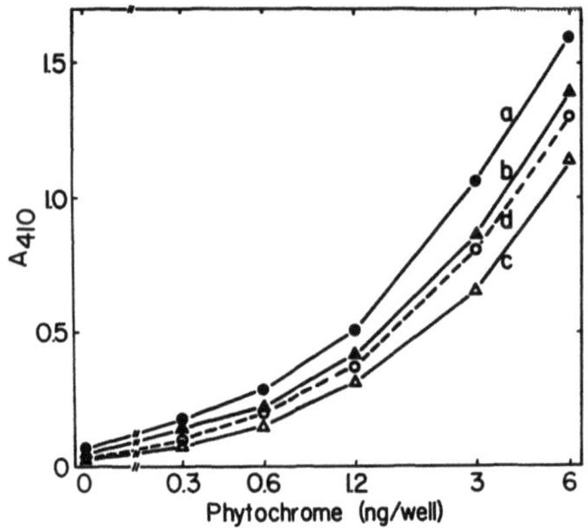

Fig. 10. Standard curves used both to correct for nonspecific interference in the ELISA and to quantitate phytochrome in crude oat extracts. *a* Immunopurified oat phytochrome was diluted with diluent and quantified by the ELISA for oat phytochrome. *b* Purified pea phytochrome was diluted with diluent and quantified by the ELISA for pea phytochrome. *c* Purified pea phytochrome was diluted with the crude oat extract that contains the phytochrome to be measured. The oat extract was used for this purpose at the same dilution as that used for determining its phytochrome content. The diluted pea phytochrome was then measured by the pea-phytochrome ELISA. *d* A corrected standard curve obtained by multiplying each absorbance value on curve *a* by the ratio of the absorbance value obtained at the same phytochrome concentration on curve *c* to that on curve *b*. (From Shimazaki et al. 1983)

concentration as that used for determining its phytochrome content. These two standard curves for pea phytochrome are prepared according to the standard ELISA protocol for oat phytochrome with two exceptions. (1) ELISA plate wells are coated with immunopurified rabbit antibodies to pea phytochrome rather than to oat phytochrome. (2) A mixture of monoclonal antibodies to pea phytochrome, which do not cross-react with oat phytochrome, is used in place of the monoclonal antibodies to oat phytochrome. Determine the extent of inhibition at each pea phytochrome concentration by dividing the absorbance value on curve c by the corresponding value on curve b. Multiply the results by the absorbance value on curve a and plot the outcome as the corrected value for the oat phytochrome standard curve (Fig. 10, curve d). This corrected standard curve is then used to estimate the phytochrome amount in the crude oat extract.

4.3 Controls

Controls are required to detect possible nonspecific adherence of any assay component, which would lead to spurious activity, and to verify the absolute accuracy of the assay.

A general control to test for possible nonspecific activity involves replacing immune antibodies, with which assay wells are coated, with nonimmune Igs at the same concentration and from the same animal species. Any nonspecific activity would be evident as an increase in reaction product.

The accuracy of the assay is verified by adding one or more known quantities of demonstrably pure antigen to aliquots of every crude extract that is assayed. If these supplemented extracts give the expected increase in ELISA activity, independently of either extent of extract dilution or the amount of pure antigen added, then one can have confidence in the accuracy of the assay. This latter control is routinely included with every unknown sample to be assayed.

4.4 Evaluation of Method

Sensitivity. With this protocol and immunopurified oat phytochrome, it is possible to prepare a standard curve for quantitation of phytochrome in crude extracts over a range from 100 pg (less than 1 fmol of monomer, which is less than 20 pM, Fig. 11) to 6 ng (Fig. 10, curve d). When quantifying phytochrome in crude extracts of etiolated oat shoots, which contain about 600 ng phytochrome in a 50-μl aliquot, the extract has to be diluted 100- to 1000-fold prior to assay. At these extreme dilutions nonspecific interference is not detected, as indicated by the close correspondence between spectrophotometrically detectable phytochrome and that detected by ELISA (Fig. 11).

Applications. The extreme sensitivity and simplicity of the ELISA has permitted the detection of less than 10 ng phytochrome per g fresh weight tissue. A number of observations, which could not have been made in any other way, thus became possible (Fig. 12). (1) Immunochemically detectable phytochrome in green oat leaves undergoes a threefold oscillation on a normal diurnal cycle. (2) Phytochrome that is accumulated during a prolonged dark incubation of light-grown seedlings is found not only in meristematic or recently meristematic tissue, but also in the mature cells of primary leaf tips (Shimazaki et al. 1983). (3) Phytochrome that reaccumulates in darkness is immediately photoresponsive (Fig. 12). (4) The herbicide norflurazone, which prevents chlorophyll accumulation, has a strong inhibitory effect on phytochrome reaccumulation during a prolonged dark period (Fig. 12).

As for any immunochemical assay, however, the ELISA described here has an inherent limitation. Namely, the assay in an absolute sense quantitates an

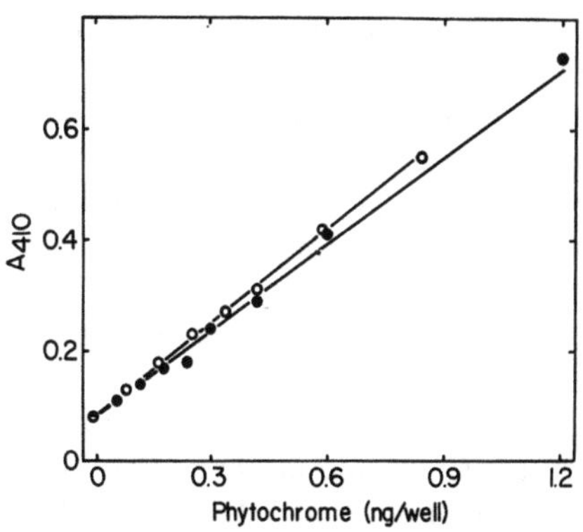

Fig. 11. ELISA standard curves obtained with immunopurified oat phytochrome (●) and with a crude extract of etiolated oats (○). Phytochrome quantity in the crude extract was determined by photoreversibility assay at 667 and 724 nm. (From Shimazaki et al. 1983)

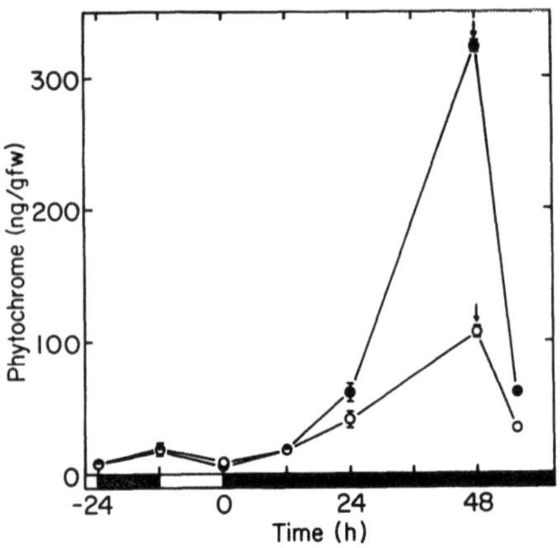

Fig. 12. Phytochrome content in crude extracts of whole shoots harvested from green (●) and norflurazon-treated (○) oat seedlings as a function of time during both a 12:12-h light: dark cycle and a subsequent prolonged dark incubation of the seedlings. Three-day-old etiolated oats were subsequently grown for 5 days under a 12:12-h light: dark cycle before being placed in continuous darkness at zero time. Light and dark periods are indicated by the *horizontal bar*: *open* = light; *closed* = dark. *Arrows* indicate time of irradiation of the plants for 5 min with Sylvania Gro-lux lamps. *Bars* are standard errors of four replicates from two independent experiments. (From Shimazaki et al. 1983)

antigen accurately only if it is immunochemically indistinguishable from that used for preparation of standard curves. This limitation can simultaneously, however, be a distinct advantage. As etiolated oat seedlings were kept for increasing time in white light, the ELISA detected a decreasing proportion of spectrophotometrically assayable phytochrome (Shimazaki et al. 1983). This inability to detect all of the phytochrome is clearly attributable to this inherent limitation of the ELISA. At the same time, however, it permitted the following tentative conclusions, which have since been confirmed. (1) Light-grown, green oats contain a pool of phytochrome that appears to be antigenically distinct from that found in etiolated oats. (2) The transition from phytochrome like that found in etiolated oats to the immunochemically distinct pool found in green oats occurs gradually following transfer of etiolated plants to continuous illumination.

Of course, the ELISA described here can be adapted to any multivalent antigen. A recent, independent application of this approach is the quantitation of a wheat germ agglutinin(WGA)-like lectin in adult wheat plants (Raikhel et al. 1984). The ELISA described here was modified by utilizing antigen-specific polyclonal antibodies obtained from both guinea pig and rabbit, rather than polyclonal rabbit and monoclonal mouse antibodies. WGA is characteristic of wheat embryos. Although a similar protein has been found in the roots of older plants, it has not been characterized in detail because of its low content in the older tissue. Application of this positive ELISA, however, permitted a number of observations, including the following. (1) The WGA-like protein from adult plants is antigenically similar, if not identical, to the embryo WGA. (2) A substantial amount of WGA-like protein is present at the base of the shoot and in the roots of adult wheat plants. (3) Wheat plants each contain up to 100 ng of WGA-like protein after the first week of growth. (4) The level of this protein fluctuates during the first 15 weeks of growth. (5) Together with other observations, it was possible to conclude that this fluctuation is attributable to de novo synthesis and that the WGA-like protein in older plants is probably not residual embryo WGA.

5 Concluding Remarks

The methods described here have been refined to take maximum advantage of the unique features of monoclonal antibodies (Goding 1983). Moreover, they have been designed to function optimally with phytochrome as antigen (Pratt 1984b). Nevertheless, they should function equally well with other polypeptide antigens and, with appropriate modification, with polyclonal antibody preparations.

It is important to emphasize that these protocols should serve only as starting points for other applications. For each method described, many variations are available. It is likely that modifications in the protocols described here will be required to optimize them for use with other antigens.

Acknowledgements. The methods presented here were developed with support provided by grants from the National Science Foundation (PCM-8022159, PCM-8315840), the Department of Energy (Contract DE-AC09-81SR10925), and the Swiss National Funds (3:292-0:82).

References

Belanger L, Sylvestre C, Dufour D (1973) Enzyme-linked immunoassay for alpha-fetoprotein by competitive and sandwich procedures. Clin Chim Acta 48:15–18

Blake MS, Johnston KH, Russell-Jones GJ, Gotschlich EC (1984) A rapid, sensitive method for detection of alkaline phosphatase-conjugated anti-antibody on Western blots. Anal Biochem 136:175–179

Briggs WR, Siegelman HW (1965) Distribution of phytochrome in etiolated seedlings. Plant Physiol (Bethesda) 40:934–941

Bullock GR, Petrusz P (eds) (1982) Techniques in immunocytochemistry, Vol 1. Academic Press, London, 306 pp

Coleman RA, Pratt LH (1974) Subcellular localization of the red-absorbing form of phytochrome by immunocytochemistry. Planta (Berl) 121:119–131

Cordonnier M-M, Smith C, Greppin H, Pratt LH (1983) Production and purification of monoclonal antibodies to *Pisum* and *Avena* phytochrome. Planta (Berl) 158:369–376

Cordonnier M-M, Greppin H, Pratt LH (1985) Monoclonal antibodies with differing affinities to the red-absorbing and far-red-absorbing forms of phytochrome. Biochemistry 24:3246–3253

Cordonnier M-M, Greppin H, Pratt LH (1986) Identification of a highly conserved domain on phytochrome from angiosperms to algae. Plant Physiol (Bethesda) 80:982–987

Daniels SM, Quail PH (1984) Monoclonal antibodies to three separate domains on 124 kilodalton phytochrome from *Avena*. Plant Physiol (Bethesda) 76:622–626

Engvall E, Perlmann P (1971) Enzyme-linked immunosorbent assay (ELISA). Quantitative assay for immunoglobulin G. Immunochemistry 8:871–874

Gershoni JM (1985) Protein blotting: developments and perspectives. Trends Biochem Sci 10:103–106

Gershoni JM, Palade GE (1983) Protein blotting: principles and applications. Anal Biochem 131:1–15
Giloh H, Sedat JW (1982) Fluorescence microscopy: reduced photobleaching of rhodamine and fluorescein protein conjugates by *n*-propyl gallate. Science 217:1252–1255
Goding JW (1983) Monoclonal antibodies: principles and practice. Academic Press, London, 276 pp
Köhler G, Milstein C (1975) Continuous cultures of fused cells secreting antibody of predefined specificity. Nature 256:495–497
Knox RB, Vithanage HIMV, Howlett BJ (1980) Botanical immunocytochemistry: a review with special reference to pollen antigens and allergens. Histochem J 12:247–272
Laemmli UK (1970) Cleavage of structural proteins during the assembly of the head of bacteriophage T_4. Nature 227:680–685
Lin W, Kasamatsu H (1983) On the electrotransfer of polypeptides from gels to nitrocellulose membranes. Anal Biochem 128:302–311
Mackenzie JM Jr, Coleman RA, Briggs WR, Pratt LH (1975) Reversible redistribution of phytochrome within the cell upon conversion to its physiologically active form. Proc Natl Acad Sci USA 72:799–803
McCurdy DW, Pratt LH (1986) Kinetics of intracellular redistribution of phytochrome in *Avena* coleoptiles after its photoconversion to the active, far-red-absorbing form. Planta (Berl) 167:330–336
Pratt LH (1984a) Phytochrome purification. In: Smith H, Holmes MG (eds) Techniques in photomorphogenesis. Academic Press, London, pp 175–200
Pratt LH (1984b) Phytochrome immunochemistry. In: Smith H, Holmes MG (eds) Techniques in photomorphogenesis. Academic Press, London, pp 201–226
Pratt LH, Coleman RA (1971) Immunocytochemical localization of phytochrome. Proc Natl Acad Sci USA 68:2431–2435
Pratt LH, Coleman RA (1974) Phytochrome distribution in etiolated grass seedlings as assayed by an indirect antibody-labeling method. Am J Bot 61:195–202
Raikhel NV, Mishkind ML, Palevitz BA (1984) Characterization of a wheat germ agglutinin-like lectin from adult wheat plants. Planta (Berl) 162:55–61
Saunders GC (1979) The art of solid-phase enzyme immunoassay including selected protocols. In: Nakamura RM, Dito WR, Tucker III ES (eds) Immunoassays in the clinical laboratory. Liss, New York, pp 99–118
Saunders MJ, Cordonnier M-M, Palevitz BA, Pratt LH (1983) Immunofluorescence visualization of phytochrome in *Pisum sativum* L. epicotyls using monoclonal antibodies. Planta (Berl) 159:545–553
Schuurs AHWM, Van Weemen BK (1977) Enzyme-immunoassay. Clin Chim Acta 81:1–40
Shimazaki Y, Cordonnier M-M, Pratt LH (1983) Phytochrome quantitation in crude extracts of *Avena* by enzyme-linked immunosorbent assay with monoclonal antibodies. Planta (Berl) 159:534–544
Sternberger LA (1979) Immunocytochemistry, 2nd edn. Wiley, New York, 354 pp
Studier FW (1973) Analysis of bacteriophage T7 early RNAs and proteins on slab gels. J Mol Biol 79:237–248
Tokuyasu KT (1980) Immunochemistry of ultrathin frozen sections. Histochem J 12:381–403
Towbin H, Gordon J (1984) Immunoblotting and dot immunobinding – current status and outlook. J Immunol Methods 72:313–340
Towbin H, Staehelin T, Gordon J (1979) Electrophoretic transfer of proteins from polyacrylamide gels to nitrocellulose sheets: procedure and some applications. Proc Natl Acad Sci USA 76:4350–4354
Van Weemen BK, Schuurs AHWM (1971) Immunoassay using antigen-enzyme conjugates. FEBS Lett 15:232–236
Voller A, Bidwell D, Bartlett A (1980) Enzyme-linked immunosorbent assay. In: Rose NR, Friedman H (eds) Manual of clinical immunology, 2nd edn. Am Soc Microbiol, Washington DC, pp 359–371

Radioimmunoassay for a Soybean Phytoalexin

H. GRISEBACH, P. MOESTA, and M. G. HAHN

Phytoalexins are low molecular weight compounds with antimicrobial properties which accumulate postinfectionally in plants from various families (Bailey and Mansfield 1982). To evaluate the significance of phytoalexins in plant disease resistance, quantitative knowledge of their exact spatial and temporal distribution within plant tissue near infection structures of invading microorganisms is of great importance.

The radioimmunoassay is a highly selective and sensitive technique that has been used to quantitate compounds such as indole-3-acetic acid (Weiler 1981), abscisic acid (Weiler 1980), and adenosine 3′:5′-cyclic monophosphate (Rosenberg et al. 1982) that occur in very low concentrations in plant tissue. In this chapter a radioimmunoassay for glyceollin I (Fig. 1) (Moesta et al. 1983a), the major phytoalexin which accumulates at infection sites in soybean (*Glycine max* L. *Merrill*) (Yoshikawa et al. 1978; Moesta and Grisebach 1981b; Hahn et al. 1985) is described.

1 Isolation of Glyceollin I

Glyceollin I is isolated from a crude mixture of glyceollin isomers I–III which is prepared from infected soybean seeds via a method described by Keen (1975) and modified by Ayers et al. (1976).

Soybean seeds (*Glycine max* cv. Harosoy 63 or another cultivar) (500 g) are soaked in distilled water for 4 h at 21 °C and sliced into quarters. The sliced seeds are then sprayed with conidia (approximately 10^7 in 20 ml H_2O) of *Cladosporium cucumerinum* (ATCC 11279) (Skipp and Bailey 1977), a nonpathogen of soybeans, and the inoculated seeds are stored in the dark at 21 °C for 72 h. The cut surfaces turn deep red during the incubation. This color is due to oxidation of glycinol (Fig. 1) or its prenylated derivatives (Zähringer et al. 1981).

The infected seeds are soaked in 500 ml of 95% ethanol for 48 h at 4 °C to extract isoflavonoids. The ethanol extract is dried under reduced pressure by rotatory evaporation at 40 °C, and the residue is dissolved in 10 ml ethanol. The solution is applied in 0.5-ml aliquots to about 20 preparative TLC plates of silica gel (Silica Gel G, F 254, Merck, Darmstadt FRG). The plates are developed with hexane-ethylacetate-methanol (60:40:5, v/v) and the glyceollin isomers are detected as a single fluorescent-quenching band at R_f 0.32. It can further be identified by its fungitoxicity in the TLC-*C. cucumerinum* assay (Keen et al. 1971). The glyceollin band is scraped from the plates, eluted from the silica with 95% ethanol, and dried by rotary evaporation under reduced pressure.

Up to 5 mg of the glyceollin mixture is dissolved in 1 ml chloroform. The solution is then applied to a SepPak silica cartridge (Waters), and nonpolar components are washed out with a total of 5 ml chloroform. The glyceollin fraction is eluted with 3 ml chloroform-2-propanol (1:1, v/v). The solvent is evaporated and the residue is taken up in 500 µl of 2-propanol. Separation of the glyceollin isomers is achieved by HPLC on either Partisil 5

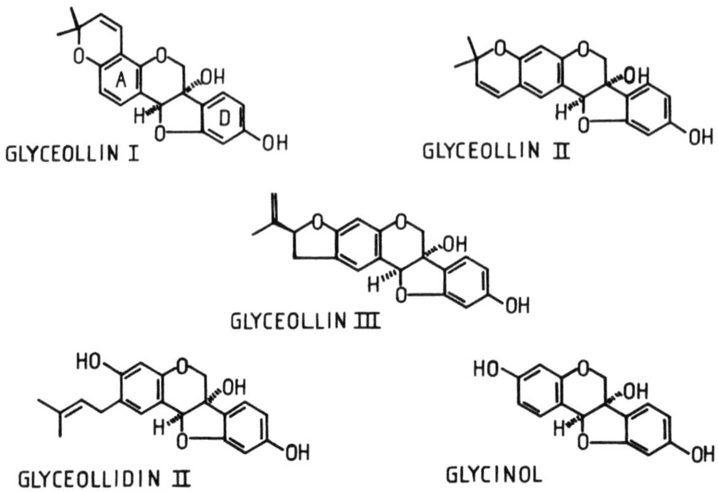

Fig. 1. Structures of several pterocarpan phytoalexins which accumulate in infected soybean tissue

Table 1. Retention times (min) on HPLC

Compound	Partisil 5	Lichrosorb RP 18
Glyceollin I	12.0	10.1
Glyceollin II	12.7	9.8
Glyceollin III	13.2	9.8
Daidzein	18.0	7.8
Glycinol	23.0	6.2

(5 µm, Whatman, Ferrieres, France) equipped with a Corasil II precolumn (Waters) and elution with n-hexane/2-propanol (90/10, v/v) at a flow rate of 2.5 ml min^{-1}, or on reversed phase (Lichrosorb RP18, 5 µm with precolumn, Merck, Darmstadt) and elution with methanol/water (75:25, v/v). Depending on the condition of the column, these solvent mixtures may have to be adjusted slightly to achieve optimum separation of the isomers. The glyceollin isomers are detected by their UV absorption at 280 nm. Retention times are listed in Table 1. The molar extinction coefficients for the glyceollin isomers are: glyceollin I E_{285} = 10300 (Ayers et al. 1976); glyceollin II, E_{286} = 8700 (Lyne et al. 1976); glyceollin III. E_{292} = 9600 (Lyne et al. 1976). The glyceollin isomers are easily oxidized in the presence of traces of air and moisture, and should be stored under nitrogen at −20 °C.

2 Synthesis of Glyceollin I-BSA Conjugate

Glyceollin I is coupled in good yield to a diazotized p-aminohippuric acid derivative of bovine serum albumin (BSA) (Fig. 2). The molar ratio of glyceollin I to BSA was determined with [^{14}C]glyceollin I and found to be 42 and 53 in two experiments (Moesta et al. 1983a).

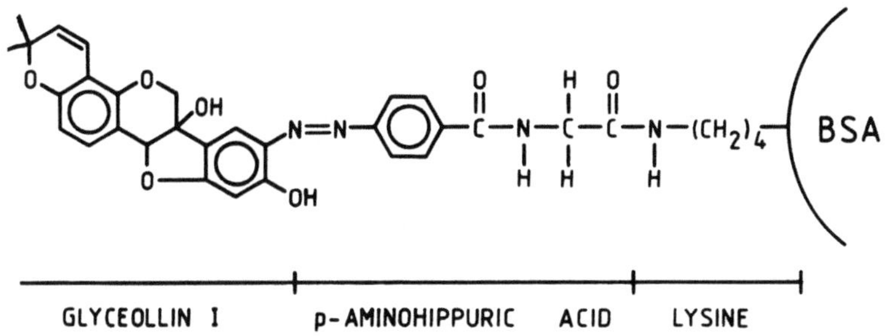

Fig. 2. Proposed structure for the glyceollin I-bovine serum albumin conjugate

2.1 Preparation of p-Aminohippuric Acid Substituted BSA (Weiler 1980)

To 200 mg p-aminohippuric acid in 120 ml water, 200 mg BSA is added and the pH of the solution is adjusted to 8.0 with 1 mol l^{-1} NaOH. The mixture is gently warmed to give a clear solution. To this solution, 200 mg 1-ethyl-3(3-dimethylaminopropyl)-carbodiimide·HCl (ECD), Merck (Darmstadt, FRG) are added and the pH is adjusted to 6.4 with 2 mol l^{-1} HCl. After stirring for 6 h at room temperature in the dark, another 100 mg ECD are added and the reaction is allowed to proceed for another 15 h. The substituted BSA is dialyzed extensively against water and the solution is then lyophilized.

2.2 Coupling of Glyceollin I to p-Aminohippuric Acid Substituted BSA (Moesta et al. 1983a)

Ten mg of the conjugate are suspended in 5 ml water and the pH adjusted to 1.5 with 1 mol l^{-1} HCl. This solution is cooled to 0 °C, and a solution of 60 mg $NaNO_2$ in 0.5 ml water is slowly added. After 5 min, excess nitrite is destroyed by addition of 30 mg ammonium sulfamate in 0.5 ml water. The diazotized protein is then added dropwise to a stirred solution of 8 mg glyceollin I in 10 ml of 50% (v/v) methanol in 0.1 mol l^{-1} borate buffer (pH 9.0) at 0 °C. After 30 min, the deep orange solution is dialized for 5 days against repeated changes of distilled water and subsequently lyophilized.

3 Generation of Antisera

Injection of the glyceollin I-BSA conjugate into rabbits leads to the production of antisera against glyceollin I (Fig. 3).

Ten mg of the glyceollin I-bovine serum albumin conjugate is suspended in 5 ml of a 0.25% (w/v) sodium bicarbonate solution, and mixed in a 1:1 ratio with complete Freund's adjuvant. Aliquots (2 ml) of this suspension are injected into the foot pads, and subcutaneously and intramuscularly at multiple sites of each

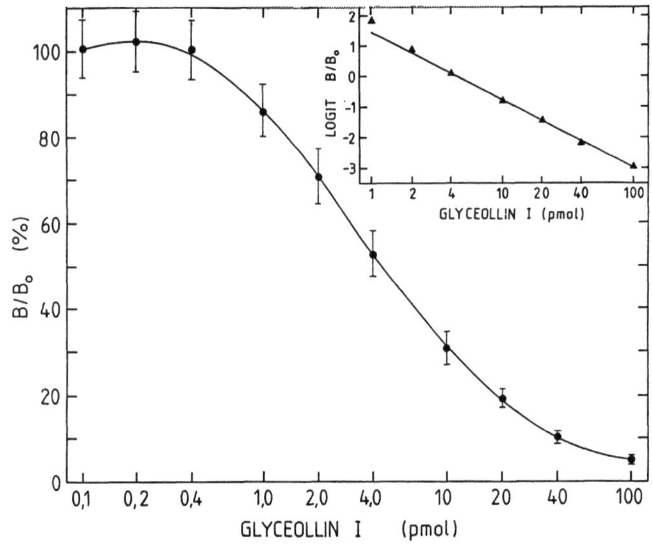

Fig. 3. Formation of antisera against a glyceollin I-bovine serum albumin conjugate in four immunized rabbits. *Arrows* indicate the times at which booster injections of the conjugate were given. The binding assay contained 100 µl of 0.01 mol l^{-1} Na-phosphate (pH 7.4) containing 0.15 mol l^{-1} NaCl and 0.1% (w/v) NaN_3 (buffer B), 100 µl of 10% (v/v) methanol in H_2O, 100 µl antiserum diluted 1:1000 in buffer B, and 50 µl [^{125}I]-glyceollin I (~10000 cpm) in 50% (v/v) methanol in buffer B. The assay was incubated and the amount of binding determined as described for the radioimmunoassay in the text

of four rabbits (New Zealand White). A booster injection is given subcutaneously and intramuscularly 5 weeks later with 2 mg conjugate per rabbit. In our experiments, sera obtained 1 week after the second injection were able to bind up to 60% of a [^{125}I]-labeled glyceollin tracer (see below) at an antiserum dilution of 1:1000. The antiserum titer was not increased following a second booster injection after 17 weeks (Fig. 3).

4 Preparation of the [^{125}I]-Glyceollin I Tracer

(Moesta et al. 1983a; Hahn et al. 1985)

In initial experiments, glyceollin I was iodinated with unlabeled sodium iodide and chloramine T, and the products analyzed by HPLC on a silica gel column [Lichrosorb Si 60 5 µm (Merck)] with a solvent mixture of n-hexane: 2-propanol, 95:5 (v/v) at a flow rate of 3 ml min^{-1}. Two iodinated glyceollin I derivatives were obtained with retention times of 9.5 min (product A) and 11.5 min (product B). The ratio of these two products depended on the molar ratio of glyceollin I to iodide in the reaction mixture. At a molar ratio of 1:1, the product A to product B ratio was 3:7. When a tenfold excess of glyceollin I was used in the iodination reaction, formation of product A could be suppressed almost completely. Since only product B behaved similarly to unlabeled glyceollin I in the radioimmunoassay (see below), the latter conditions were chosen for the preparation of [^{125}I]-glyceollin I.

Purified glyceollin I is labeled with [^{125}I] using essentially the method of Greenwood et al. (1963). To a glass test tube the following are added in rapid succession: 10 μl of 0.1 mol l^{-1} sodium phosphate (pH 7.5), 2.8 μl Na[^{125}I] (0.148 nmol: 481–629 MBq μg^{-1}), 20 μl glyceollin I (1.48 nmol) dissolved in methanol, and 10 μl chloramine T (2.8 mg ml^{-1} in 0.1 mol l^{-1} sodium phosphate, pH 7.5). This solution is mixed gently and allowed to stand at room temperature for 2 min. Subsequently, 100 μl cysteine (5 mg ml^{-1}) and 100 μl KI (1 mg ml^{-1}), each dissolved in 0.1 sodium phosphate (pH 7.5) are added, the solution again gently mixed, and then extracted three times with 0.5 ml chloroform. The combined chloroform phase is dried over anhydrous sodium sulfate, applied to a SepPak silica cartridge (Waters, Königstein, FRG), and the iodinated glyceollin I eluted from the cartridge with 8 ml chloroform. Uniodinated glyceollin I remains bound to the cartridge. The eluate is evaporated to dryness and redissolved in 50% (v/v) methanol in 0.05 mol l^{-1} sodium acetate (pH 6.2) (buffer A). An average yield of 82.6% was calculated based on the recovery of ^{125}I in labeled product in three experiments. The labeled glyceollin I is stored at −20 °C until needed, and is diluted to about 2×10^5 cpm ml^{-1} with 50% (v/v) methanol in buffer A for use in the radioimmunoassay.

5 Radioimmunoassay for Glyceollin I

Radioimmunoassays are carried out in Eppendorf vials (1.5 ml capacity). Each vial contains: 100 μl buffer A, 100 μl of sample in 10% (v/v) methanol in water, 50 μl [^{125}I]-glyceollin I tracer (~10^4 cpm) in 50% (v/v) methanol in buffer A, and 100 μl of antiserum diluted 1:1000 [1] in buffer A, added in that order and then mixed. The final titer of the antiserum in the immunoassay is therefore 1:3500. After incubation for 15 h at 8 °C, 100 μl normal rabbit serum diluted 1:10 in buffer A is added, followed by 0.5 ml of a saturated ammonium sulfate solution. The final solutions, after mixing, are 53% saturated with respect to ammonium sulfate. The solutions are incubated for 30 min at room temperature and then centrifuged at 8800 × g for 4 min. The supernatants are carefully decanted and discarded. The pellets are washed once by suspension in 0.75 ml of a saturated ammonium sulfate solution and centrifugation at 8800 × g. The supernatant is again discarded, and the vials containing the pellets are counted in a gamma counter. A complete standard curve with glyceollin I, as well as a blank control containing no antiserum and a sample containing only tracer, are included in each assay.

The standard curve for the assay of glyceollin I is shown in Fig. 4. A logit plot of the standard curve gives a straight line in the range of 1.0 to 100 pmol (0.34 to 34 ng) of glyceollin I. The assay is therefore more than 1000-fold more sensitive than any other technique available for the quantitation of glyceollin I.

[1] Dilution depends on the antiserum titer.

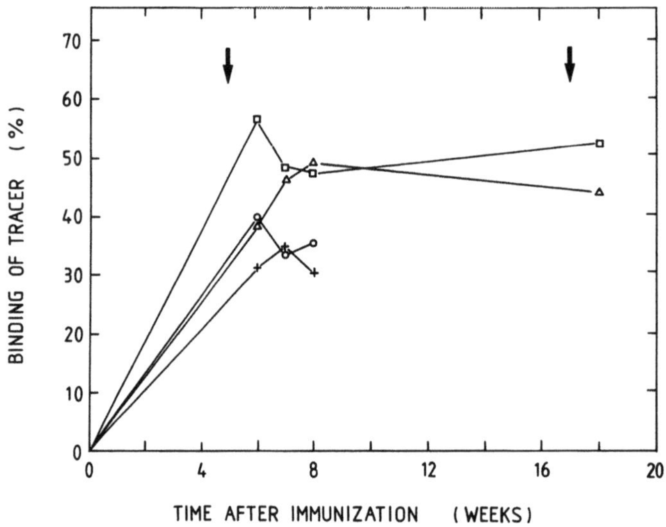

Fig. 4. Standard curve for the radioimmunoassay of glyceollin I. Each *point* represents the average, with standard deviations, of 31 independent assays with duplicates for each value. B_0 = cpm of tracer bound in the absence of exogenous glyceollin I, B = cpm of tracer bound in the sample. Logit $B/B_0 = \ln [(B/B_0)/(100\text{-}B/B_0)]$, where B/B_0 is expressed as a percentage.

The [^{125}I]-glyceollin I tracer is not very water soluble, and significant amounts are absorbed from aqueous solutions onto the surface of the vials used for the radioimmunoassay. Several organic solvent additives (methanol, ethanol, acetone, Triton X-100) were tested to overcome this problem. Of these, methanol was most effective in preventing surface adsorption without significantly affecting the assay. A methanol content of 10% (v/v) in the final assay mixture decreases the maximal amount of tracer bound by the antibody by only 10%.

The assay blank is very sensitive to the protein used as the carrier to insure complete precipitation of the antigen-antibody complex with ammonium sulfate. For example, the use of normal bovine serum or rabbit gamma globulin as carrier gives unacceptably high blank values (as great as 33% of the total counts). Normal rabbit serum at a 1:10 dilution consistently yields low blank values (<4% of the total counts).

6 Specificity of the Radioimmunoassay

The radioimmunoassay is very specific for glyceollin I (Table 2). Glyceollin II and III, two structural isomers of glyceollin I (Fig. 1), show cross-reactivity only at much higher concentrations (20- to 100-fold). The same is true for two biosynthetic precursors of the glyceollins, glycinol and glyceollidin II. The related isoflavones, daidzein and genistein, show no cross-reactivity even when assayed at

Table 2. Cross-reactivity of pterocarpans and isoflavones in the radioimmunoassay for glyceollin I

Compound	% Cross-reactivity[a]
Glyceollin I	100
Iodo-glyceollin I (Product B)	100
Glyceollin II	5.7
Glyceollin III	1.0
Glyceollidin II	0.56
Glyinol	1.9
Daidzein	< 0.84
Genistein	< 1.1

[a] Cross-reactivities were compared at 50% displacement of the tracer.

7.2 nmol and 5.3 nmol, respectively. Thus, the presence of these structurally and biosynthetically related molecules will not interfere significantly with the detection of glyceollin I in tissue extracts using the radioimmunoassay.

Product B, obtained upon iodination of glyceollin I with a tenfold excess of glyceollin I (see above), shows a displacement curve that is indistinguishable from that of uniodinated glyceollin I, whereas product A does not cross-react when assayed at 10 pmol. Since synthesis of the radioactive tracer is carried out under conditions leading almost exclusively to the formation of product B, it is certain that the iodination does not alter the binding of glyceollin I to the antibody.

The cross-reactivity data suggest that binding to the antibody is strongly influenced by the substituents on ring A of glyceollin I. The D ring is identical for all of the pterocarpans tested. These results further imply that the diazo-coupling of glyceollin I to p-aminohippuric acid substituted bovine serum albumin occurs ortho to the phenolic hydroxyl group in ring D (Fig. 2). Iodination of glyceollin I to form product B probably occurs at the same location.

7 Application of the Radioimmunoassay for Glyceollin I to the Quantitation of Phytoalexins in Infected Soybean Tissue

Since the radioimmunoassay detects only glyceollin I, the relative amounts of the pterocarpan phytoalexins produced in infected soybean tissue must first be determined in the respective tissue by HPLC (see Sect. 1). The predominant phytoalexins that accumulate in infected soybean hypocotyls are the glyceollin isomers I–III, which accumulate in a ratio of 8:1:1 (Moesta and Grisebach 1981b). The amounts of the other pterocarpan phytoalexins produced in hypocotyls, glycinol, and the glyceollidins, are insignificant (Moesta and Grisebach 1981b; Zähringer et al. 1981).

In soybean roots, glyceollin I constitutes approximately 90% of the pterocarpan phytoalexins (Morandi et al. 1984; Hahn et al. 1985). Only small amounts (<2%) of glyceollins II and III and glycinol are present.

In soybean cotyledons, the glyceollin isomers I–III accumulate in a ratio of 3:1:1 (Moesta et al. 1982). Furthermore, glycinol, which does not accumulate in hypocotyl or root tissue, accounts for about half of the phytoalexins produced in the cotyledons (Moesta and Grisebach 1981a; Weinstein et al. 1981). In soybean leaves, the glyceollin isomers I–III accumulate in a ratio of 1:3:6 (Ingham et al. 1981). Considerable amounts of glyceofuran and small amounts of glyceollidin II also are produced. No glycinol was detected. Thus, use of the radioimmunoassay for glyceollin I would result in an underestimation of total phytoalexin content by 20%, 70%, and 92% in soybean hypocotyls, cotyledons, and leaves, respectively. Only in soybean roots does this radioimmunoassay yield a good quantitation of total phytoalexin content.

The radioimmunoassay was applied to the quantitation of glyceollin I in soybean hypocotyl tissue (Moesta et al. 1983a) and, in particular, in soybean roots (cv. Harosoy 63) infected with either race 1 (incompatible; plant resistant) or race 3 (compatible; plant susceptible) of *P. megasperma* f.sp. *glycinea* (Pmg) (Hahn et al. 1985).

For quantitation of glyceollin I in infected soybean roots, root segments were embedded vertically upside down in Tissue Tek II O.C. T. compound (Miles Laboratories, Naperville, Illinois, USA) and were cut into 14 µm thick sections using a cryotome (Reichert-Jung, Heidelberg) at −20 °C. Every fourth section from dip-inoculated roots (Hahn et al. 1985) were affixed to microscope slides that had been coated with Haupt's adhesive (Gerlach 1977). The areas of these sections were determined under a microscope and the volumes of the sections calculated. Subsequently, the sections were examined immunohistochemically for the presence of fungal hyphae (Moesta et al. 1983b). The three intervening sections were combined and extracted with 250 µl of 10% (v/v) methanol in distilled water. The amounts of glyceollin I present in these extracts was determined directly using the radioimmunoassay. The concentrations of glyceollin I present in each thin section were calculated using a computer program written by the authors. The radioimmunoassay was able to measure glyceollin I concentrations in the sections as low as 0.01 $\mu mol\ ml^{-1}$ tissue volume.

The recovery of glyceollin I from cryotome sections was estimated by adding known amounts of purified glyceollin I to cryotome cross-sections (3 × 14 µm thick) from uninoculated soybean roots. Average recoveries of 85% from 3 × 14 µm thick sections were obtained.

The quantitative distribution of glyceollin I along the axis of soybean roots at differing times after infection with the incompatible race 1 of Pmg is shown in Fig. 5. Glyceollin I was first quantifiable in cross-sections 5 h after infection, being present in concentrations approaching the EC_{50} of this phytoalexin (0.17 $\mu mol\ ml^{-1}$) measured in vitro against *P. megasperma* (Lazarovits and Ward 1982). At later times, the phytoalexin concentrations in resistant roots reached or exceeded the in vitro EC_{90} (0.6 $\mu mol\ ml^{-1}$; Lazarovitz and Ward 1982). The glyceollin I concentrations dropped sharply at the leading edge of the infection. In no resistant root were fungal hyphae seen in advance of measurable

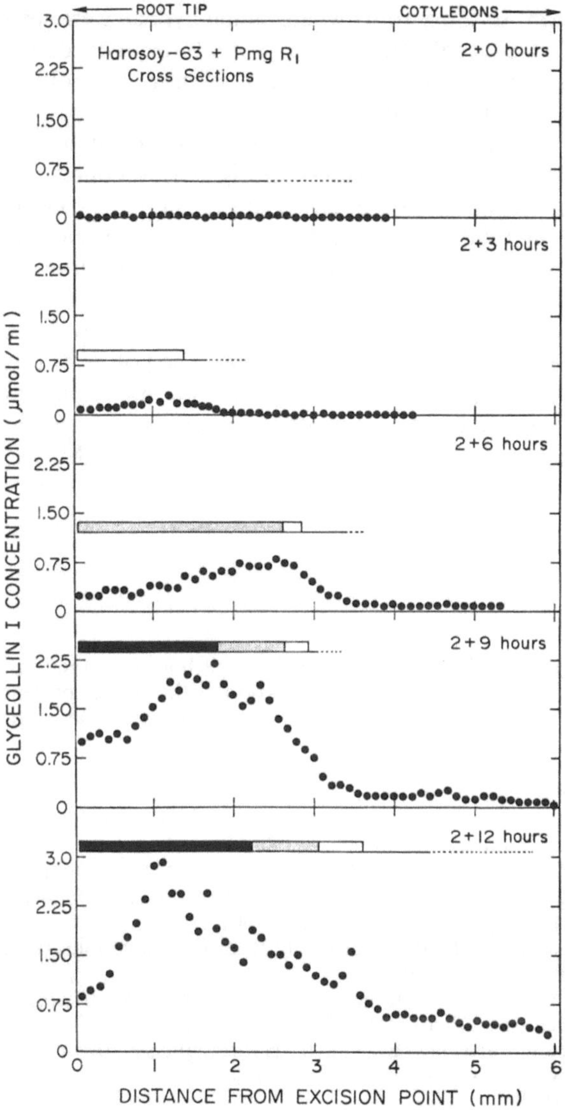

Fig. 5. Spatial and temporal course of glyceollin I accumulation along the axis of soybean roots after infection with race 1 of *P. megasperma* f.sp. *glycinea*. Seedlings (2-days-old) were dip-inoculated with about 10^4 zoospores. Glyceollin I was quantitated in extracts of cryotome cross-sections (3×14 µm thick) by radioimmunoassay, and each *point* is the average of two determinations. Data are shown for every other triple section. The in vitro EC_{90} of glyceollin I against *P. megasperma* is 0.6 µmol ml^{-1}. The extent of fungal colonization, determined immunohistochemically, is indicated by the *bars*: (▬) >90% of cross-section colonized; (▬) 50–90% colonized; (▭) 25–50% colonized; (———) a few hyphae; (---) single hyphae

concentrations of the phytoalexin. In those roots where single hyphae were seen throughout the root, glyceollin I concentrations were at inhibitory concentrations in all cross-sections. Occasionally, significant concentrations of glyceollin I were detected in advance of the fungal hyphae.

Cryotome cross-sections were also prepared from roots dip-inoculated with the compatible race 3 at each of the time points shown in Fig. 5. Low concentrations (<0.08 μmol ml^{-1}) of glyceollin I were detected in occasional cryotome cross-sections of roots 11 and 14 h following inoculation, but not earlier. No glyceollin I was detected in cross-sections of control roots and any time following inoculation.

These results demonstrate that the use of a radio-immunoassay provides a very sensitive technique for the quantitation of phytoalexins. The power of this technique is shown by the ability to quantitate phytoalexins in a few cell layers of sites of infection. Such studies could be complemented by analysis of glyceollin content at the cellular level using laser microprobe mass analysis (LAMMA) (Moesta et al. 1982).

References

Ayers AR, Ebel J, Finelli F, Berger N, Albersheim P (1976) Host-pathogen interactions. IX. Quantitative assays of elicitor activity and characterization of the elicitor present in the extracellular medium of cultures of *Phytophthora megasperma* var. *sojae*. Plant Physiol (Bethesda) 57:751–759

Bailey JA, Mansfield JW (eds) (1982) Phytoalexins. Blackie, Glasgow

Gerlach D (1977) Botanische Mikrotechnik, 1st edn. Thieme, Stuttgart, p 108

Greenwood FC, Hunter WM, Glover JS (1963) The preparation of ^{131}I-labeled human growth hormone of high specific radioactivity. Biochem J 89:114–123

Hahn MG, Bonhoff A, Grisebach H (1985) Quantitative localization of the phytoalexin glyceollin I in relation to fungal hyphae in soybean roots infected with *Phytophthora megasperma* f. sp. *glycinea*. Plant Physiol (Bethesda) 77:591–601

Ingham JL, Keen NT, Mulheirn LJ, Lyne RL (1981) Inducibly-formed isoflavonoids from leaves of soybean. Phytochemistry 20:795–798

Keen NT (1975) The isolation of phytoalexin from germinating seeds of *Cicer arietinum, Vigna sinensis, Arachis hypogaea*, and other plants. Phytopathology 65:91–92

Keen NT, Sims JJ, Erwin DC, Rice E, Partridge JE (1971) 6a-Hydroxyphaseollin: an antifungal chemical induced in soybean hypocotyls by *Phytophthora megasperma* var. *sojae*. Phytopathology 61:1084–1089

Lazarovits G, Ward EWB (1982) Relationship between localized glyceollin accumulation and metalaxyl treatment in the control of *Phytophthora* rot in soybean hypocotyls. Phytopathology 72:1217–1221

Lyne RL, Mulheirn LJ, Leworthy DP (1976) New pterocarpinoid phytoalexins of soybean. J Chem Soc Chem Commun 1976:497–498

Moesta P, Grisebach H (1981a) Investigation of the mechanism of phytoalexin accumulation in soybean induced by glucan or mercuric cloride. Arch Biochem Biophys 211:39–43

Moesta P, Grisebach H (1981b) Investigation of the mechanism of glyceollin accumulation in soybean infected by *Phytophthora megasperma* f. sp. *glycinea*. Arch Biochem Biophys 212:462–467

Moesta P, Seydel U, Lindner B, Grisebach H (1982) Detection of glyceollin on the cellular level in infected soybean by laser microprobe mass analysis. Z Naturforsch Sect C Biosci 37:748–751

Moesta P, Hahn MG, Grisebach H (1983a) Development of a radioimmunoassay for the soybean phytoalexin glyceollin I. Plant Physiol (Bethesda) 73:233–237
Moesta P, Grisebach H, Ziegler E (1983b) Immunohistochemical detection of *Phytophthora megasperma* f. sp. *glycinea*. Eur J Cell Biol 31:167–169
Morandi D, Bailey JA, Gianinazzi-Pearson V (1984) Isoflavonoid accumulation in soybean roots infected with vesicular-arbuscular mycorrhizal fungi. Physiol Plant Pathol 24:357–364
Rosenberg N, Pines M, Sela I (1982) Adenosine 3′:5′-cyclic monophosphate – Its release in a higher plant by an exogenous stimulus as detected by radioimmunoassay. FEBS Lett 137:105–107
Skipp RA, Bailey JA (1977) The fungitoxicity of isoflavonoid phytoalexins measured using different types of bioassay. Physiol Plant Pathol 11:101–112
Weiler EW (1980) Radioimmunoassays for the differential and direct analysis of free and conjugated abscisic acid in plant extract. Planta (Berl) 148:262–272
Weiler EW (1981) Radioimmunoassay for pmol-quantities of indol-3-acetic acid for use with highly stable [^{125}I] and [^{3}H]-IAA derivatives as radiotracers. Planta (Berl) 153:319–325
Weinstein LI, Hahn MG, Albersheim P (1981) Host-pathogen interactions XVIII. Isolation and biological activity of glycinol, a pterocarpan phytoalexin synthesized by soybeans. Plant Physiol (Bethesda) 68:358–363
Yoshikawa M, Yamachi K, Masago H (1978) Glyceollin: its role in restricting fungal growth in resistant soybean hypocotyls infected with *Phytophthora megasperma* var *sojae*. Physiol Plant Pathol 12:73–82
Zähringer U, Schaller E, Grisebach H (1981) Induction of phytoalexin synthesis in soybean. Structure and reactions of naturally occurring and enzymatically prepared prenylated pterocarpans from elicitor-treated cotyledons and cell cultures of soybean. Z Naturforsch Sect C Biosci 36:234–241

The Measurement of Low-Molecular-Weight, Non-Immunogenic Compounds by Immunoassay

R. J. Robins

1 Introduction

The analysis of compounds by immunoassay requires an antibody preparation active against the compound of interest. Small molecules (< 1000 daltons) do not themselves stimulate antibody formation. In order to produce the required antibodies, the small molecule (the hapten) must be covalently linked to a larger molecule (usually a protein) that is immunogenic and this conjugate used to immunise an animal. The antiserum thus generated will contain a mixture of antibodies, some of which will be capable of recognising the hapten even when free in solution. The application of such methods to the analysis of mammalian steroids resulted in the successful development of a number of immunoassays and revolutionised their determination (Abraham 1974). To date, immunological methods have been developed and applied in three major areas of plant research:

1. The assay of plant growth regulators;
2. The assay of plant secondary products;
3. The determination of toxic residues associated with material of plant origin.

1.1 The Analytical Role of the Immunoassay

An immunoassay offers a number of major advantages over more conventional forms of assay such as TLC, HPLC, and GC and those utilising enzymatic or biological properties. These tend to require considerable work-up and purification of the sample prior to analysis and may require the analyte to be concentrated to achieve a detectable limit. Furthermore, such methods may be of limited sensitivity, may require expensive equipment and chemicals and are ill-suited to handle routinely a large number of samples. In contrast, an immunoassay can be extremely sensitive and specific and seldom requires significant sample work-up. Several hundred samples may be analysed simultaneously and it is possible to perform an immunoassay with very simple equipment. The major disadvantage of an immunoassay is that an assay needs to be developed de novo for each compound of interest. Such development involves several months work and requires special facilities, particularly for the experimental animals. It is therefore important to consider the objectives of the experimental programme under consideration when evaluating the applicability of an immunoassay for the determination of a particular small molecule. In any instance where there occurs a regular analytical requirement for the quantitative determination of a single molecular species, particularly when that species occurs at very low concentration in the tissue, then an immunoassay is the method of choice.

1.2 Types of Immunoassay

There are many variations on the possible ways in which an immunoassay may be performed. These all depend on mixing samples containing unknown amounts of hapten with antiserum. Quantification of unknown hapten is by reference to the behaviour under identical conditions of standard amounts of hapten. This reaction of antiserum with hapten is most commonly brought about by one of four basic methods:

1. The radioimmunoassay (RIA), in which standard amount of radioactive hapten is mixed with the sample (Fig. 1 a). Radiolabelled and "cold" hapten compete for a limited number of antibody-binding sites. The extent to which the binding of radiolabelled hapten is distributed between "antibody-bound" and "antibody-free" fractions is related to the amount of cold hapten present. All the reagents in the assay are in free solution. For maximum sensitivity the binding should be allowed to come to equilibrium but this is rarely tested. Separation of the "bound" and "free" phases is conveniently achieved either by the rapid adsorption of the hapten or by precipitation of the protein, although a number of other methods have been developed (see Hunter and Corrie 1983), and the amount of radioactivity in one or other phase determined.

Competition immunoassays have also been developed using fluorescent- (Smith et al. 1981) or chemiluminescent- (Weeks and Woodhead 1984) labelled hapten and are performed in much the same way.

2. The direct, competitive enzyme-linked immunosorbent assay (EIA) (Voller 1978). In this method (Fig. 1 b) hapten is labelled with enzyme rather than radioactivity. The antibody, which again provides a limited number of binding sites, is bound to a solid phase, usually polystyrene tubes or the wells of a microtitration plate. Competition for the immobilised binding sites occurs between the free hapten and the enzyme-linked hapten, and the "free" phase is readily separated from the "bound" phase by rapidly removing the liquid from the vessel coated with immobilised antibody. The differing amounts of enzyme that have been bound by the immobilised antibodies in different tubes or wells are detected by addition of substrate.

3. The double-antibody, indirect enzyme-linked immunosorbent assay (ELISA) (Voller et al. 1979; Morgan et al. 1983b). Here, hapten is conjugated to a protein to form the immobilised phase and adsorbed to a support, most commonly the polystyrene wells of a microtitration plate (Fig. 1 c). Free hapten and antiserum are introduced, and antiserum is distributed between free hapten and the immobilised hapten-protein conjugate. "Bound" and "free" phases are again rapidly separated by decanting liquid from tubes or wells. A second antibody, specific to the IgG fraction from the animal in which anti-hapten antiserum was raised and covalently labelled with enzyme, is introduced in excess. This antibody binds to the primary antibody retained on the solid support. After the liquid

(a) RIA

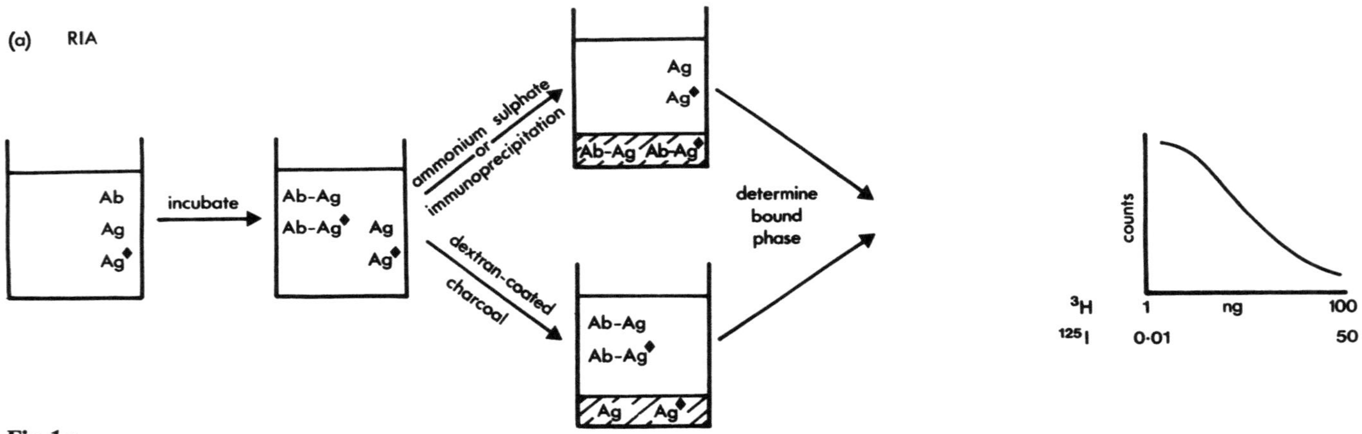

Fig. 1a

Fig. 1a–e. The basic techniques for performing an immunoassay. For details, see text. **a** Radioimmunoassay (RIA); **b** the direct enzyme-linked immunosorbent assay (EIA); **c** the indirect enzyme-linked immunosorbent assay (ELISA); **d** the fluorescent polarisation immunoassay (FPIA); **e** the enzyme-multiplied immunoassay (EMIT). *Abbreviations* used: Antibody, *Ab*; antigen, *Ag*; radio-labelled antigen, *Ag*◆; antibody-antigen complex, *Ab-Ag*; antibody-radiolabelled antigen complex, *Ab-Ab*◆; enzyme-labelled antigen, *Ag-E*; antibody-enzyme-labelled antigen complex, *Ab-Ag-E*; substrate, *S*; product, *P*; protein-antigen conjugate, *B-Ag*; protein-antigen-conjugate complexed to antibody, *B-Ag-Ab*; enzyme-labelled second antibody *2Ab-E*; protein-antigen conjugate complexed to antibody and further complexed to enzyme-labelled second antibody *B-Ag-Ab-2Ab-E*; fluorescent-labelled antigen, *Ag-F*; antibody-fluorescent-labelled antigen complex, *Ab-Ag-F*; antibody-enzyme-labelled antigen complex in which enzyme activity is suppressed, *Ag-E-Ab*

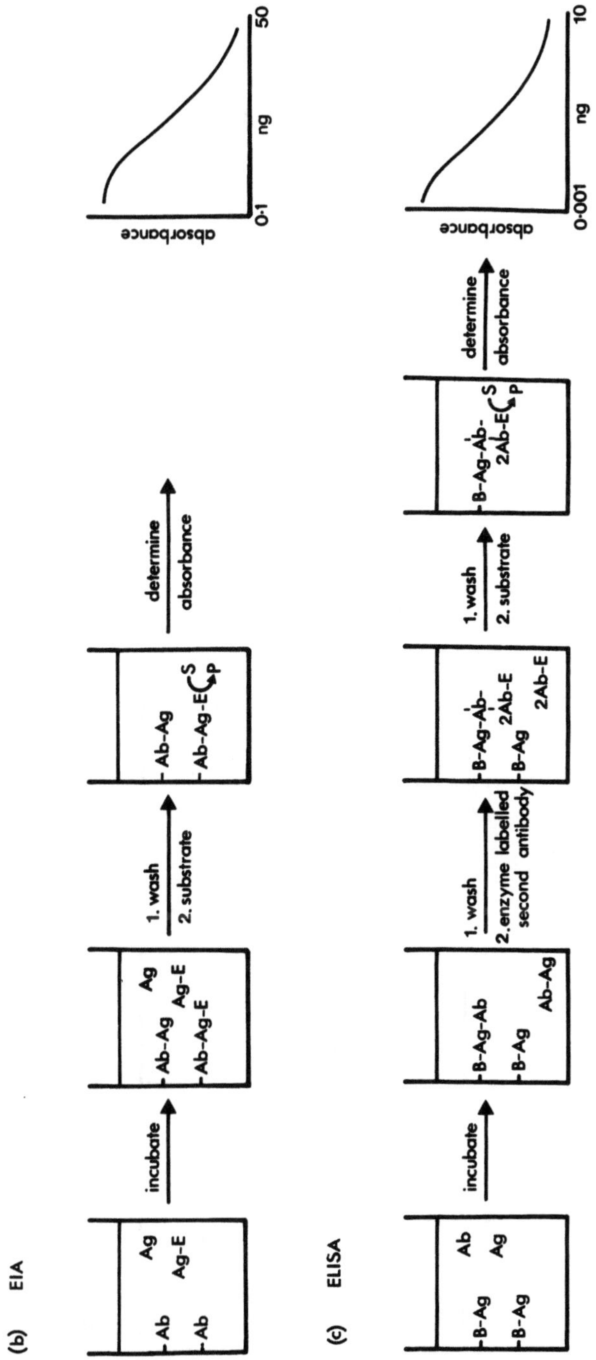
(b) EIA
Ab Ag Ag-E Ab
incubate
Ab-Ag Ag Ag-E Ab-Ag-E
1. wash
2. substrate
Ab-Ag Ab-Ag-E S P
determine absorbance
absorbance
0·1 ng 50
(c) ELISA
B-Ag Ab Ag B-Ag
incubate
B-Ag-Ab B-Ag Ab-Ag
1. wash
2. enzyme labelled second antibody
B-Ag-Ab-2Ab-E B-Ag 2Ab-E
1. wash
2. substrate
B-Ag-Ab-2Ab-E S P
determine absorbance
absorbance
0·001 ng 10

Fig. 1 b, c

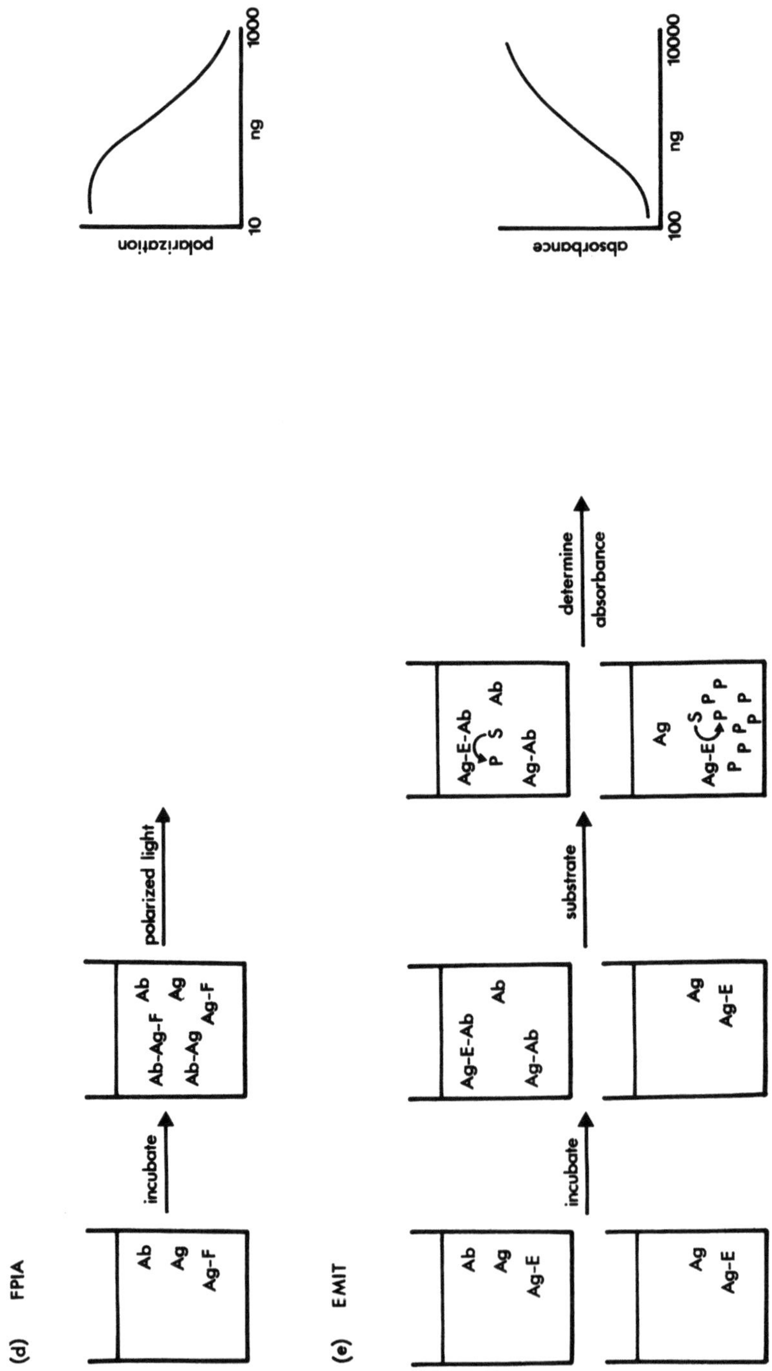

(d) FPIA
Ab
Ag
Ag-F
incubate
Ab-Ag-F
Ab
Ab-Ag
Ag
Ag-F
polarized light
polarization
10
ng
1000
(e) EMIT
Ab
Ag
Ag-E
Ag
Ag-E
incubate
Ag-E-Ab
Ab
Ag-Ab
Ag
Ag-E
substrate
Ag-E-Ab
P S
Ab
Ag-Ab
Ag
Ag-E
S
P P P P P P P P
determine absorbance
absorbance
100
ng
10000

Fig. 1 d, e

phase has again been removed, the amount of second antibody bound is quantified with a substrate appropriate to the enzyme present.

4. The homogeneous immunoassay. In this type of assay the whole process is carried out without having to separate bound and free phases. Instead, it relies on observing the change in property of a molecule that results from the formation of a hapten-antibody complex.

In the fluorescence-polarization immunoassay (FPIA) the hapten is coupled to fluorescein (Dandliker and Saussure 1970), which is excited to fluoresce by a beam of polarized blue light (Fig. 1 d). In the free state the hapten-fluorescein conjugate rotates freely and the degree of polarization of the emitted light is small. When bound to antibody, however, the rotational freedom is restricted and the degree of polarization is increased. The amount of hapten in a sample is thus quantified by the extent to which it displaces the polarization away from the maximum.

In another form (Fig. 1 e), the enzyme-multiplied immunoassay (EMIT), enzyme-labelled hapten is used (Rowley et al. 1975). Hapten is coupled to enzyme in such a way that when the hapten portion binds to antibody, enzyme activity is reduced due to occlusion of the active site preventing access of substrate. Thus, the extent to which enzyme activity is enhanced by the presence of free hapten in a sample gives a measure of the amount present.

Each type of assay has its advantages and disadvantages. The RIA requires the synthesis of a radiolabelled hapten, as few are available commercially for plant products (see Appendix). With many natural products this may prove difficult although a number of satisfactory radiolabelling methods have been developed (Sect. 2.3.1). Expensive equipment is also required, most often a centrifuge capable of spinning 100–200 sample tubes simultaneously and a liquid scintillation counter [^{3}H] or γ-emission counter [^{125}I]. One great advantage with [^{3}H] is that the tracer species can be made so that its molecular structure differs from the analyte only in isotopic composition. The use of fluorescent- or chemiluminescent-labelled haptens, while avoiding the use of radioisotopes, loses this advantage. As expensive dedicated equipment is required for their use it remains to be seen whether these recently developed techniques will find wide application.

By using an enzyme-labelled hapten and an immobilised phase the EIA simplifies the analytical procedure as phase separation is easier and quantification only requires a colorimetric recorder. It does, however, require the synthesis of enzyme-labelled hapten of a particular hapten:enzyme ratio for maximal performance. Furthermore, it uses the most valuable component of the assay – the antiserum – inefficiently. This is because adsorption is incomplete and will mask some binding sites. Another disadvantage is that the enzyme and the crude extract-containing sample are incubated together, which may lead to problems, such as enzyme inactivation, the severity of which will vary from sample to sample. The homogeneous assays also suffer from this problem and, additionally, are much less sensitive. Furthermore, the synthesis of enzyme-antigen conjugates of the required properties can prove very difficult.

The ELISA avoids many of these disadvantages. The synthesis of a coating conjugate is no more difficult than the synthesis of immunogen and the protein is not required to retain enzymatic activity. The antiserum is used at much greater dilution, which favours high affinity, high specificity antibodies, tending to provide an assay of superior sensitivity and specificity (but see Lehtonen and Eerola 1982). Although it involves an extra incubation step, enzyme-labelled second antibodies to IgG from a number of species are readily available commercially (see Appendix) and the use of a second antibody can theoretically improve sensitivity (Ekins 1980) and provide wider working ranges for the standard curve. This method is rapidly gaining in popularity and for most purposes is the best method currently available. By quantifying the ELISA with a radiometric substrate, as little as 10^{-16} g (6×10^2 molecules) can be detected (Harris et al. 1979).

1.3 Automation

The major advantage of the homogeneous assays is that by performing the assay in a single phase it is readily automated, making it particularly attractive for routine analysis with a very large sample through-put. Although rather more complex to achieve, however, it is also possible to automate the EIA (Carlier et al. 1979) or ELISA and process 900–1000 samples/day. For screening a very large number of samples semi-quantitatively, Weiler and Zenk (1979) have developed an autoradiographic immunoassay which they claim can be used to analyse 10000 samples/day and have used this to screen for digoxin production (Weiler 1977).

At the other extreme, it is possible to perform an EIA or ELISA under field conditions, using a portable, battery-operated photometer (Rook and Cameron 1981), making the technique very versatile.

2 Development of an Immunoassay

2.1 Synthesis of Immunogen

To synthesise a suitable immunogen for a hapten it is necessary to conjugate the hapten covalently to a larger molecule such as a protein. It has long been known that antibody specificity is directed to sites on the hapten distal to the point of conjugation (Landsteiner 1945). This is because regions of the molecule are occluded due to proximity to the polypeptide chain. It is important, therefore, to consider very carefully the features of the molecule which are desired to be involved in determining the specifity of the antiserum. For example, if it is intended to assay for a compound when an hydroxylated product also occurs in the tissue, then this group should be distal to the point of conjugation, ensuring that it is exposed to the immune system. Conversely, if it is desirable to assay both the compound and its hydroxylated product then by conjugating through the hydroxyl group the difference is destroyed, generating an antiserum capable of recognising both molecular species to the same degree. Hence, by the careful choice of different linkage moieties it is possible to generate antiserum of predicted selectivity in the assay (Robins et al. 1985). The use of antisera made by different coupling methods to obtain different specificities is well demonstrated for abscisic acid (Weiler 1980).

Table 1. Summary of the conjugation methods and their application

Directed towards	Method	Application	Ref.
-COOH	Mixed-anhydride	Ajmalicine	Arens et al. (1978)
		Quassin	Robins et al. (1984a)
		Ochratoxin A	Morgan et al. (1983a)
	Carbodiimide	Cotinine	Langone et al. (1973)
		Vinblastine	Langone et al. (1979)
		Vincristine	Langone et al. (1979)
		Serpentine	Arens et al. (1978)
		Quassin	Robins et al. (1984a)
		Sennoside	Atzorn et al. (1981)
		Colchicine	Boudene et al. (1975)
		Ochratoxin A	Morgan et al. (1983a)
-OH	Succinic anhydride	Nicotine	Langone et al. (1973)
		Vindoline	Kutney et al. (1980)
		Quinine	Robins et al. (1984c)
		Quinidine	Morgan et al. (1985c)
		Bruceantin	Fong et al. (1980)
		Quassin	Robins et al. (1984b)
		Digitoxin	Oliver et al. (1968)
		Solanidine	Vallejo and Ercegovich (1979)
		Chloramphenicol	Campbell et al. (1984)
	Glutaric anhydride	Vindoline	Westekemper et al. (1980)
	Sebacoyl dichloride	Sterigmatocystin	Morgan et al. (1986a)
	Trans-1,4-cyclohexan-dicarbonyl dichloride	Sterigmatocystin	Morgan et al. (1986a)
 -OH	Carboxymethylation	Morphine	Spector and Parker (1970)
		Morphine	Vunakis et al. (1972)
		Morphine	Adler and Liu (1971)
	Diazotisation	Naringin	Jourdan et al. (1982)
		Chloramphenicol	Hamburger (1966)
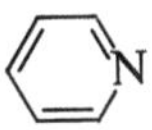	6-Bromohexanoic acid	Paraquat	Niewola et al. (1983)
$-NH_2$	Mannich reaction	Vinblastine	Teal et al. (1977)
		Vinblastine	Hacker et al. (1984)
	Acrylic acid	Scopolamine	Weiler et al. (1981)
>=O	Amino-oxyacetic acid	Limonin	Mansell and Weiler (1980)
		Naringin	Jourdan et al. (1983)
		Aflatoxin B_1	Morgan et al. (1985b, c)
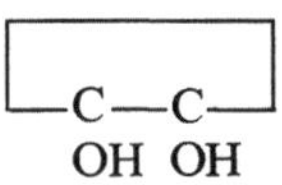	Periodate cleavage	Digoxin	Butler and Chen (1967)
		Solanine	Morgan et al. (1983b)
		Solasodine	Weiler et al. (1980)
	6′-Carboxylic acid formation	Loganin	Tanahashi et al. (1984)

The range of functional groups present on most natural products provides a number of possible modes for conjugation. If no satisfactory group is present then it is often possible to use a derivative of the molecule to be assayed in which a functional group occurs at a satisfactory position to give the desired exposure of the hapten in the conjugate.

2.1.1 Methods of Conjugation

A number of methods have been successfully used for conjugation, linkage through a carboxylic acid or alcohol being the most common. Frequently a two-stage synthesis is applied, in which the first step involves a bifunctional reagent and generates a derivative containing a free carboxylic acid. Table 1 summarises the various methods and their application to a range of compounds. In the following sections details of the methodology are given for a particular compound, which is indicated in parenthesis with the relevant reference at the end of each sub-section. When a method is applied to a new compound minor changes in the conditions may be required and the progress of the reaction should be monitored by HPLC or TLC and the structure of intermediates confirmed by reference to standard compounds or by MS and NMR spectroscopy.

2.1.1.1 Carboxylic Acid-Directed Reagents

(a) The mixed anhydride method (Erlanger et al. 1959). Dissolve protein (80 mg) in water (4.3 ml) and adjust to pH 9.5 with NaOH. Add 1,4-dioxan (2.6 ml), cool to 10 °C, and filter to remove undissolved protein if necessary. Dissolve compound (25 mg) in 1,4-dioxan (2.1 ml), cool to 10 °C, add tri-n-butylamine (0.05 ml), leave 20 min then add iso-butylchlorocarbonate (0.02 ml) and leave a further 20 min. Mix solution of activated compound with protein solution, rapidly adjust pH to 7–8 and stir for 4 h at 4 °C. Lyophilise after extensive dialysis against distilled water. (Quassin: Robins et al. 1984a.)

(b) The carbodiimide method (Meisner and Meisner 1981). The disadvantage of this method is that it frequently leads to insoluble products. Dissolve protein (60 mg) in 0.01 M NaCl and remove undissolved material by filtration. Dissolve compound (8 mg) in EtOH (0.12 ml) and add to 0.01 M phosphate buffer pH 7.0 (3.0 ml). Mix with protein solution, add 1-ethyl-3(dimethylaminopropyl) carbodiimide (28 mg) and stir for 24 h at 20 °C. Dialyse extensively against water and lyophilise. (Quassin: Robins et al. 1984a.)

2.1.1.2 Alcohol-Directed Reagents

(a) Succinic anhydride. Dissolve compound (40 mg) and succinic anhydride (200 mg) in pyridine (1.0 ml). Flush vessel with N_2, stopper and leave at room temperature. After 2–4 days, or when the reaction has been shown to go to completion, remove the pyridine by rotary evaporation, dissolve product in a suitable solvent and wash with water to destroy and remove residual anhydride. Collect the organic phase containing the hemisuccinate derivative and dry by rotary evaporation. Conjugate to protein through the carboxylic acid group (Sect. 2.1.1.1). (Quassin: Robins et al. 1984b.)

(b) Glutaric anhydride. Proceed as in (a) but substitute glutaric anhydride (250 mg) for succinic anhydride. (Vindoline: Westekemper et al. 1980.)

(c) Sebacoyl dichloride. Weigh compound (20 mg) into a dry vessel and add sebacoyl dichloride (200 µl) dropwise. Heat at 100 °C for 5 min then decrease temperature to 50 °C for 10 min. Cool and distill off excess sebacoyl dichloride in vacuo for 16 h. Dissolve product in 1,4-dioxan (6 ml). Dissolve protein (200 mg) in 10 ml 0.1 M NaCl, add 0.5 ml 0.1 M NaOH, and cool on ice. Mix product and protein solutions and stir for several hours at 4 °C. Dialyse extensively against water and lyophilise conjugate. (Sterigmatocystin: Morgan et al. 1986a.)

(d) The *trans*-1,4-cyclohexan-dicarbonyldichloride method. Dissolve 1,4-benzene dicarboxylic acid (18 mg) in the minimum amount of tetrahydrofuran. Add thionylchloride (20 mg) dropwise and reflux overnight. Cool and distill off excess thionylchloride and solvent in vacuo, leaving a brown liquid. Proceed as described for sebacoyl dichloride. (Sterigmatocystin: Morgan et al. 1986a.)

2.1.1.3 Phenolic-Directed Reagents

(a) Carboxymethylation. The free phenolic group is reacted with sodium-β-chloroacetate in absolute ethanol. (Morphine: Spector and Parker 1970.)

(b) Diazotization. The presence of a phenolic group on an aromatic ring activates the position *ortho* to the substituent and coupling may be achieved in mild conditions by diazotization to an aromatic amine group on the protein. Dissolve compound (18 mg) in MeOH (0.9 ml), mix with 0.1 M borate buffer pH 9.0 (30 ml) and cool on ice. Dissolve 4-aminohippurate-BSA (Sect. 2.1.2.3) (87 mg) in water (15 ml) and adjust pH to 1.5 with 1 M HCl, cool on ice and, while gently stirring, add dropwise sodium nitrite (180 mg in 1.5 ml water). After 4 min add ammonium sulfamate (90 mg in 1.5 ml water). Leave 1 min and add compound solution, stirring vigorously for 5 min then gently for 1.5 h. Dialyse and lyophilise product. (Naringin: Jourdan et al. 1982.)

2.1.1.4 N-in-Aromatic-Ring-Directed Reagents

6-Bromohexanoic acid. Dissolve compound (0.5 g) in propan-2-ol, add 6-bromohexanoic acid (1.0 g), mix, and reflux for 22 h. Filter off precipitate and recrystallise from propan-2-ol. Couple to protein using the carboxylic acid group (Sect. 2.1.1.1). (Paraquat: Niewola et al. 1983.)

2.1.1.5 Amine-Directed Reagents

(a) Mannich reaction. Dissolve compound (30 mg) in water (1.5 ml). Dissolve BSA (100 mg) in water (2.0 ml), add 3 M sodium acetate (0.67 ml) followed by 7.5% (w/v) formaldehyde (1.2 ml) and mix with solution of compound. Leave stirring overnight at room temperature, dialyse extensively and lyophilise conjugate. (Vinblastine: Teal et al. 1977.)

(b) Acrylic acid. Dissolve compound (2.0 g) in EtOH (4.0 ml) and dissolve acrylic acid (580 mg) in EtOH (1.0 ml). Mix reagents and incubate at 70 °C for 5.5. h. Add a further aliquot of acrylic acid (385 mg) in EtOH (0.65 ml) and after

2 h at 70 °C reduce temperature to 20 °C for 43 h or until reaction has gone to completion. Reduce temperature to 4 °C for 20 h and harvest crystals of propionic acid derivative. Couple to protein using the carboxylic acid group (Sect. 2.1.1.1). (Scopolamine: Weiler et al. 1981.)

2.1.1.6 Ketone-Directed Reagents

(a) Amino-oxyacetic acid. Dissolve compound (500 mg) and amino-oxyacetic acid (300 mg) in 20 ml dry EtOH. Add to 20 ml dry pyridine and reflux for 3 h. Remove all traces of pyridine by evaporation in vacuo (35 °C), dissolve the oily residue in the minimum quantity of warm water and store at 4 °C. Filter off crystals of carboxymethyl oxime product. Couple to protein using the carboxylic acid group (Sect. 2.1.1.1). (Naringin: Jourdan et al. 1983.)

2.1.1.7 Sugar-Directed Reagents

(a) The periodate cleavage method (Butler and Chen 1967). Mix compound (150 mg) with water (15 ml) for 2 h at room temperature. (It is not essential to dissolve all the substrate as the products are much more soluble.) Add 15 mM sodium periodate (15 ml) and leave 16 h, in the dark at room temperature. Add 1 M ethylene glycol (2.7 ml) and leave 15 min. Dissolve protein (150 mg) in water (30 ml) and adjust to pH 9.5 with 40 mM potassium carbonate. Mix protein solution with "activated" compound and stir for 1 h at room temperature. Add sodium borohydride (450 mg) and leave overnight in the dark. Dialyse extensively against water and lyophilise product. (Solanine: Morgan et al. 1983b.)

(b) Conversion to the 6′-carboxylic acid. Dissolve compound (1.0 g) in pyridine (8 ml), add trityl chloride (1.4 g), leave 4 days and add acetic anhydride (6 ml). Leave overnight and separate acetylated 6′-O-trityl product by chromatography. Dissolve 6′-O-trityl product (565 mg) in 80% acetic acid (18 ml) and heat at 80 °C for 40 min. Cool, dilute with water and extract with chloroform (3 × 50 ml). Wash organic layers with saturated sodium hydrogen carbonate solution and water and dry in vacuo. Dissolve residue in acetone (15 ml), treat with Jones' reagent for 1 h, add propan-2-ol, dilute with water and extract with chloroform (3 × 50 ml). Concentrate chloroform in vacuo to 50 ml and extract with saturated sodium hydrogen carbonate (5 × 30 ml). Retain aqueous phase, slightly acidify with 1 M HCl, extract with chloroform and dry organic phase in vacuo. Dissolve residue (150 mg) in methanol (30 ml), add a saturated methanolic barium hydroxide solution (1.0 ml) and a few drops water. Stir at room temperature for 4 h, neutralise with Amberlite IR-120 (H^+), filter off resin and dry filtrate to obtain product. Conjugate using the free acidic group (Sect. 2.1.1.1). (Loganin: Tanahashi et al. 1984.)

2.1.2 Bridge Conjugates

Some of the conjugation methods described in Sect. 2.1.1 insert a short spacer molecule between the protein and the hapten. In some cases in which the spacer is only a C_2 or C_3 unit, conjugates so constructed have failed to raise an antiserum able to recognise free hapten. In these instances it may prove valuable (see for example Sect. 3.1.4) to introduce a spacer molecule of a further C_5 or C_6 unit, which presumably improves the presentation of hapten for antibody recognition.

2.1.2.1 The 6-Amino-N-Hexanoic Acid Method

The compound to be conjugated (25 mg) is "activated" by a suitable method (Sect. 2.1.1) to leave a free carboxylic acid group to react with the amino function. Dissolve 6-amino-N-hexanoic acid (10 mg) in water (4.3 ml), adjust the pH to 9.5 with NaOH, add 1,4-dioxan (2.6 ml) and cool to 12 °C. Mix with the solution of activated compound, rapidly adjust the pH to 7–8 and stir for 4 h at 4 °C. Conjugate the remaining free carboxylic acid to protein (Sect. 2.1.1.1). (Quinidine: Morgan et al. 1985c.)

2.1.2.2 Alcohol-Directed Reagents

The substitution of sebacoyl dichloride for succinic or glutaric anhydride will increase the length of the bridge when an alcoholic moiety is being used for conjugation (Sect. 2.1.1.2). Alternatively, if a more rigid bridge is required, use *trans*-1,4-cyclohexan-dicarboxydichloride instead (Sect. 2.1.1.2).

2.1.2.3 The 4-Aminohippuric Acid Method

In this method the bridge is bound to the conjugating protein. Dissolve 4-aminohippuric acid (400 mg) in water (240 ml) made to about pH 8.0. Add BSA (400 mg), readjust the pH to 8.0, add 1-ethyl-3-(3-dimethyl-aminopropyl) carbodiimide, adjust to pH 6.4 with 2 M HCl and stir at room temperature for 6 h. Add further carbodiimide (200 mg), readjust pH to 6.4 and leave for 16 h. Dialyse and lyophilise. The BSA is now activated with a side chain terminating in an aromatic amine group. (Naringin: Jourdan et al. 1982.)

2.1.3 Choice of Protein

In the majority of methods the final linkage to a protein involves a Schiff base reaction between the activated hapten and the amine groups of the protein. The availability of lysine groups is therefore important. Although polylysine can be used (Langone et al. 1973; Vunakis et al. 1972), it does not appear to provide a superior carrier to proteins such as bovine (BSA) and human (HSA) serum albumins or keyhole limpet hemocyanin (KHLH), which are the most frequently used protein carriers for the hapten. In the ELISA, where conjugation to two proteins is required, it is important that these are unrelated proteins and that the animal will not have encountered them previously.

Although not strictly necessary, it is common practice to pass the crude conjugate through a column of Sephadex G-25 prior to final lyophilisation and storage. This ensures that all unreacted compound and reagents used in excess still left after dialysis are removed from the final product. It is also normal to ascertain the molar ratio of the conjugate. It seems that both very high and very low densities of hapten should be avoided for optimal antisera quality. For BSA, ratios of between 10:1 and 25:1 (hapten:protein) should be attained – though successful antisera have been raised to conjugates of lower and higher ratios.

While it is frequently possible to obtain an estimate of the molar ratio by spectroscopy, usually UV or fluorescence, in some instances it may be necessary to disrupt chemically the conjugate and assay the free hapten by chromatographic techniques (Morgan et al. 1983b). Although tedious, this ensures that several months work is not wasted by injecting an inferior immunogen, which may lead to a very poor response.

2.2 Production of Antiserum

For an immunoassay, antiserum may be prepared in any experimental animal, although normally rabbits, guinea pigs, mice or rats are used. Larger animals, such as sheep, have the advantage of providing a greater blood volume. Monoclonal antibodies, which once identified are available in unlimited supply, will not be discussed in this chapter as their application in this area has been very limited to date. If, however, it is envisaged that a large quantity of antiserum is to be required then mice or rats must be used as, currently, suitable hybridomas are only available for these species. For a recent review of the generation of monoclonal antibodies see Galfre and Milstein (1981).

Immunisation may be intramuscular, subcutaneous, intravenous, intraperitoneal or a combination of these (Herbert 1978). The objective is to introduce the immunogen in a way that ensures its slow release into the circulatory system, leading to maximal exposure of the immune system to the foreign protein, thus generating a strong response. The responsivity is enhanced by introducing the immunogen homogeneously mixed with Freund's complete adjuvant, a water-in-oil emulsion containing killed *Mycobacterium*, and a non-ionic detergent such as Tween 80. Booster injections are administered at periodic intervals using a similar mixture with Freund's incomplete adjuvant, which lacks the mycobacteria. As a number of alternative routines are available, only those used in our laboratory will be given in detail.

Whatever the animal used, a sample of pre-immune serum should be taken prior to administering immunogen to check for non-specific binding.

2.2.1 Immunisation Procedure

Blend to homogeneity immunogen (15 mg dissolved in 5 ml 0.15 M NaCl), Freund's complete adjuvant (5 ml) and Tween 80 [5 ml of a 1% (w/v) solution]. Inject rabbits subcutaneously into the back (4 × 0.5 ml) and intramuscularly in each hind leg (1.0 ml). Inject mice or rats intraperitoneally (0.2 ml). Administer booster injections at monthly intervals in which Freund's incomplete adjuvant is used. At 8, 11 and 14 days after the second booster injection and each booster thereafter, collect about 10 ml blood from the marginal ear vein of rabbits or 0.5 ml blood from the tail of mice or rats into heparinized tubes, centrifuge at 3000 rpm for 15 min (4 °C) and store at −40 °C.

An alternative procedure for rabbits using reagents as above except an immunogen concentration of only 100–500 μg ml^{-1}, is to shave the back of the animal and inject approximately 0.05 ml into 20 sites intradermally (a total of 1 ml per rabbit). The booster injections and bleeds are performed at the same times as for the former method. This method requires very much less conjugate.

Fully illustrated details of these procedures are give in Herbert (1978). They should result in antibody production against a good hapten-protein conjugate in sufficient amount and of sufficient avidity to allow the serum to be used in an assay for hapten at dilutions of greater that 1000-fold. Such high dilution is important for continuity of work, for the reduction of non-specific binding effects, and to allow wide dissemination of sera. There are, of course, many reports of serum being used at lower dilutions than 1:1000.

2.3 Synthesis of Labelled Tracers

2.3.1 Radiolabelled Tracers

The method of synthesis of a radiolabelled tracer is dependent on the precise chemistry of the compound to be assayed, so general methods cannot be described. There are, however, two approaches to the incorporation of a radiolabel which can be applied in a number of cases.

2.3.1.1 Aromatic Ring Substitution

The protons on an aromatic ring can be exchanged under acidic conditions. In practice, great care is required to establish precisely the right conditions to avoid undesirable side reactions, in particular acylation of the aromatic ring, and the method needs careful development for each molecule. For very labile molecules, such as vincristine, it is possible to use trifluoroacetic acid, tritiated in the acidic group, and achieve specific tritiation of the aromatic ring in the presence of a Pt-C catalyst (Owellen and Donigian 1972).

2.3.1.2 Incorporation at the Point of Conjugation

It is sometimes possible to synthesise a useable radiolabelled tracer by the incorporation of a small radiolabelled substituent of high specific activity at the point of conjugation. Problems can occur because some antibody preparations recognise a radiolabelled tracer made in this way with greater affinity than "cold" hapten. Fortunately, this phenomenon, known as bridge binding (Corrie 1983), is not always present.

(a) Acetylation. Alcohol groups can readily be acetylated using [^{3}H]-acetic anhydride. Dissolve compound (5 mg) in pyridine (0.5 ml) and introduce into the upper chamber of a sealed [^{3}H]-acetic anhydride tube (25 mCi) cooled in dry ice. Stopper firmly, break seal, and run pyridine into lower chamber. Allow to warm to room temperature and leave 1–2 h or longer until reaction is complete (as determined by a "cold" reaction). Separate [^{3}H]-acetyl derivative from unreacted substrate by, for example, TLC. (Quinine: Robins et al. 1984c.)

(b) Alkylation. Carboxylic acids can readily be alkylated using [^{3}H]-diazomethane. Purchase N-[^{3}H]-methyl-N'-nitro-N-nitrosoguanidine (from e.g. Amersham) and an MNNG-diazomethane generation apparatus, millimole size (Aldrich Chemical Co.) and generate [^{3}H]-diazomethane as per instructions, collecting in ether. Add, dropwise, [^{3}H]-diazomethane solution to a solution of compound until solution just stays yellow. Work up product. Both diazomethane and the precursor are highly carcinogenic.

(c) N-methylation. Secondary amines will react with ^{3}[H]-methyl iodide. The reaction should be carried out in vacuo in an apparatus as described by Werner and Mohammad (1966). Dissolve compound (85 mg) in ether/methanol (4:1, 2 ml) and combine with 21 mg [^{3}H]-methyl iodide. Heat at 40 °C for 170 h. Cool and dry off solvent. Wash product with water (4.5 ml), dry over sodium sulphate in a desiccator and work-up. (Scopolamine: Weiler et al. 1981.)

2.3.1.3 Reduction with Sodium [^{3}H]-Borohydride

Dissolve compound (3.3 mg) in methanol (0.3 ml) and add to sodium [^{3}H]-borohydride (0.36 mg). If solution decolourises within 30 min add compound (3.3 mg) and leave 22 h. Dilute with chloroform (0.2 ml) and work-up by, for example, TLC (Ajmalicine: Arens et al. 1978).

2.3.1.4 [^{125}I]-Incorporation

The method described by Greenwood et al. (1963) for the iodination of growth hormone is widely used. Purchase sodium [^{125}I]-iodide (2–4 mCi in 0.05 ml solution) in a rubber-capped vial. Add the following by injection, mixing after each addition: 0.5 M sodium phosphate buffer pH 7.5 (0.025 ml); sample (5 µg in 0.025 ml 0.05 M sodium phosphate buffer, pH 7.5); fresh chloramine-T (100 µg in 0.025 ml of 0.05 M sodium phosphate, pH 7.5); sodium metabisulphite (240 µg in 0.1 ml of 0.05 M sodium phosphate buffer, pH 7.5). The iodinated product then requires appropriate work-up to provide a labelled tracer.

[^{125}I]-Incorporation offers some advantages over [^{3}H]. Tracers are of higher specific activity, leading to greater potential assay sensitivity; the equipment and materials for its detection are cheaper and γ-isotope counting is more rapid than [^{3}H] determination. In most cases, however, these are outweighed by the disadvantages. The isotope has a short half-life, making it necessary to resynthesise fresh tracer every few months; radio-induced decay of the samples may be greater than with [^{3}H]; greater precautions are required for the handling of γ-emitters; the structure of the tracer molecule cannot be as close to the unlabelled form as with [^{3}H]. Usually, [^{3}H] is now the tracer of choice for small molecules but [^{125}I] is still used in certain circumstances, such as when the synthesis of a [^{3}H]-labelled tracer is unsuccessful or when the laboratory is only equipped for the determination of one type of emitter.

2.3.2 Enzyme-Linked Tracers

The preparation of an enzyme-linked tracer for EIA may be carried out by substitution of enzyme for the protein carrier in the procedure used to make the immunogen, as described in Sect. 2.2. Care must be taken to avoid loss of enzyme activity and to produce a conjugate of particular hapten:enzyme ratio in order to realise maximum assay sensitivity. Bridge binding (Sect. 2.3.1.2) may also be a problem, necessitating use of an alternative coupling procedure. The use of pairs of bifunctional reagents, such as the straight-chained sebacoyl dichloride and *trans*-1,4-cyclohexandicarbonyldichloride (Sect. 2.1.1.2), may prove particularly useful in this context (Morgan et al. 1986a).

Alkaline phosphatase (EC.3.1.3.1), horseradish peroxidase (EC.1.11.1.7) and β-galactosidase (EC.3.2.1.23) are the most commonly used enzymes because of their cost and availability in pure form, their stability in a variety of coupling procedures, their suitable specific activities and the ease of their colorimetric assays.

Enzyme-linked second antibodies are commercially available (see Appendix), an advantage of using the ELISA.

2.4 ELISA Procedure

As the ELISA is the method of choice only this technique will be described in detail. The RIA is described elsewhere in this volume (page 88, Fig. 1a) and the EIA by Voller (1978). For a successful ELISA a number of inter-related parameters need to be varied to define a satisfactory system, and this will be exemplified using the immunoassay for quassin (Robins et al. 1984a).

2.4.1 Microtitration Plates

A number of different makes of microtitration plates are available (see Appendix) and, although ostensibly interchangeable, some assays work better in one plate type than another. Routinely, we use NUNC (with Certificate) or Sterilin plates. It is advisable to try several makes.

2.4.2 Buffers

(a) Coating buffer: 0.05 M carbonate/hydrogen carbonate, pH 9.6 (formula: sodium carbonate, 1.59 g; sodium hydrogen carbonate, 2.93 g; sodium azide 0.2 g; water, 1.0 litre;

(b) rinse buffer (PBS-Tween): phosphate-buffered saline pH 7.4, containing 0.5% (v/v) Tween 20 (formula: sodium chloride, 8 g; potassium dihydrogen phosphate, 0.2 g; disodium hydrogen phosphate duodecahydrate 2.9 g; potassium chloride, 0.2 g; sodium azide, 0.2 g; Tween 20, 0.5 ml; water, 1.0 litre).

2.4.3 Initial Titre Measurement

To confirm the presence of antibodies in the serum a dilution curve is required. In a RIA this is achieved by looking at the binding of radiolabelled hapten but in an ELISA it can be a little more difficult as it is dependent on the coating concentration and second antibody dilutions.

Coat plates at 1 μg ml^{-1} (see Sect. 2.4.4). Dilute antiserum over a range 1:10 to 1:10^6 in ten fold steps. Place, in triplicate, 0.2 ml into wells and leave overnight

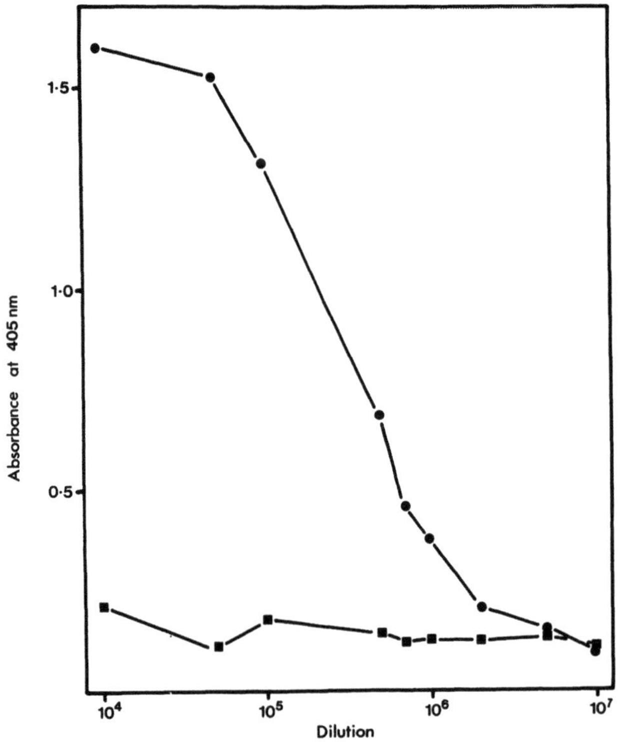

Fig. 2. Dilution curve for anti-iso-quassinic acid antiserum obtained by ELISA showing the effect of increasing the dilution of primary antiserum in the assay. For details see text. Serum obtained prior to immunisation (■); serum obtained after immunisation and three booster injections administered at monthly intervals (●)

at 4 °C in a saturated atmosphere. Wash wells 5 × 0.3 ml PBS-Tween and introduce 0.2 ml enzyme-labelled second antibody diluted 1:750 in buffer B. Incubate at 37 °C for 2 h and wash 5 × 0.3 ml PBS-Tween. Introduce 0.2 ml substrate solution (see Sect. 2.4.8) into each well and incubate at 37 °C for 1–2 h. An approximation to the dilution at which the antiserum should be used will be given by the dilution causing a 50% decrease in optical density compared to a pre-immune serum blank (Fig. 2).

2.4.4 Coating Concentration

Dissolve coating conjugate in buffer (A) at ten-fold increments of dilution over the range 10–0.01 μg ml^{-1}. After stirring 1–2 h at room temperature, filter and pipette 0.3 ml into each well. Leave overnight in a saturated atmosphere at 4 °C. Empty the wells, wash with 5 × 0.3 ml PBS-Tween and blot the plates to remove residual droplets of buffer. Perform a standard curve (Sect. 2.4.5) in plates of each coating density, preferably using antiserum at two or three intermediate dilutions covering a ten-fold range around the 50% decrease value already obtained. If the

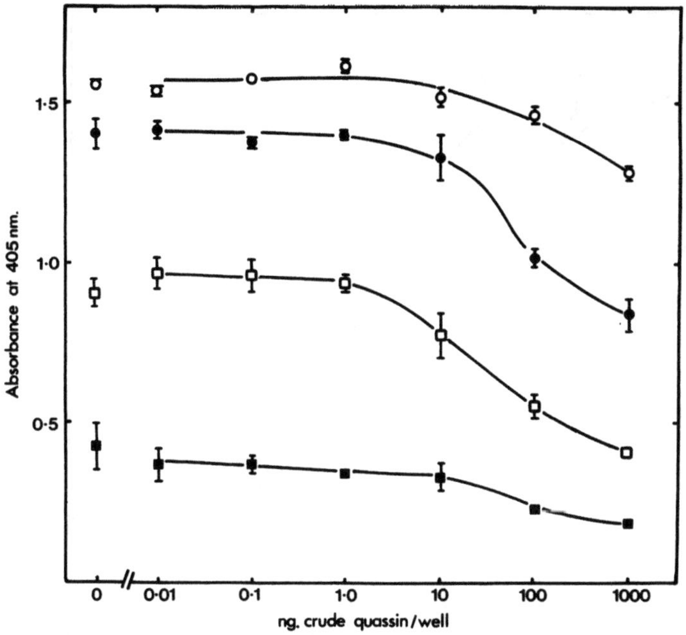

Fig. 3. The effect of varying the coating concentration of the ELISA standard curve for "crude" quassin (i.e. the mixture of quassin and derivatives as purchased from Koch Light Ltd.: see Robins et al. 1984a). Microtritration plates coated with iso-quassinic acid-KHLH conjugate at 10 μg ml^{-1} (○), 1 μg ml^{-1} (●,□), and 0.1 μg ml^{-1} (■) were used to prepare standard curves and developed with antiserum diluted either $1{:}10^4$ (*closed symbols*) or $1{:}5 \times 10^4$ (*open symbols*) as described in the text. The mean ± SE for triplicates of each condition is given. [Reproduced from Robins et al. (1984a)]

coating concentration is too low (Fig. 3), then insufficient antiserum will bind to develop a dose-response curve even using concentrated antiserum. In contrast, if the coating concentration is too high then too much binding occurs. Intermediate values will show displacement curves. Once a suitable value is attained, batches of plates can be prepared, washed with water (5 × 0.3 ml) after coating and stored indefinitely in a desiccator. Wash with PBS-Tween (5 × 0.3 ml) prior to use.

2.4.5 Standard Curve

An estimate of the range and sensitivity of the assay should now be performed. Make dilutions of hapten in buffer B over the range 1 μg to 0.1 pg in ten-fold steps. A concentrated stock can satisfactorily be held in methanol or ethanol but it is best to make fresh dilutions each time an assay is performed. Introduce, in triplicate, 0.1 ml of each dilution into wells followed by 0.1 ml antiserum at the dilution determined in Sect. 2.4.3. Incubate overnight at 4 °C, wash and develop with second antibody as described in Sect. 2.4.3. A standard curve for quassin is shown in Fig. 4.

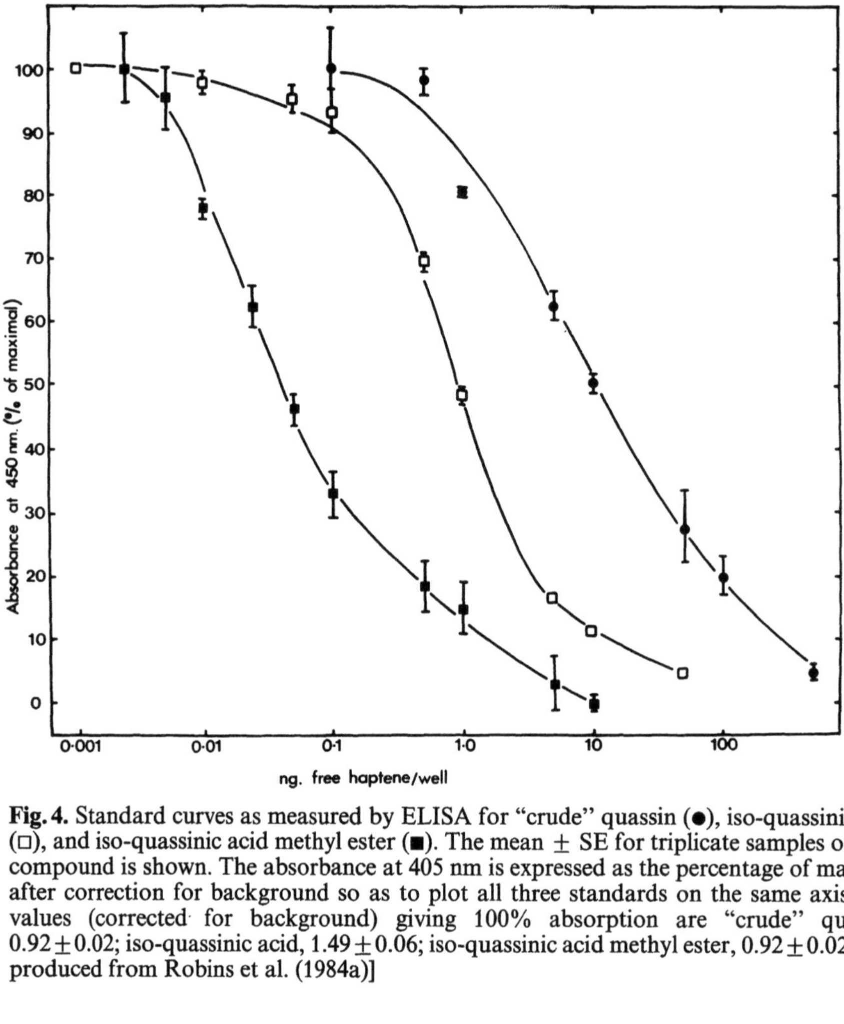

Fig. 4. Standard curves as measured by ELISA for "crude" quassin (●), iso-quassinic acid (□), and iso-quassinic acid methyl ester (■). The mean ± SE for triplicate samples of each compound is shown. The absorbance at 405 nm is expressed as the percentage of maximal after correction for background so as to plot all three standards on the same axis. The values (corrected for background) giving 100% absorption are "crude" quassin, 0.92 ± 0.02; iso-quassinic acid, 1.49 ± 0.06; iso-quassinic acid methyl ester, 0.92 ± 0.02. [Reproduced from Robins et al. (1984a)]

2.4.6 Serum Dilution

To improve the standard curve, the inter-relationship of the three parameters serum dilution, second antibody dilution and the duration of the incubations need to be studied. In practice a rigorous optimisation is not usually profitable. The normal aim is for a final incubation with substrate in which an optical density reading of 1.0–1.5 is attained within about 1 h. Figure 5 shows standard curves obtained using different final dilutions of antiserum and a constant amount of second antibody. Frame d shows that if the antiserum dilution is too great ($1:64 \times 10^4$), then a very shallow slope is obtained and this effect is not prevented even by using a long period of incubation with substrate. Under these conditions the assay is insensitive to differences in hapten concentration within the range of detection. In contrast, frame a shows that at a lower antiserum dilution ($1:8 \times 10^4$)

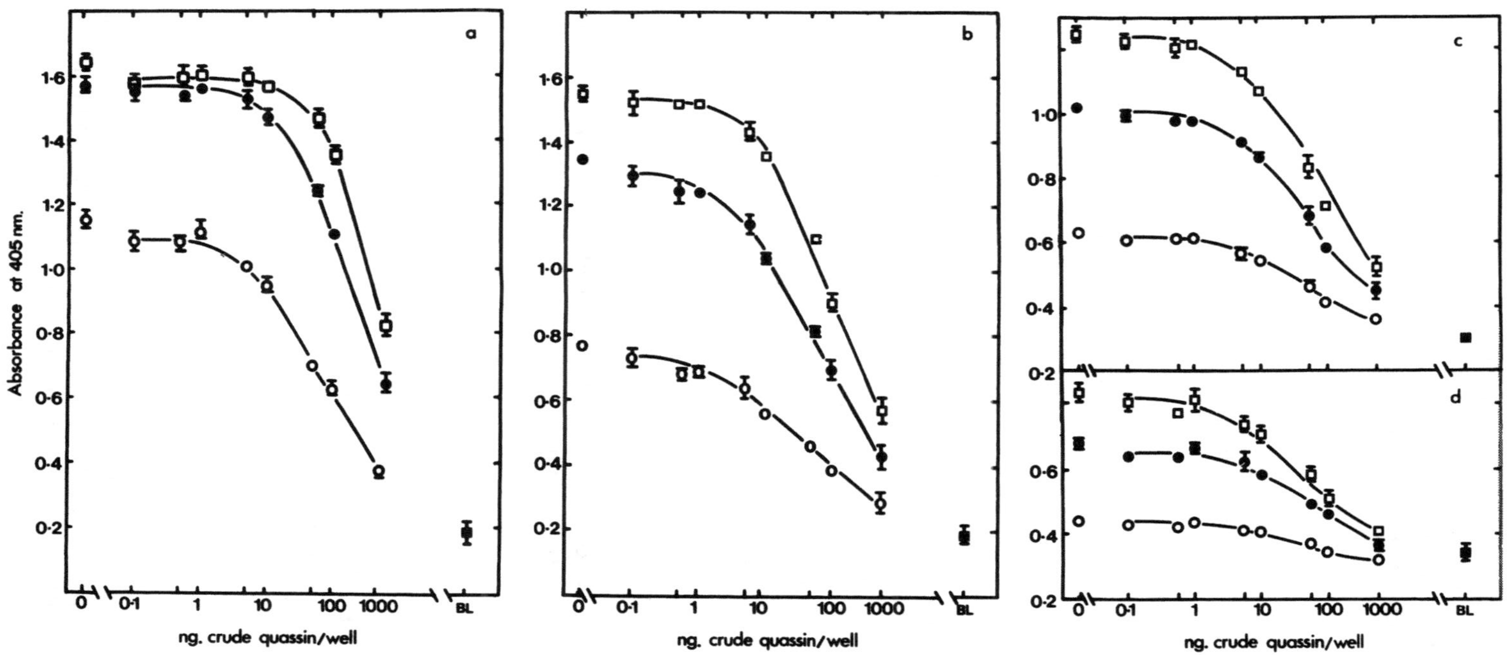

Fig. 5 a–d. The effect of different dilutions of antiserum on the development of ELISA standard curve for "crude" quassin. Plates, coated with 1 μg ml^{-1} iso-quassinic acid-KHLH conjugate, were incubated overnight with "crude" quassin and antiserum diluted **a** 1.8×10^4; **b** $1{:}16 \times 10^4$; **c** $1{:}32 \times 10^4$; **d** $1{:}64 \times 10^4$ times. After development (see text) the absorbance at 405 nm was read after 1.5 h (○), 3.0 h (●), and 4.5 (□). Values given as mean ± SE for triplicates at each condition. *BL*, "no antibody" blank. [Reproduced from Robins et al. (1984a)]

too long a period of incubation with substrate leads to a flattened curve at low antigen dilutions and a less sensitive assay. In this example curves of similar shape were obtained either at $1{:}8 \times 10^4$ with a 1.5 h incubation, or $1{:}16 \times 10^4$ with a 3 h incubation, the former being used for convenience.

2.4.7 Second Antibody Dilution

The effect of varying this parameter is usually only checked at the serum dilution and incubation time selected in Sect. 2.4.6. Using, therefore those conditions, develop with dilutions of second antibody in PBS-Tween from 1:100 to $1{:}10^4$. The effect is similar to that seen by varying the primary antibody. Insufficient dilution leads to rapid colour development and a high background absorbance; excessive dilution leads to poor colour development and shallow, insensitive, standard curves.

2.4.8 Second Antibody Development

The most commonly used second antisera are linked to either alkaline phosphatase (EC.3.1.3.1) or to horseradish peroxidase (EC.1.11.1.7), although other enzymes such as β-galactosidase (EC.3.2.1.23) are also available. These enzymes have high activity against substrates which produce coloured or fluorescent products. This enables their simple quantification within a period of 1–2 h, during which a high optical density (1.0–1.5 units) can be obtained.

2.4.8.1 Alkaline Phosphatase

Dissolve 4-nitrophenylphosphate at 1 mg ml^{-1} in buffer (0.05 M sodium carbonate/hydrogen carbonate, pH 9.6, containing 0.5 mM magnesium chloride) immediately prior to use. Introduce 0.2 ml per well and read the optical density at 405–410 nm after optimal incubation time.

2.4.8.2 Peroxidase

A number of substrates for peroxidase are available but they all tend to be toxic. Dissolve 2,2′-azino-bis (3-ethylbenzthiazoline)-6-sulphonic acid, diammonium salt (50 mg) in buffer (100 ml of 0.1 M citrate-phosphate buffer, pH 4.0). Add 30% (v/v) hydrogen peroxide solution (10 μl per 100 ml) immediately before use and introduce 0.2 ml per well. Read optical density at 415 nm after optimal incubation time.

2.4.8.3 β-Galactosidase

Dissolve 2-nitrophenyl β-galactopyranoside (50 mg) in buffer (100 ml of 0.1 M sodium dihydrogen phosphate, pH 7.0). Introduce 0.2 ml per well and read the optical density at 410 nm after optimal incubation time.

2.4.9 Assay Validation

Having used antisera to produce an immunoassay standard curve, the method must be validated for each application. Closely related metabolites must be tested to determine the extent to which they are recognised by the antibody preparation (the cross-reaction). In this way the degree of specificity of the assay is assessed for actual and potential interfering compounds. General, non-specific interferences may occur due to compounds present in the sample matrix and can only be avoided by selecting a suitable sample preparation procedure. Positive samples should show linearity to the volume assayed, and this should be identical to that obtained from both spiked negative samples and standards in assay buffer. (If only the third response is not superimposable then it may be possible to perform standard curves in negative sample matrix rather than assay buffer – though this usually entails loss of assay sensitivity.) Analytical recovery of spiked samples should show little variation, and inter- and intra-assay variation should be checked. Since sample work-up prior to immunoassays is minimal, recovery is often complete. If possible, immunoassay results should be correlated with an alternative method.

Since the antibody-hapten interaction is robust, it has proved possible to perform the assay in the presence of ethanol or methanol at 10% or even 20% (v/v) concentration without affecting sensitivity (Weiler and Zenk 1976; Weiler et al. 1981; Robins et al. 1984a–c). This means that these solvents can be used to extract tissues and need not be removed prior to analysis. Immiscible solvents, however, must be dried off and samples redissolved in buffer B for introduction to the microtitration plate (e.g. Morgan et al. 1983b).

2.4.10 Future Developments

Because, on theoretical grounds, the sensitivity of a reagent-excess assay like the ELISA is dependent on the specific activity of the labelled antibody (Ekins 1981), greater sensitivity can be generated by improving the method for quantifying the amount of second antibody bound to the microtitration plate. This may be done either by using fluorescent or chemiluminescent methods or by a radiometric enzyme assay. As yet, none of these techniques have been applied to the molecular species covered in this chapter.

2.4.10.1 Fluorescent Enzyme Assays

By using a fluorogenic substrate the sensitivity of the assay used to detect binding in EIA or ELISA can be greatly enhanced. Particularly good substrates for both β-galactosidase and alkaline phosphatase have been developed by synthesis of the appropriate derivatives of 4-methylumbelliferyl, the β-D-galactoside (Leaback and Walker 1961) and the phosphate (Fernley and Walker 1965) respectively. Peroxidase may be detected using 3,5-diacetyl-1,4-dihydrolutidine (Godicke and Godicke 1973). All these substrates and suitable measuring apparatus are available commercially (see Appendix).

2.4.10.2 The Biotin-Streptavidin System

The extremely high affinity interaction ($k = 10^{-15}\ M^{-1}$) between biotin and the protein streptavidin has been exploited (Amersham 1984) to develop an amplification system. The extent to which antibody has bound to the immobilised antigen is quantified using a second antibody (see Sect. 2.4.8) to which biotin has been linked. Streptavidin is then introduced and the extent to which this binds is quantified using a biotin-labelled enzyme. Because both the second antibody and the enzyme may be biotinylated with more than one molecule, amplification may be achieved while maintaining a very high signal-to-noise ratio.

2.4.10.3 The Enzymatic Radioimmunoassay

The technique of Harris et al. (1979) employs alkaline-phosphatase-labelled second antibody, and [^{3}H]-AMP as substrate. The production of [^{3}H]-adenosine is monitored and as little as 10^6 molecules of analyte can be determined. The method is not suitable for large-scale operations as the [^{3}H]-adenosine has to be separated by ion-exchange chromatography from the substrate.

2.4.10.4 Chemiluminescent-Labelled Antibodies

Instead of labelling the second antibody with an enzyme it is possible to prepare IgG to which a chemiluminescent molecule is attached. The use of both luminol (Simpson et al. 1979) and acridinium esters (Weeks et al. 1983) has been described. The extent to which such techniques may enhance sensitivity is yet to be determined for small molecules of plant origin. Potentially, as little as 10^{-18} mol might be detectable.

2.4.11 Speed of Assay

The development of an ELISA as described in Sects. 2.4.1 through 2.4.10 should lead to an assay of high sensitivity but which takes 18–24 h to obtain a result. For many clinical or routine analytical applications it is desirable to perform the analysis within a few hours at most. Short analytical periods of only 2–4 h can be attained with both EIA (Hacker et al. 1984) and ELISA (Morgan, personal communication). Although the saving in time may lead to some loss of sensitivity, the assay range of a good ELISA is so low that for many applications the sensitivity is still more than sufficient.

Most of the parameters of the assay need to be reassessed for a more rapid analysis. Coating conjugate concentration should be increased, both primary and secondary antiserum dilutions decreased and incubations performed at 37 °C. Sample, primary antibody and secondary antibody can be introduced together, cutting down the number of incubations required. Incubation with substrate for 15–30 min is often sufficient. Thus, much of the time taken for the assay is saved.

3 Immunoassays

The range of compounds for which assay kits are available is steadily increasing and it is likely that as immunoassays become established as a routine technique, a very much larger number of these will become available. The commercial production of kits is simplified by the ease with which the EIA and ELISA can be marketed in this form. For the time-being, as immune serum for the assays described in this section is often produced in excess, it may be possible to obtain or purchase a small supply on application to the relevant authors. Antisera to a number of important drugs (Marrero 1984) are already commercially available.

3.1 Alkaloids

3.1.1 Nicotiana Alkaloids

RIAs are available for nicotine and its principal metabolite in mammals, cotinine (Langone et al. 1973). The technique was developed to assay this interconversion in vitro using liver extracts and to determine the levels of these analytes in the physiological fluids of smokers.

3.1.1.1 Nicotine

For immunogen synthesis 3′-hydroxymethyl nicotine was activated with succinic anhydride (Sect. 2.1.1.2) in benzene to make *trans*-3′-succinylmethylnicotine, which was conjugated to HSA, KHLH and poly-L-lysine by the carbodiimide method (Sect. 2.1.1.1). [^{3}H]-Nicotine was prepared by random catalytic [^{3}H] exchange on the pyridine ring. In the RIA procedure "bound" and "free" phases were separated using goat anti-rabbit IgG as immunoprecipitant, radioactivity being determined in the bound fraction. Nicotine could be determined over the range 0.35–7 ng and significant cross-reaction was only obtained with *dl*-2-aminonicotine (2.4%). Cotinine below 50 ng could not be detected.

3.1.1.2 Cotinine

Immunogen was prepared directly from *trans*-4-carboxycotinine, by carbodiimide (Sect. 2.1.1.1) conjugation to HSA, KHLH and poly-L-lysine. A label for RIA was prepared by [^{125}I]-incorporation (Sect. 2.3.1.4) into N-(4-hydroxyphenethyl)-*trans*-cotinine carboxamide and RIA conducted as for nicotine. The assay proved rather more sensitive, detecting cotinine over the range 0.01–1.0 ng and only cross-reacting significantly with *dl*-desmethylcotinine (0.9%). Nicotine below 70 ng could not be detected.

Application (Sects. 3.1.1.1 and 3.1.1.2). The time-course for production of cotinine from nicotine by rabbit liver extracts was determined. Complete oxidation was not achieved, only 37% of the nicotine being converted in 10 h. The two assays were also used to determine nicotine and cotinine levels in the body fluids of smokers and the effect on these of ceasing smoking. Cotinine was shown to be the major metabolite present in smokers' serum (73–650 ng ml^{-1}) and urine (2.0–6.1 mg 24 h^{-1}) but levels decayed rapidly within 8 days of stopping inhaling tobacco smoke. No significant amounts were found in fluids from non-smokers. No application was made to plant tissues.

3.1.2 Indole Alkaloids of *Catharanthus roseus*

The immense pharmaceutical importance of the alkaloidal products of this species has led to the development of several immunoassays for their determination. The great demand for these drugs has stimulated extensive investigation of ways of improving their production both by plant breeding and bio-technologically through the use of plant cell cultures.

3.1.2.1 Vincristine and Vinblastine

These bisindole alkaloids, although minor components of the total alkaloid spectrum of *C. roseus*, are among the most valuable products derived from plants and are used extensively as anti-neoplastic agents. Because of their extreme potency as well as side effects such as myelosuppression and neurotoxicity their clinical administration requires careful monitoring, a task for which the immunoassay is ideally suited.

The Assays. Three groups have developed antiserum to vinblastine and vincristine, all of which used a different conjugation method. Teal et al. (1977) used the Mannich reaction (Sect. 2.1.1.5) to conjugate vinblastine directly to BSA via the secondary amine groups. Although the assay titre was poor (1:20) good sensitivity was obtained (Table 2) in a RIA using [^{3}H]-vinblastine (Amersham) but the assay could not differentiate between the two drugs. This was achieved, however, by Langone et al. (1979) who successfully raised antisera with differential recognition of vinblastine and vincristine. Both drugs were separately converted to the 6,7-dicarboxylic acid derivatives, coupled to HSA by the carbodiimide method (Sect. 2.1.1.1), and injected into rabbits. [^{3}H]-Vinblastine (Amersham) and [^{3}H]-vincristine (Moravek) were used as tracers in a RIA method in which "bound" and "free" phases were separated using goat anti-rabbit IgG and the bound phase determined. One animal provided an antiserum (Table 2) which was totally specific for vincristine, vinblastine only cross-reacting at 0.05%. Antiserum from another rabbit was moderately selective for vinblastine, vincristine cross-reacting at 25%. These properties make it the most useful antiserum available, in particular for selection of plant strains of particular properties.

Eli-Lilly Co. Ltd have prepared a commercially available antiserum by converting vinblastine to 4-deacetyl vinblastine C-3-carboxazide and conjugating this to BSA (details unpublished). The antiserum has been used for both a RIA (Sethi et al. 1980) and an EIA (Hacker et al. 1984). The RIA used [^{3}H]-labelled drugs and a system of sequential saturation, where unlabelled analyte is pre-incubated with immune serum prior to addition of radio-tracer. This increased sensitivity about ten fold (Table 2). A far greater improvement in sensitivity was achieved

Table 2. Comparison of assay parameters for determination of vinblastine and vincristine

Assay parameter	Assay[a]				
	1	2	3	4a[b]	4b[b]
Type of assay	RIA	RIA	RIA	RIA	EIA
Conjugation method	Mannich	Carbodiimide	Carbodiimide	Carboxazide	Carboxazide
Immunisation conjugate	Vinblastine-BSA	Vinblastine-HSA	Vincristine-HSA	Vinblastine-BSA	Vinblastine-BSA
Tracer	[^{3}H]-vinblastine	[^{3}H]-vinblastine	[^{3}H]-vincristine	[^{3}H]-vinblastine	Vinblastine-alkaline phosphatase
Antiserum titre	1:20	–	–	–	–
Range (ng)	0.5–2	1–10	1–15	0.2–30	0.01–0.1
Detection limit (ng)	0.1	0.4	0.5	0.2	0.005
Cross-reaction with (%):					
Vinblastine	100	100	0.05	100	100
Vincristine	100	25	100	100	100
Bleomycin	0	–	–	0.1	0
Catharanthine	–	0	0	–	–
Vindoline	–	0	0	–	–

[a] *1* Teal et al. (1977); *2, 3* Langone et al. (1979); *4a* Sethi et al. (1980); *4b* Hacker et al. (1984).
[b] Different types of assay using the same antiserum.

with the EIA (Hacker et al. 1984) in which vinblastine was labelled with alkaline phosphatase at 155:1 using the Mannich reaction (Sect. 2.1.1.5). EIA was conducted in microtitration plates coated with antibody (1:10000) and maximum sensitivity (Table 2) could be obtained after only a 2 h incubation, the limit of detection at 5 pg being 100-fold lower than with the RIA methods.

Application. All three assays were used to look at the clearance of drug from serum following intra-venous injection and showed a rapid phase of loss to only 10% in 30–60 min. The drugs are accummulated differentially by various tissues, as shown both by [^{3}H]-tracer studies and by RIA analysis (Owellen et al. 1977, 1981). Only Langone et al. (1979) examined plant tissue. Because the concentration of vincristine is only 1–2% that of vinblastine it will not contribute significantly to vinblastine determination. They showed leaves to contain the most alkaloid (119 μg/plant vinblastine; 1.6 μg/plant vincristine) although the highest concentration was in flowers. They further showed by HPLC analysis of leaf extracts that while the antisera reactivity correctly identified the vincristine or vinblastine peaks respectively, they both reacted strongly to a more polar fraction, as yet unidentified. This was not vindoline or catharanthine, as no cross-reaction occurred with these alkaloids but, surprisingly, they did not examine either serpentine or ajmalicine, also major alkaloids of *C. roseus*, for cross-reactivity of HPLC elution.

3.1.2.2 Vindoline

This monomeric indole alkaloid, a biosynthetic precursor of vinblastine, is of major importance as it can be economically used to synthesise chemically vinblastine (Mangeney et al. 1979). Although a very major component of the total alkaloid of *C. roseus* it would be advantageous to breed plants with increased yield and two groups have developed RIAs for this compound, although only one application to plants is reported.

The Assays. Westekemper et al. (1980) activated 17-deacetyl vindoline with glutaric anhydride (Sect. 2.1.1.2) and then coupled the resulting hemisuccinate to BSA by the carbodiimide method (Sect. 2.1.1.1). Antiserum rasied in rabbits was used in a RIA with [^{3}H]-vindoline (Amersham) using ammonium sulphate precipitation to prepare the "bound" phase. Kutney et al. (1980) also synthesised an immunogen from 17-deacetyl vindoline and BSA but used the succinic anhydride (Sect. 2.1.1.2)/mixed anhydride (Sect. 2.1.1.1) route for conjugation and performed the RIA in a similar way to Westkemper et al. (1980). The assay of Kutney et al. (1980) was the less sensitive, though both assays show satisfactory specificity (Table 3), in particular showing no cross-reaction with other *Catharanthus* alkaloids. Both assays could tolerate methanol in the competition stage.

Application. Westekemper et al. (1980) showed that the only immunoreactive material on a TLC plate corresponded to a vindoline standard. They showed (Fig. 6) that in 14-week-old plants, although average vindoline content is 0.4% dry weight, freshly matured green leaves may contain up to 7% alkaloid but that the

Table 3. Comparison of the RIA methods for vindoline

Assay parameter	Assay[a]	
	1	2
Precipitant	Ammonium sulphate (50%)	Ammonium sulphate (satd.)
Serum titre	1:170	1:80
Tracer	[^{3}H]	[^{3}H]
Tracer/assay (pmol)	1.64	11.4
Specific activity (Ci $mmol^{-1}$)	14.3	0.91
Non-specific binding (%)	2	4
Range (ng)	0.2–45	10–1000
Detection limit (ng)	0.05	5
Maximum affinity constant ($M\,l^{-1}$)	1.7×10^{-10}	0.4×10^{-10}
Cross-reaction with (%):		
Ajmalicine	0	0
Serpentine	0	0
Vinblastine	0	0
Catharanthine	0	0

[a] *1* Westekemper et al. (1980); *2* Kutney et al. (1980)

content decreases rapidly as leaves mature. Within this tissue alkaloid was evenly distributed across the lamella, leaf veins having a much lower concentration. Flowers, seeds and roots contained very little alkaloid; so did calli and suspension cultures initiated from high-vindoline containing leaves (0.002%). The vindoline content of a number of other *Catharanthus* species was examined but none was superior to *C. roseus*. A population of plants of this species was therefore examined and high-yielding individuals (1–2%) identified for a breeding programme. The RIA was also used to examine the synthesis of vindoline from tryptamine and secologanin by cell-free extracts of leaf and it was demonstrated that no product was made, in contrast to a previous report (Stuart et al. 1978).

3.1.2.3 Ajmalicine and Serpentine

Ajmalicine, also a major product of *Catharanthus* and *Rauwolfia* species, is used extensively in the treatment of circulatory diseases, in particular cerebral obstruction and hypersensitivity. Serpentine, a closely related alkaloid is not used directly but converted to ajmalicine by chemical reduction. Because it is desirable to select for plants with a high ajmalicine:serpentine ratio, separate antisera have been raised to each alkaloid.

The Assays. (Arens et al. 1978) (a) Ajmalicine. Ajmalinic acid was generated by alkaline hydrolysis of ajmalicine and coupled to HSA by a modified, mixed-anhydride method. [^{3}H]-Ajmalicine was prepared by the reduction of serpentine with sodium [^{3}H]-borohydride (Sect. 2.3.1.3) and used in a RIA with antiserum at a titre of 1:630 in which "bound" and "free" were separated by ammonium sulphate precipitation, the bound phase being determined. The assay could deter-

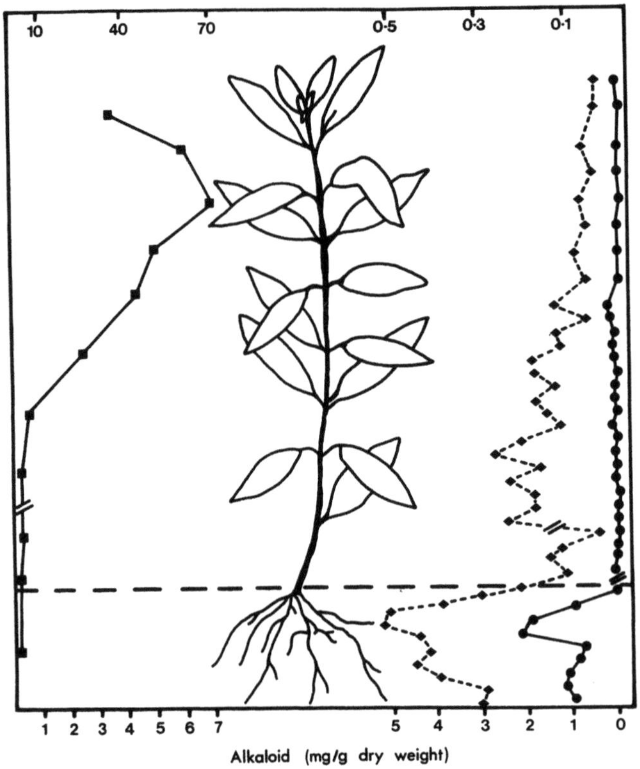

Fig. 6. The distribution of the indole alkaloids ajmalicine (●) and serpentine (◆) within a mature plant and vindoline (■) within a 14-week-old plant of *Catharanthus roseus* as determined by RIAs. Alkaloids were extracted with hot methanol and assayed directly after suitable dilution in buffer. In each distribution analysis the break-point (-//-) indicates wheather the values of alkaloid content should be read against the scales below or above the break. [Redrawn from Arens et al. (1978) and Westkemper et al. (1980)]

mine ajmalicine over the range 1–20 ng with a detection limit of 0.1 ng. The antiserum was specific for ajmalicine (100%) and 19-epi-ajmalicine (100%), cross-reacting only with tetrahydroalstonine (11%). No cross-reaction occurs with other structurally related alkaloids including serpentine, vinblastine, vincamine and vindoline.

(b) Serpentine. Serpentinic acid was generated by alkaline hydrolysis of serpentine and coupled to HSA via the N-hydroxysuccinimide ester derivative using a modified carbodiimide method. RIA was performed as in (a). [^{3}H]-Serpentine was prepared by Pd/C oxidation of [^{3}H]-ajmalicine, sufficient label being retained in the C-ring. Serpentine was detected over the range 2–50 ng with a detection limit of 0.5 ng and only cross-reacted with alstonine (8%).

Application. The assays were shown to correlate well to fluorometric determination (r = 0.95) and both assays showed good recovery of alkaloids from ethanolic plant extracts. For plant screening, the distribution of the alkaloids within plants was examined (Fig. 6) and it was found that in total contrast to vindoline, maximal concentration of both alkaloids occurs in the roots just below the soil surface, serpentine being at about five-fold greater concentration than ajmalicine. Using this tissue, with an average content of 0.15% dry weight, ten individuals with greater than 0.5% ajmalicine and 1.2% serpentine were identified, selfed and 750 plants of the S_1-generation examined. In this single generation a 1.5- to 2-fold increase in average alkaloid content occurred and individuals with greater than 0.9% ajmalicine and 1.4% serpentine were identified for further improvement. The RIA has also been used to screen for alkaloid production (Zenk et al. 1977) in plant cell suspension cultures, in which an improvement can be obtained more rapidly as sexual stages are not required for propagation. Cell cultures with over 1.8% dry weight alkaloids have been established and their exploitation for the commercial production of these alkaloids is being investigated.

These assays have also proved valuable in the study of the pathway of ajmalicine production. Treimer and Zenk (1978) demonstrated, by detecting the alkaloid with the ajmalicine-RIA, that ajmalicine could be made from tryptamine and secologanin by cell-free systems derived from *C. roseus* cell cultures. Because of the specificity of the assay they were able to perform a full investigation of this enzymic process, despite the product only being made at the nanomolar level. Cathenamine synthesis by a cell-free extract was also demonstrated using this antiserum by Westekemper et al. (1980).

3.1.3 Tropane Alkaloids

Scopolamine, produced by species of *Datura* and *Duboisia* (Solanaceae), is used pharmaceutically as a parasympatholytic, anti-emetic and anti-colinergic agent. In plant tissue it usually is found associated with larger quantities of hyoscyamine, its biosynthetic precursor. It may be determined by a RIA able to distinguish these two alkaloids.

The Assay (Weiler et al. 1981). When scopolamine was coupled to HSA via the side-chain hydroxyl group using the glutaric anhydride (Sect. 2.1.1.2)/mixed anhydride (Sect. 2.1.1.1) method and used as an immunogen, no anti-hapten activity was found. A successful immunogen was synthesised by reacting nor-(-)-scopolamine with acrylic acid (Sect. 2.1.1.5), forming the N-β-propionic acid derivative, and coupling this to HSA with the mixed anhydride or carbodiimide (Sect. 2.1.1.1) method. N-[^{3}H-Me]-scopolamine was prepared from nor-scopolamine by [^{3}H]-methyl iodide treatment (Sect. 2.3.1.2). Antiserum was raised in rabbits and used at a titre of 1:2250, the bound fraction being precipitated with ammonium sulphate for [^{3}H]-determination. The assay showed good sensitivity, detecting 0.5–50 ng scopolamine, with a limit of 0.2 ng. The serum is specific to scopolamine (100%) and nor-scopolamine (160%), showing no recognition if only the tropane moiety is present. It only cross-reacts 0.8% to hyoscyamine, and 10.4% to 6-hydroxyhyoscyamine, which is present in minute amounts in plant tissue.

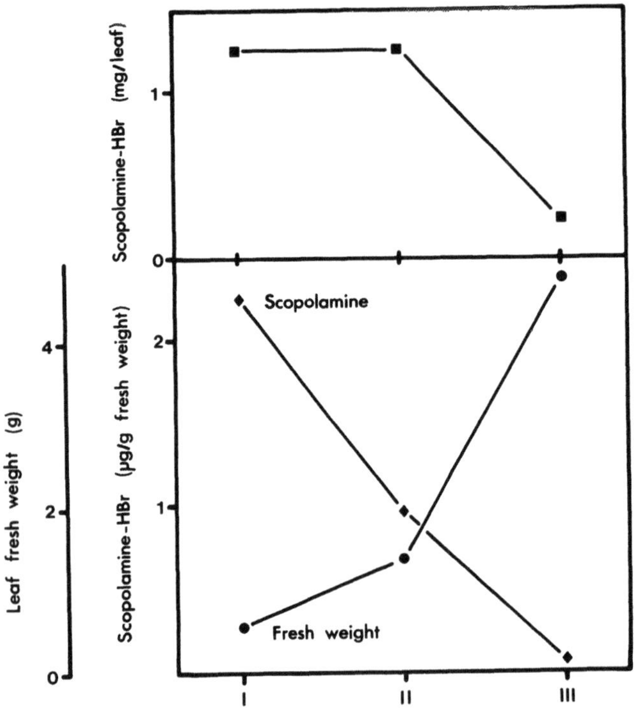

Fig. 7. Concentration and absolute amounts of scopolamine in leaves of increasing age from a *Datura sanguinea* plant (age: 6 months) as determined by RIA. Alkaloid was extracted with hot methanol and assayed directly after suitable dilution in buffer. [Reproduced from Weiler et al. (1981)]

Application. Excellent correlation was shown between the RIA and TLC quantification of scopolamine (r = 0.96) in which the RIA was done directly in crude methanolic extracts. The distribution of alkaloid within the tissues of both *Datura innoxia* and *D. sanguinea* was examined and the majority of alkaloid (83%) shown to be deposited in the leaves, although the highest concentration is found in unripe seeds. An important observation (Fig. 7) was that scopolamine levels decreased about 20-fold in leaves as they matured, showing that improved yield would be obtained by harvesting only young leaves. The RIA was also used to examine the range of alkaloid productivity. Of 528 plants examined most contained 0.8–1.2% dry weight scopolamine, but a few individuals containing in excess of 3% were identified.

3.1.4 Quinoline Alkaloids

Several species of the genus *Cinchona* (Rubiaceae) produce the isomeric alkaloids quinine, used as an anti-malarial and bittering agent, and quinidine, a powerful anti-arrhythmic drug. A number of assays for these compounds have been developed for both clinical or plant-selection application.

Table 4. Comparison of assay parameters for quinine and quinidine determination

Assay parameter	Quinine		Quinidine	
	1[a]	2	3	4
Assay type	RIA	ELISA	RIA	ELISA
Serum titre	1:10000	1:100000	1:5000	1:50000
Coating concentration (μg ml^{-1})	–	1.0	–	1.0
Non-specific binding (%)	5	3	–	–
Range (ng)	1–50	0.01–1	1–1000	0.1–10
Detection limit (ng)	0.05	0.01	1.0	0.1
Cross-reaction with (%):				
Quinine	100	100	0	0
10,11-Dihydroquinine	35	2.7	–	0
Cinchonidine	14	1.2	–	0
Cupreine	7.3	0.8	–	0
10,11-Dihydrocupreine	3.2	0.7	–	0
Quinidinone	0.6	1.6	–	4
Cinchoninone	0	0	–	0.1
Quinidine	0	0	–	100
10,11-Dihydroquinidine	0	0	–	0
Cinchonine	0	0	–	0
Cupreidine	0	0	–	0.15
10,11-Dihydrocupreidine	0	0	–	0

[a] *1* Robins et al. (1984c); *2,3,4* Morgan et al. (1985c).

3.1.4.1 Quinine

At FRI we have developed both a RIA (Robins et al. 1984c) and an ELISA (Morgan et al. 1985c) for quinine, using the same antiserum. Immunogen was synthesised by activating quinine with succinic anhydride (Sect. 2.1.1.2) and coupling to BSA by the mixed anhydride method (Sect. 2.1.1.1), and antiserum raised in rabbits. In the RIA, dextran-coated charcoal was used to separate "bound" and "free" phases. 9-[^{3}H-acetyl]-Quinine, synthesised as in Sect. 2.3.1.2, was used as tracer. Details of the assay are given in Table 4. For the ELISA, coating protein was synthesised by carbodiimide conjugation (Sect. 2.1.1.1) of 9-hemisuccinylquinine to KHLH and the bound antiserum quantified using alkaline phosphatase-labelled goat anti-rabbit IgG at a dilution of 1:750. As shown from the data in Table 4 the ELISA is both more sensitive and more specific to quinine.

3.1.4.2 Quinidine

Commercial assays are available for the pharmacological determination of quinidine. Beckman has developed a rate nephelometric inhibition immunoassay (Quattrone et al. 1984) which determines both quinidine and a number of pharmacologically active mammalian metabolites. Syva Corporation has developed an EMIT which is more specific to quinidine (Pape 1981). Both methods require specialised measuring apparatus, use large amounts of antiserum, and are of low sensitivity (1–10 μg ml^{-1}). Their total automation makes them very suitable for clinical application but neither has been applied to plant material.

Quinidine may also be assayed by RIA or ELISA (Morgan et al. 1985c). Immunogen synthesised as for quinine (Sect. 3.1.4.1) failed to elicit a suitable response in rabbits. Success was achieved by inserting a 6-amino-N-hexanoic acid bridge (Sect. 2.1.2.1) between the 9-hemisuccinylquinidine and BSA. RIA and ELISA methodology and coating conjugate synthesis for ELISA were the same as described for quinine. From Table 4 it is clear that the ELISA is again a more sensitive and more selective technique.

Application (Sects. 3.1.4.1 and 3.1.4.2). The RIA for quinine has been used to screen clonal shoot cultures of *Cinchona ledgeriana* for their quinine content (Robins et al. 1984c), the highest yielding material being used to initiate a plant-cell culture line. The same assay has also been used to screen 100 microcalli derived from individual seeds (Robins, Payne and Rhodes, unpublished results). A wide range of quinine content was found (0–600 μg g^{-1} fresh weight) and the highest yielding individuals were retained for a selection programme to improve quinine production.

3.1.5 Papaver Alkaloids

The benzylisoquinoline alkaloids morphine and codeine, extracted from *Papaver somniferum* (Papaveraceae) latex are very widely used narcotic analgesics. Because of the habit-forming nature of morphine, careful monitoring of clinical application is desirable. Furthermore, there has been limited breeding and strain improvement of *P. somniferum*, a task greatly simplified by using immunoassays.

3.1.5.1 Morphine and Codeine

Two RIAs (Spector and Parker 1970; Spector 1971; Vunakis et al. 1972) and one hemagglutination-inhibition assay (Adler and Liu 1971) have been developed. In all cases, immunogen was made by carboxymethylation (Sect. 2.1.1.3) at the 3-hydroxyl of morphine and carbodiimide (Sect. 2.1.1.1) conjugation to protein.

The Assays. Antiserum was raised both in rabbits and guinea pigs. Vunakis et al. (1972) prepared [^{125}I]-labelled tracer by iodination (Sect. 2.3.1.4) of the tyrosine residues in a copolymer of L-glu, L-lys, L-ala and L-tyr in which 3-O-carboxymethylmorphine was conjugated to the L-lysine residues. As shown in Table 5 the most sensitive and specific assay was developed from the guinea pig serum, although this has the disadvantages of requiring an [^{125}I]-labelled tracer and is unable to distinguish between morphine and codeine.

Application. Spector (1971) examined the binding of antiserum to a range of synthetic opiates. None of those tested were recognised at more than 0.25%. He also examined the biological half-life of morphine in rat serum, finding two main phases with $t_{0.5}=0.85$ h and 6.8 h, the latter presumably representing the release of morphine from tissues. Vunakis et al. (1972) analysed sera from 36 "sudden death" cases and showed a good correlation of the assay with conventional techniques. Using the RIA they were able to detect morphine in a number of samples in which it was not detected by normal methods.

Table 5. Comparison of the assays for morphine

Assay parameter	Assay[a]			
	1	2	3	4
Assay type	RIA	Hemagglutination	RIA	RIA
Animal	Rabbit	Rabbit	Rabbit	Guinea pig
Immunogen protein	BSA	BSA	Polylysine	Polylysine
Tracer	[^{3}H-7,8]-Dihydro-morphine	Sheep erythrocytes	[^{125}I-L-tyr]-carboxy-methylmorphine	
Specific activity (μCi μg^{-1})	1.4	–	75	75
Precipitant	Ammonium sulphate	Carboxymethyl-morphine-rabbit serum albumin	Goat-anti-rabbit	Goat-anti-guinea-pig
Non-Specific binding (%)	15	–	15	10
Titre	1:400	1:51000	1:100000	1:10000
Range (ng)	0.1–0.8	–	2–60	0.05–5
Detection limit (ng)	0.1	0.025	1.0	0.01
Cross-reaction with (%):				
Morphine	100	100	100	100
7,8-Dihydro-morphine	–	–	3.4	1
Codeine	100	–	583	167
Dihydrocodeine	–	–	194	–
Nor-morphine	100	–	–	–
Heroine	100	–	27	3.8
Morphine-3-mono-glucuronide	0.3	–	–	–
Nalorphine	0.25	–	0.5	0.2
Methadone	0.25	0.1	–	<0.001

[a] *1* Spector and Parker (1970); Spector (1971); *2* Adler and Liu (1971); *3,4* Vunakis et al. (1972).

3.1.5.2 The Benzylisoquinoline Pathway

Recently, Wieczorek et al. (1984) have reported (in abstract only) the development of six RIAs for the determination of (S)-reticuline, (R)-reticuline, salutaridine, thebaine, codeine and morphine using [^{3}H]-tracers. The antisera are all highly specific with a range of detection of 0.01–10 ng. They provide a powerful tool for the study of this important biosynthetic pathway, and for the selection of plants or cell cultures with enhanced production of alkaloid and are currently being used for these purposes.

3.2 Terpenoids

3.2.1 Limonin

This bitter triterpene is concentrated in the seeds of many types of *Citrus* (Rutaceae) and, when fruit is processed to extract juice, may be released, making the product unacceptably bitter. As the amount of limonin in fruit varies considerably each batch of produce needs to be checked because the intensity of bitterness develops sharply over the range 0.1–1 μg. This requirement has led to the development of three immunoassays, which have been assessed for their suitability in quality control.

The Assays. Two RIAs, one using [^{125}I]-tracer (Mansell and Weiler 1980) and the other using [^{3}H]-tracer (Weiler and Mansell 1980), and an EIA (Jourdan et al. 1984) have been described for limonin. All the assays use the same antiserum, raised in rabbits to a limonin-BSA conjugate synthesised as described in Sect. 2.1.1.6. The [^{125}I]-tracer was synthesised by iodination of the tyrosine in limonin-7-(O-carboxymethyl) oxime-tyrosine methyl ester (Sect. 2.3.1.4) and the [^{3}H]-tracer by the reduction of limonin to limonol (Sect. 2.3.1.3). For the EIA limonin-7-(O-carboxymethyl) oxime was conjugated with alkaline phosphatase by the carbodiimide method (Sect. 2.1.1.1). Table 6 shows the characteristics of each assay. While greater sensitivity is obtained with the RIAs, the [^{125}I]-tracer method is the least satisfactory as it requires the regular synthesis of tracer and a γ-counter. The EIA, however, uses a stable tracer and for routine use quantitative results may be obtained within 1 h.

Application. Using the [^{125}I]-tracer assay the distribution of limonin within tissue of *Citrus paradisi* (White Marsh) was examined and found to agree with previous results, the major concentration being found in the seeds (Fig. 8). In the analysis

Table 6. Comparison of assays for limonin in which the same antiserum was used

Assay parameter	Assay[a]		
	1	2	3
Assay time	RIA	RIA	EIA
Tracer	[^{125}I]	[^{3}H]	Alkaline phosphatase
Antiserum titre	1:27000	1:1350	1:1650
Range (ng)	0.1–100	0.5–100	1–100
Detection limit (ng)	0.07	0.22	1.0
Non-specific binding (%)	27	1.1	–
Intra-assay variation (%)	7.5	9.0	8.5
Inter-assay variation (30 day) (%)	16.6	–	13.7
Cross-reaction with (%):			
Deoxylimonin	100	27	25
Deacetylnomalin	–	6.6	3.4
Nomalin	1.2	0.9	0.5

[a] *1* Mansell and Weiler (1980); *2* Weiler and Mansell (1980); *3* Jourdan et al. (1984).

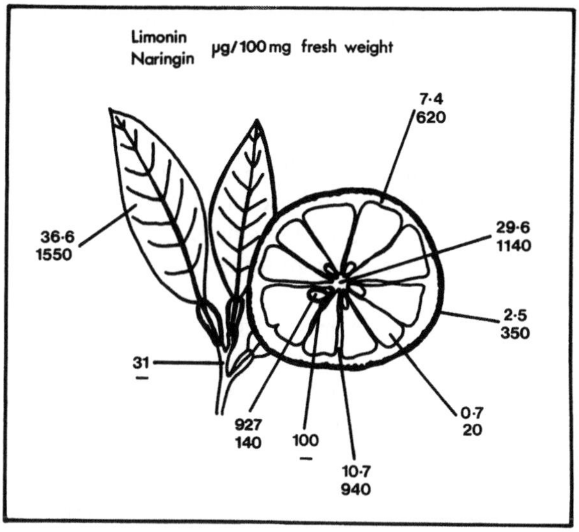

Fig. 8. Distribution of limonin (*upper value*) and naringin (*lower value*) in a fruit and vegetative parts of a grapefruit, *Citrus paradisi* as determined by RIAs. Limonin was extracted with hot acetone from the variety White Marsh, acidified and the extract assayed after dilution with water. Naringin was extracted with hot methanol from variety Ruby Red and assayed after suitable dilution. – = not determined. [Redrawn respectively from Mansell and Weiler (1980) and Jourdan et al. (1983)]

of 30 juice samples a good correlation (r = 0.68) was obtained between EIA and HPLC, the latter method requiring much greater sample preparation and time for analysis.

3.2.2 Bruceantin

This seco-triterpene, obtained from *Brucea antidysenterica* (Simaroubaceae), shows a broad spectrum of anti-tumour activity. As with other drugs, the availability of an immunoassay greatly facilitates pharmacological studies.

The Assay. For a RIA to bruceantin (Fong et al. 1980), antiserum was raised in rabbits to a bruceantin-BSA conjugate prepared by the succinic anhydride method (Sect. 2.1.1.2) and a tracer of [^{3}H]-acetylbruceantin by acetylation with [^{3}H]-acetic anhydride (Sect. 2.3.1.2). The level of displacement was determined on the "bound" phase, "free" [^{3}H]-bruceantin being removed by adsorption onto dextran-coated charcoal. Antiserum was used at a final titre of 1:125 and bruceantin could be detected over the range 0.5–50 ng. No cross-reactions were measured.

Application. The assay was used (Fong et al. 1980) to examine the tissue distribution of bruceantin in mice and the urinary and bilary clearance of bruceantin in dogs. In both animals the drug appears to be rapidly metabolised to a product not recognised by the immunoassay. Unmetabolised drug is accumulated in liver and the majority of excreted bruceantin is found in bile. No application to plant tissue was undertaken.

3.2.3 Quassin

This intensely bitter seco-triterpene is synthesised by *Quassia amara* and related members of the Simaroubaceae. It is of interest as a bittering agent in the food industry, and has a useful insecticidal action. Two ELISAs have been developed for use in quality control and clone-improvement programmes.

The Assays. (a) (Robins et al. 1984a) A free acidic group was generated from the δ-lactone of quassin and stabilised by epimerisation to isoquassinic acid. This was conjugated to BSA by the mixed-anhydride method (Sect. 2.1.1.1) and antiserum raised in rabbits. Coating conjugate was made using KHLH, linked by the carbodiimide method (Sect. 2.1.1.1). Assays were performed in microtitration plates and satisfactory conditions (see Sect. 2.4) were obtained using coating conjugate at 1 $\mu g\ ml^{-1}$, antiserum at a titre of 1:80000 and alkaline-phosphatase-labelled anti-rabbit IgG at 1:750. It was established that samples could be assayed in the presence of 10–20% MeOH in the well without loss of sensitivity.

The assay was relatively insensitive to quassin, which could be detected over the range 2–200 ng, but was extremely sensitive to the methyl ester of isoquassinic acid, which was detected over the range 5–100 pg. Considerable cross-reaction occurred with the closely related quassinoids of *Q. amara*, neoquassin (100%), 14,15-dehydroquassin (81%), 12-hydroxyquassin (100%), 18-hydroxyquassin (250%) and the nigakilactones (11–33%) and picrasin A (13.5%) of *Picrasma excelsa*.

(b) (Robins et al. 1984b) 18-Hydroxyquassin was conjugated to BSA by the succinic anhydride method (Sect. 2.1.1.2) and used as immunogen in rabbits. Coating conjugate was prepared by the same technique using KHLH. Assays were performed in microtitration plates with coating conjugate at 1 $\mu g\ ml^{-1}$, antiserum at a titre of 1:50000 and alkaline-phosphatase-labelled anti-rabbit IgG at 1:1500. Quassin and 18-hydroxyquassin were detected over the range 5–100 pg. Cross-reactions were similar to (a) except for 14,15-dehydroquassin (10.7%), and picrasin A (0.4%). No loss of sensitivity ensued from a final concentration of 10% MeOH in the wells.

Application. Assay (b) was used to examine the distribution of immunoreactivity among the tissues of *Q. amara*, *Q. indica* and *P. excelsa*. The amount of quassin present varied considerably both between different tissues and between species, the wood of *Q. amara* containing the highest levels (Fig. 9). No application to soft drinks or other beverages that might contain quassin was performed.

3.2.4 Iridoid Glucosides

The monoterpene glucosides loganin and secologanin are key intermediates in the biosynthesis of many important alkaloids of the indole and ipecac series. Prior to condensation with tryptamine, loganin is converted by a unique ring cleavage

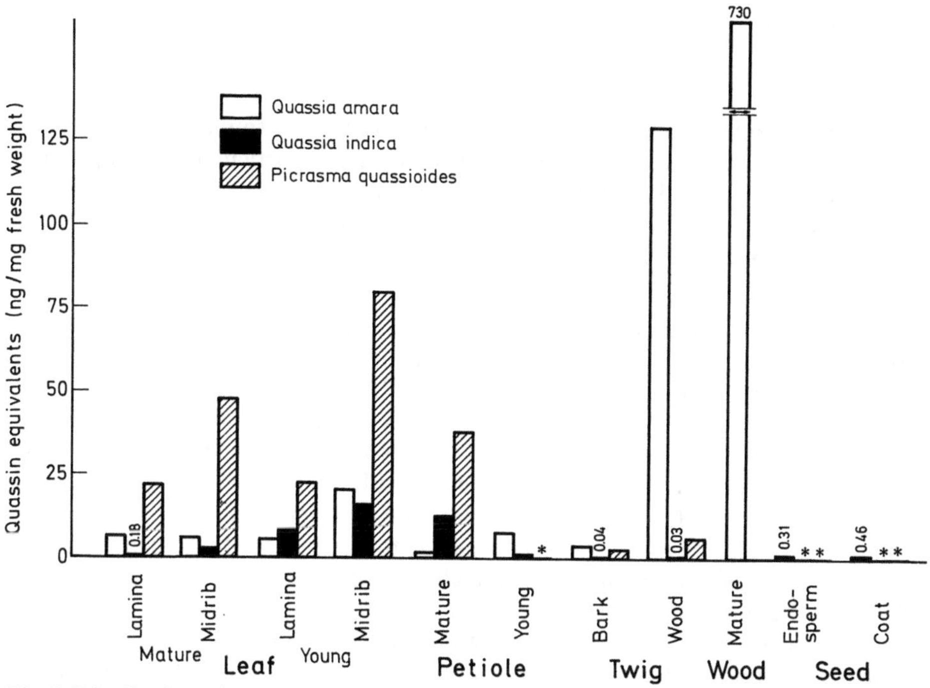

Fig. 9. Distribution of quassin in various tissues of three species of Simaroubaceae as determined by ELISA. Tissue was extracted with hot methanol and quassin assayed directly after suitable dilution in buffer. * = Material unavailable. [Reproduced from Robins et al. (1984b)]

to secologanin, a reaction of key interest about which little is known. In order to study this process RIAs have been developed for loganin (Tanahashi et al. 1984) and secologanin (Weiler and Zenk, unpublished).

The Assays. To ensure suitable specificity loganin was conjugated to BSA through the sugar by formation of the 6′-carboxyl derivative (Sect. 2.1.1.7) and carbodiimide coupling (Sect. 2.1.1.1). A [^{3}H]-tracer was formed by periodate cleavage (Sect. 2.1.1.7) and sodium [^{3}H]-borohydride reduction (Sect. 2.3.1.3). Serum, raised in rabbits, was used at a final titre of 1:450 and binding determined by ammonium sulphate precipitation to obtain the bound fraction. The assay detected loganin over the range 0.5–100 ng with a limit of sensitivity of 0.3 ng. The serum was specific to loganin (100%), the only natural product also recognised significantly being the aglycone (30%). Secologanin only cross-reacted at 0.07%.

Application. Representative species of a number of genera of Caparifoliaceae, known to contain substantial amounts of secologanin in the intact plant, have been examined in cell culture for loganin formation. Even with the RIA no loganin was detected, although cultures of both *Weigelia gigantiflora* and *Lonicera tatarica* readily absorb added loganin and transform it into secologanin, none of which is released from the cells.

3.3 Flavonoids

3.3.1 Naringin

A number of bitter flavonoid-7-*O*-neohesperidosides are found in *Citrus* (Rutaceae) fruits. The principal of these is naringin, which, particularly in grapefruit (*Citrus paradisi*), can, like limonin, occur at levels that impart excessive bitterness to both the fruits and juice obtained from them. Two RIAs have been developed to provide a rapid analytical approach to quality control.

The Assays. (a) (Jourdan et al. 1982) Naringin was coupled to BSA by diazotization (Sect. 2.1.1.3), the protein having been substituted with 4-aminohippuric acid at the free amino residues (Sect. 2.1.2.3). An [^{125}I]-labelled tracer, with the [^{125}I] in the same position as used for conjugation, was synthesised as described in Sect. 2.3.1.4. In the RIA procedure antiserum was used at a titre of 1:900, "bound" and "free" phases were separated by precipitation with ammonium sulphate and the [^{125}I] in the bound phase determined. Naringin was detected over the range 5–200 ng (Table 7). The assay was very sensitive to the nature of the sugar substituents at the 7-position, but insensitive to changes in the aromatic ring system, all flavonoid-7-*O*-neohesperidosides being recognised to the same extent.

(b) (Jourdan et al. 1983) Naringin was coupled to BSA by the mixed-anhydride method (Sect. 2.1.1.1) after activation at the 4-keto group with amino-oxyacetic acid (Sect. 2.1.1.6). An [^{3}H]-labelled tracer was prepared by condensing naringin-(4-O-carboxymethyl) oxime with 5-aminolevulinic acid, using the mixed-anhydride reaction (Sect. 2.1.1.1) and reducing the side chain with sodium [^{3}H]-borohydride (Sect. 2.3.1.3). The RIA was conducted as in (a). The alternative conjugation routine resulted in a ten-fold improvement in the sensitivity of

Table 7. Comparison of assay parameters for naringin determination by RIA

Assay parameter	Assay[a]	
	1	2
Tracer	[^{125}I]	[^{3}H]
Coupling mode	Diazotisation/hippuric acid	Amino-oxyacetic acid
Antiserum titre	1:900	1:900
Range (ng)	5–200	0.5–50
Detection limit (ng)	2	0.2
Non-specific binding (%)	7	1.5
Intra-assay variation (%)	3.9	2.3
Inter-assay variation (%)	7.3	5.0
Cross-reaction with (%):		
Rhoifolin	95	27
Poncirin	100	71
Neohesperidin	100	57
Fortunellin	100	8

[a] *1* Jourdan et al. (1982); *2* Jourdan et al. (1983).

the assay (Table 7) and a decrease in the recognition of other flavonoid-7-*O*-neohesperidosides.

Application. Both assays were used to determine the flavonoid-7-*O*-neohesperidoside content of grapefruit juice and the distribution of immunoreactivity in grapefruit tissues. The only immunoreactant found in juice was naringin and it was confirmed that in the intact fruit this is primarily localised to the pith, the albedo and the membranes (Fig. 8). Assay (b) was applied to the large-scale analysis of grapefruit at a major juice-processing plant. It was shown that the range of naringin levels is very wide (6–2120 ppm:mean = 412:N=6685) and, as the upper limit of acceptance is about 700 ppm, clearly demonstrates the usefulness of such techniques in quality control. An excellent correlation was found between this assay and HPLC (r=0.92).

3.4 Anthrones

3.4.1 Sennosides

The dianthrone glucosides, sennosides A and B, products of a number of *Cassia* species (Caesalpinaceae), are powerful laxatives, and are one of the most widely used pharmaceuticals of plant origin. Uncertainty as to the precise origin of the dimeric molecules led to the development of a RIA designed only to detect the dimeric glucosides.

The Assay (Atzorn et al. 1981). Antiserum was raised in rabbits to a conjugate made by coupling sennoside B directly to BSA by the carbodiimide method (Sect. 2.1.1.1). [^{3}H]-Tracer was made by reduction (Sect. 2.3.1.3) of sennoside B with sodium [^{3}H]-borohydride to form [^{3}H]-8-glucosidorheinanthrone. (Convertion of this back to sennosides A and B by Pd/O_2 condensation resulted in about 80% loss of [^{3}H]. Separation of "bound" and "free" phases was achieved by precipitation with ammonium sulphate, the bound fraction being determined. Antiserum showed a high affinity constant for tracer (4×10^8 l/mol) and was used at a final titre of 1:1350. Sennoside B was detected over the range 0.4–8 ng with no interference from other materials in a crude leaf extract. The antiserum showed no cross-reaction with a range of anthraquinones and a strong requirement both for the 8-O-glucose moitey and for a *meso* linkage in the dimeric anthrone glucosides. Thus only sennosides A (75%) and C (100%) and the tracer material 8-glucosidorheinanthrone (55%) cross-reacted significantly, sennidines A (25%) and B (10%) showing some recognition.

Application. Previously, it has not proved possible to detect the dimeric sennosides in fresh plant material and it has been suggested that they are derivatives produced during the drying process. This was confirmed by using the RIA to examine the immunoreactivity of fractions from a TLC separation of dry and fresh leaves. In the latter only 8-glucosidorheinanthrone was detected (minute traces of sennoside A and B may have derived from work-up), whereas dry leaves con-

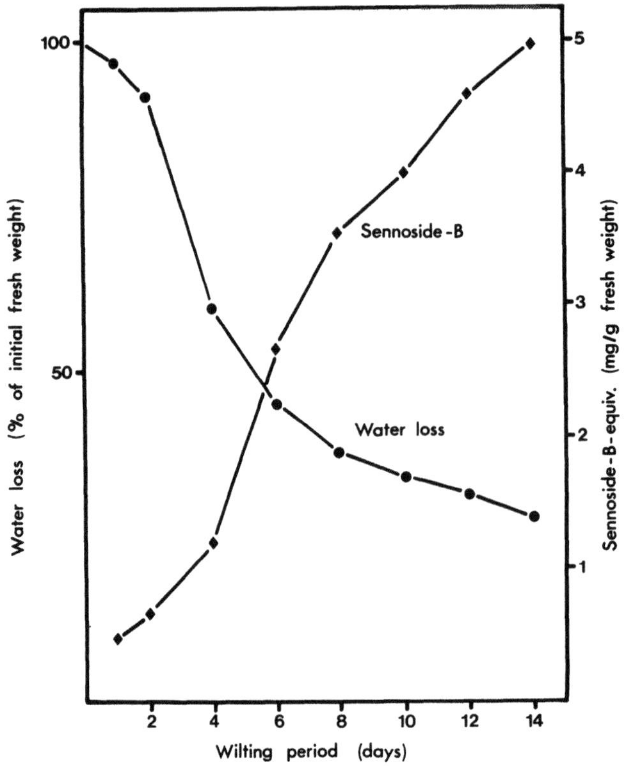

Fig. 10. Relationship between the amount of sennoside B formed, as determined by RIA, and the wilting of leaf material of *Cassia augustifolia*. Tissue was frozen in liquid nitrogen then extracted with hot water, the extract being diluted with water prior to assay. [Reproduced from Atzorn et al. (1981)]

tained in addition sennosides A and B at about 6 mg g^{-1} initial fresh weight. It was further demonstrated that sennoside B accumulated steadily in excised leaves left to desiccate over a 14-day-period (Fig. 10). As seedlings mature following germination the sennoside-equivalents/g fresh weight increases steadily. In mature tissue the influorescence contains the most material (4.3% dry weight), though leaves (2.8%) and seeds (2.4%) are also good sources. By examining fruits from 100 plants it was found that the sennoside content varied between 1% and 4% (mean = 2.4%). Thus, selection of high-yielding strains may prove profitable using this technique for rapid screening.

3.5 Steroids

3.5.1 Digitalis Cardenolides

The important myocardial stimulants digoxin and digitoxin (= 12-hydroxydigoxin) are extracted from *Digitalis lanata* (Scrophulariaceae) where they and a number of structurally closely related glycosides and aglycones (principally digoxigenin) accumulate in the leaves.

3.5.1.1 Digoxin

Following the pioneering work of Butler and Chen (1967), RIAs have been developed for clinical application (Smith et al. 1970) and for use in plant-strain improvement (Weiler and Zenk 1976; Weiler 1977; Weiler and Westekemper 1979).

The Assays. Both groups used essentially the same assay procedure. Antisera were raised in rabbits to a conjugate of digoxin with HSA coupled by the periodate cleavage method (Sect. 2.1.1.7). G-[^{3}H]-digoxin (NEN) was used as labelled tracer, though Weiler and Zenk (1976) also developed an assay using digoxin-[^{125}I]-tyrosine methyl ester. "Bound" and "free" phases were separated with dextran-coated charcoal, the antigen present in the bound phase being determined. The assays (Table 8) were very similar, except that the antiserum of Smith et al. (1970) had considerably greater sensitivity to the key 12-hydroxyl function. Neither antiserum recognised common steroids likely to be found in clinical samples or plant extracts. Weiler and Zenk (1976) examined the cross-reactivity with a very large number of related structures and showed that the serum would determine the sum of the digitoxigenin glycosides in crude extracts of *D. lanata* but would not recognise other related compounds.

Application. Smith et al. (1970) showed that serum levels of digoxin as low as 3×10^{-13} mol ml^{-1} could be determined. The distribution of digoxin-equivalents

Table 8. Comparison of RIAs for determination of *Digitalis* cardenolides

Assay parameter	Digoxin			Digitoxin
	1[a]	2	3	4
Tracer	[^{3}H]	[^{3}H]	[^{125}I]	[^{125}I]
Antiserum titre	1:63000	1:9000	1:36000	1:3000
Range (ng)	0.8–40	0.5–5	0.1–1	0.2–20
Detection limit (ng)	0.4	0.1	0.02	0.1
Non-specific binding (%)	–	4	9	1
Cross-reaction with (%):				
Digoxin	100	100	100	10
Digitoxin	4	100	100	100
Deslanoside	100	–	–	–
Digitoxigenin	–	75	59	–
Lanatoside C	–	113	133	–
Digoxoside	–	134	188	–
Digitoxigenin-bis-digitoxoside	–	114	128	–
Digitoxigenin-di-digitoxoside	–	71	103	–

[a] *1* Smith et al. (1970); *2,3* Weiler and Zenk (1970); *4* Oliver et al. (1970).

within the tissues of *D. lanata* was examined by Weiler and Zenk (1976) and is also discussed in Weiler (1977). They confirmed that the majority of product is present in leaf tissue and showed that the distribution throughout a leaf is uniform, with the exception of a lower concentration along the midrib. The concentration of digoxin in leaves was shown to rise over the first 8–12 weeks of growth but subsequently to remain static. Of commercial importance was their demonstration that by analysing a 3 mm-diam disc excised from a single leaf an exact measurement of the digoxin content ($r = 0.97$) of the dried leaf could be obtained. Thus it is possible to determine the optimal age for harvesting leaves. Using such techniques Weiler and Westekemper (1979) screened 50000 field-grown and 15000 phytotron-grown plants from which they selected 100 high-yielding individuals for a breeding programme. Within two selection steps the average digoxin content of the population was increased 2.5-fold to 0.6% dry weight with a few individuals producing 0.9–1% dry weight as digoxin. Such rapid improvement again demonstrates powerfully the effectiveness with which immunoassays can be applied to a strain-improvement programme.

Nickel and Staba (1977) raised antiserum and developed an assay after the method of Smith et al. (1970). They used this to study alterations in digoxin content of *Digitalis lanata* and *D. purpurea* plants from seed to maturity and the subcellular distribution of digoxin in different parts of plants during their development. Highest digoxin was always found in the supernatant fraction, whatever the tissue. Mature leaf contained considerable levels of digoxin but the highest concentration was in the seeds. Tissue cultures of *D. lanata* were apparently devoid of *Digitalis* cardenolides.

3.5.1.2 Digitoxin

Oliver et al. (1968) raised antisera in rabbits to a conjugate of digitoxigenin (the aglycone) with BSA made by 3-*O*-succinylation (Sect. 2.1.1.2) and carbodiimide condensation (Sect. 2.1.1.1). A radio-tracer was prepared by iodination (Sect. 2.3.1.4) of 3-*O*-succinyl digitoxigenin tyrosine methyl ester, made by the mixed-anhydride reaction (Sect. 2.1.1.1). The RIA (Table 8) was performed using goat anti-rabbit IgG to separate "bound" and "free" phases, the radioactivity in the bound phase being determined. The antiserum cross-reacted about 10% with digoxin but no further cardenolides were tested, despite the known metabolism of digitoxin in mammals. Physiological levels of several steroids were not recognised, although moderate inhibition occurred with 500 ng progesterone or testosterone.

Application. It was shown that patients receiving digitoxin therapy contained between 4–60 ng ml^{-1} serum, whereas patients not being treated or being administered digoxin caused no displacement of the radiolabel. No application to plants was made.

3.5.2 Solanum Steroidal Glycoalkaloids

The genus *Solanum* (Solanaceae) produces a number of glycoalkaloids of steroidal origin. Interest has focused on two groups of these:

1. Potato tubers (*Solanum tuberosum*) contain a number of toxic, bitter derivatives of solanidine (largely α-solanine and α-chaconine). There is no problem at the levels found normally in commercial varieties but it is vital to ensure that in breeding programmes using wild species any tendency towards higher glycoalkaloid production is identified at an early stage. Furthermore, poor storage of tubers or growth under stressed conditions can greatly increase the glycoalkaloid content of potatoes. Thus, the routine examination of samples is required.

2. The alkaloid solasodine, accumulated in the leaves of a number of species (e.g. *S. laciniatum* and *S. aviculare*) is of great interest as a potential starting material for the synthesis of steroidal pharmaceuticals.

Although all these compounds are alkaloids, they are grouped in this section on biosynthetic grounds.

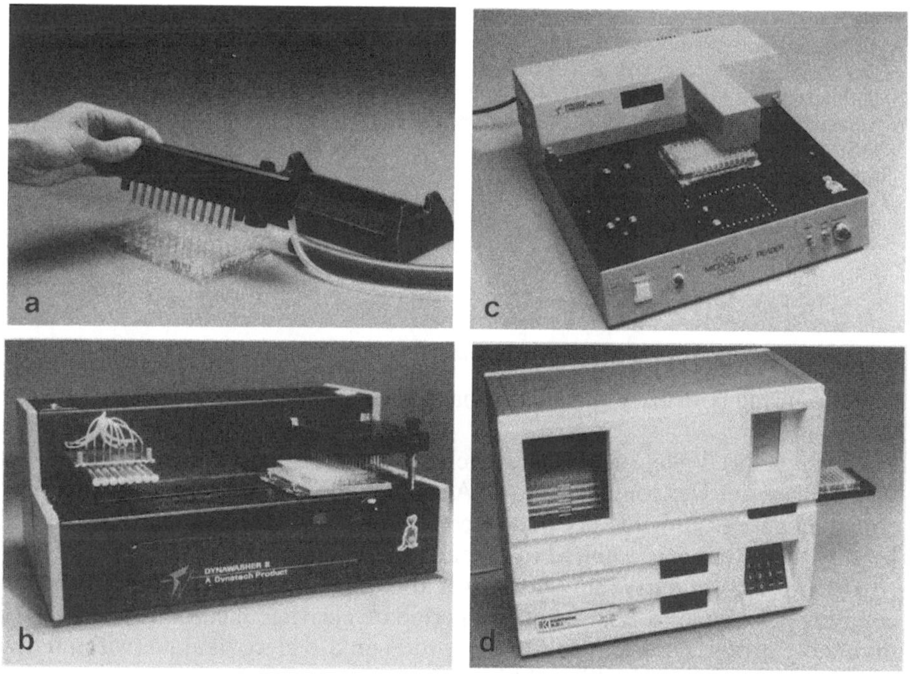

Fig. 11 a–d. A selection of equipment to facilitate the performance of a microtitration plate ELISA. **a** The NUNC-Immuno Wash 12. A hand-held, inexpensive plate washer capable of filling and aspirating 12 wells in a single operation (supplied by A/S Nunc). **b** The Dynawasher II a virtually automated plate washer capable of filling and aspirating 96 wells simultaneously (supplied by Dynatech Labs Inc.). **c** A simple plate reader in which the plate is moved by hand and the optical density determined for each well. More recent versions incorporate a printer or computer interface (supplied by Dynatech Labs Inc.). **d** A virtually automated plate reader, capable of determining the optical density of all wells in a plate in a single operation and of reading up to 10 plates without attention. Incorporates a printer with limited calculating capacity but can be interfaced to a computer for which data-handling programmes are available (supplied by Kontron AG)

Table 9. Comparison of the assays for determination of solanidine glycoalkaloids

Assay parameter	Assay[a]		
	1	2	3
Assay type	RIA	RIA	ELISA
Tracer	[^{3}H]	[^{3}H]	Alkaline phosphatase
Coupling method	Hemisuccinate/ mixed-anhydride	Periodate	Periodate
Antiserum titre	1:25	1:3000	1:20000
Range (ng)	0.2–50	0.2–8	0.01–1
Detection limit (ng)	0.15	0.2	0.005
Non-specific binding (%)	–	–	5
Intra-assay variation (%)	7.2	17.8	6.5
Inter-assay variation (%)	–	18.8	7.3
Cross-reaction with (%):			
Solanidine	100	100	100
α-Solanine	28	100	100
α-Chaconine	–	100	100
Demissidine	–	100	100
Rubijervine	–	4	4
Solasodine	–	< 0.1	0
β-Sitosterol	0	< 0.1	0
Stigmasterol	0	< 0.1	0

[a] *1* Vallejo and Ercegovich (1979); *2* Matthew et al. (1980); *3* Morgan et al. (1983b).

3.5.2.1 Solanidine-Derived Glycoalkaloids

Methods using both RIA and ELISA have been developed.

The Assays. (a) Vallejo and Ercegovich (1979) developed a RIA (Table 9) in which solanidine was conjugated to BSA through the 3-hydroxyl group. This generated an antiserum with very poor recognition of the glycosylated forms and a lengthy hydrolysis was required to prepare the samples for analysis.

(b) (Morgan et al. 1983b; Matthew et al. 1983) Immunogen was prepared by coupling α-solanine to BSA using the periodate cleavage method (Sect. 2.1.1.7) in order to produce antisera able to assay total potato glycoalkaloids without hydrolysis. Antiserum, raised in rabbits, proved not to differentiate between solanidine, α-solanine and α-chaconine whether used in RIA or ELISA (Table 9). It is therefore ideal for the determination of total glycoalkaloids in tubers and to study the distribution of ingested glycoalkaloids in human body fluids. Matthew et al. (1983) used [16,22-^{3}H]-solanidine as tracer in a RIA for solanidine (the aglycone being separated from the more polar derivatives by chloroform extraction). "Bound" and "free" phases were separated with dextran-coated charcoal. The ELISA (Morgan et al. 1983b) was performed in microtitration plates (see Sect. 2.4) coated at 20 μg ml^{-1} with solanidine-KHLH (made as for the immunogen) and developed with alkaline phosphatase-labelled goat anti-rabbit IgG.

Application. Matthew et al. (1983) and Harvey et al. (1985) examined serum samples extracted with chloroform and dried directly into assay vials. They demonstrated for the first time that solanidine is detectable in human plasma and that the level present (0.33–22.5 ng ml^{-1} serum) correlates ($r = 0.88$) with the dietary intake of potato. Abstaining from potato in the diet caused a significant decrease in serum solanidine levels.

Because of the importance of the routine testing of potato tubers for glycoalkaloid content and the difficulty of the conventional analyses, a rigorous comparison of the ELISA with chemical methods of glycoalkaloid determination was undertaken (Morgan et al. 1985a, b). For the ELISA, potato tissue (5 g) is extracted with methanol:water:acetic acid (94:6:1, 20 ml) and an aliquot diluted appropriately in PBS-Tween. Excellent correlation was achieved between all the techniques employed, of which the simplest, quickest, most direct and least susceptible to technical error was the ELISA. Such a trial demonstrates the suitability of ELISAs for routine quality control and it is anticipated that their use in this role will expand rapidly in the near future.

3.5.2.2 Solasodine-Derived Glycoalkaloids

(Weiler et al. 1980) Immunogen was synthesised from HSA by periodate coupling (Sect. 2.1.1.7) of a mixed preparation of solasonine and solamargine (40:60) and used to raise antiserum in rabbits. For RIA, a [^{3}H]-radio-tracer of undefined structure but of high immunoreactivity was prepared from the mixed glycoalkaloids by periodate treatment (Sect. 2.1.1.7) and reduction of the purified aldehydes with sodium [^{3}H]-borohydride (Sect. 2.3.1.3). Samples were prepared by methanolic extraction of dried plant material, the presence of methanol up to 10% having no effect on the assay. "Bound" and "free" phases were separated with dextran-coated charcoal and the bound [^{3}H]-antigen quantified. The assay was performed at a final titre of 1:4500 and would detect solasodine glycosides over the range 1–100 ng with a limit of detection of 0.7 ng. Cross-reactivity was assessed with a wide range of steroids, only the glycoalkaloids of the solasodine-type (solasonine, 100%; solamargine, 110%) or the closely related tomatidine-type (tomatine, 104%) being recognised strongly, the aglycones solasodine (7%), soladulcidine (5.4%) and tomatidine (20.7%) also being recognised.

Application. In a survey of *S. laciniatum* tissues solasodine was found in all parts of the plant, the highest concentration being in fully developed green fruits (1.1%) but the greatest extractable quantity being in the leaves (85%). As found with digoxin (Sect. 3.5.1.1), the alkaloid is distributed evenly throughout the lamella of the leaf but only at low concentration in the midrib, enabling a similar screening and selection procedure to be conducted. Leaf discs (3 mm) from *S. laciniatum* and *S. aviculare* were analysed and, although the mean content on a dry weight basis was 1.48% (N = 3213) and 1.54% (N = 353) respectively, a few individuals of greater than 2% dry weight were identified to be used in breeding experiments. In addition, over 250 species of *Solanum* were investigated for their solasodine content using small pieces of herbarium material, some over 100 years old. *S. laciniatum* proved the best accumulator of this alkaloid.

3.6 Miscellaneous

In this section a brief survey is given of a number of immunoassays for compounds which are not endogenous to plant tissue but which it may be desirable to determine at very low levels in plant material. These fall into two categories:

1. Materials that have physiological activity when applied exogenously to plants. Using an immunoassay it may prove possible to examine features of this activity such as the internal concentration of analyte, its conversion to a different form and its sub-cellular distribution;
2. Undesirable compounds in plant material destined for food use generated as a result of microbial spoilage.

3.6.1 Physiologically Active Metabolites

None of the assays presented herein have been applied to plant material, having been developed for the determination of these toxic compounds in human serum.

3.6.1.1 Colchicine

This fungal metabolite is extensively used in the study of the plant cytoskeleton. Boudene et al. (1975) have developed a RIA which could prove useful in cytoskeletal studies. The free amino group of N-desacetylthiocolchicine was coupled to BSA by the carbodiimide method (Sect. 2.1.1.1). [^{3}H]-Colchicine was used as tracer with immune serum at a titre of 1:100, "bound" and "free" phases being separated by immunoprecipitation. Colchicine could be satisfactorily determined over the range 5–50 ng but surprisingly both N-desacetylthiocolchicine (used for immunogen synthesis) and N-desacetyl-N-methylcolchine (altered at the point of conjugation) only showed 8% cross-reaction.

3.6.1.2 Chloramphenicol

This antibiotic has been widely applied in plants as a result of its specific inhibition of chloroplastic protein synthesis. Although Hamburger (1966) prepared antibodies to chloramphenicol-bovine γ-globulin by diazotisation (Sect. 2.1.1.3) the assay was not developed. Recently Campbell et al. (1984) have described an ELISA in which antiserum was raised in rabbits to a chloramphenicol-KHLH conjugate made by succinic anhydride (Sect. 2.1.1.2)/carbodiimide (Sect. 2.1.1.1) coupling. The assay employs polystyrene tubes coated with a chloramphenicol-BSA conjugate [made by the mixed-anhydride method (Sect. 2.1.1.1)] at 20 ng ml^{-1}. Sample is incubated with antiserum at a titre of 1:300 and the assay is developed using peroxidase-labelled goat anti-rabbit IgG at 1:500. The ELISA is fully developed, coating concentration, serum dilutions, incubation time and second antiserum dilution all having been varied and the analyte can be assayed over the range 0.4–40 ng. A wide range of other antibiotics were assessed for cross-reactivity and the antiserum was shown to be very specific, not even recognising known metabolic products of chloramphenicol.

3.6.1.3 Paraquat

A potent herbicide, of considerable toxicity to mammals, paraquat may be determined both by RIA (Levitt 1977; Fatori and Hunter 1980) and ELISA (Niewola et al. 1983). As the method of choice is the ELISA, only this will be described. Immunogen was synthesised from monoquat by activation with 6-bromohexanoic acid (Sect. 2.1.1.4) and carbodiimide condensation (Sect. 2.1.1.1) with BSA. Microtitration plates were coated at 1 $\mu g\ ml^{-1}$ with a monoquat-KHLH conjugate synthesised by the same methods. Samples were incubated with antiserum diluted 1:40000 and the assay quantified using peroxidase-labelled goat anti-rabbit IgG diluted 1:800. Paraquat was detected satisfactorily over the range 0.03–1 ng and only diethyl paraquat and monoquat showed significant cross-reaction. The ELISA was shown to give excellent correlation with the previous RIA (Levitt 1977) and to be more sensitive than the [^{125}I]-tracer assay (0.1–10 ng) of Fatori and Hunter (1980).

3.6.2 Contamination of Food and Feedstuffs

Mycotoxins are highly toxic and sometimes carcinogenic secondary metabolites produced under particular environmental conditions by certain moulds, notably *Aspergillus* and *Penicillium* species. Particular effort has been put into developing immunoassays for these compounds for several reasons:

1. Contamination is often non-uniform, requiring the assay of large numbers of sub-samples;
2. Assay of toxin levels as low as 1 in 10^{12} ($pg\ g^{-1}$) is sometimes required;
3. Methodology needs to be specific because of the presence of many other chemically related compounds.

Sadly, despite more than fifty reports concerning mycotoxin immunoassay since 1976, very little routine analysis is being done by this method. The reasons for this relate to either the poor quality of antiserum generated (in terms of amount, avidity and specificity) or insufficient assay validation. Recently, however, methods have been developed at the FRI using antiserum of high quality, available in large amounts. These are already being applied in other laboratories.

3.6.2.1 Aflatoxin B_1

This toxin, one of the most carcinogenic compounds known, is a product of *Aspergillus*. Both monoclonal (Morgan et al. 1986b) and polyclonal (Morgan et al. 1986c) antisera to this compound have been produced by linking the ketone group of the toxin to amino-oxyacetic acid (Sect. 2.1.1.6) and coupling the product to BSA by the mixed-anhydride method (Sect. 2.1.1.1). Both polyclonal and monoclonal sera were of high affinities, giving limits of detection in an ELISA of 0.1 pg per well. It was notable that though different monoclonal antibodies gave a range of specifities these were also reflected in the polyclonal sera obtained from different rabbits. Both types of sera have been validated for application to cereals and food material such as peanut butter.

3.6.2.2 *Sterigmatocystin*

Another toxic metabolite of *Aspergillus*, sterigmatocystin is a potent hepatocarcinogen. To avoid potential solubility problems, it was decided to assay a hydroxylated derivative of the toxin. Therefore, the BSA-sterigmatocystin conjugate (see Sect. 2.1.2.2) was converted to the hemiacetal prior to immunisation of rabbits (Kang et al. 1984; Morgan et al. 1986a). The antiserum obtained showed excellent sensitivity in an ELISA, detecting sterigmatocystin hemiacetal over the range 0.001–1 ng and not cross-reacting with other aflatoxins likely to be present in samples. Conversion of sterigmatocystin extracted from cereal samples was readily and quantitatively achieved.

3.6.2.3 *Ochratoxin A*

This potent kidney toxin frequently occurs in stored cereals in temperate climates. By linking through a carboxylic acid function (Sect. 2.1.1.1) distal to the chlorine moiety of ochratoxin A an antiserum was obtained capable of detecting only 10 pg ochratoxin A. It showed cross-reaction of only 0.5% to ochratoxin B which is non-toxic because of the absence of the chlorine (Morgan et al. 1983a). Using this assay, Anderson et al. (1984) examined 417 kidneys from pigs fed stored cereals and found one to have a high level (59 $\mu g\ kg^{-1}$) of toxin. Overall they concluded that none of the animals examined represented a potential health hazard to humans.

Acknowledgements. I am extremely grateful to Dr. Mike Morgan (Head, Biospecific Analysis Group, FRI, Norwich) for expert advice and generous help with the preparation of this manuscript, and, in the longer term, for having taught me much of what I know about immunoassays. My thanks are also warmly extended to Dr. Heather Kemp and Angray Kang for information, to Judy Furze for preparing the figures and to Anne Christopher for so efficiently transforming the manuscript into a presentable typescript.

Appendix: Sources of Materials and Equipment for ELISA

Clearly, it is impractical to provide a complete list of suppliers in this section. The aim is to direct first-time users towards sources which between them offer a range of the necessary equipment and chemicals for performing an ELISA. Suppliers are listed in alphabetical order and the types of material they provide coded as follows:

A: Microtitration plates
B: Primary antiserum (towards hapten in parenthesis)
C: Diagnostic kits (towards hapten in parenthesis)
D: Enzyme-labelled species-specific anti-IgG preparations
E: Conjugation proteins and enzyme substrates
F: Amplification systems (type in parenthesis)
G: Reagent dispensors
H: Plate washers
I: Plate readers (other than colorimetric in parenthesis)
J: Computerised data handling available for ELISA analysis

A selection of equipment designed to automate the processing of ELISAs is illustrated in Fig. 11.

ABBOTT Labs Ltd., Diagnostics Division, Brighton Hill Parade, Basingstoke RG22 4EH, U.K.	C (chloramphenicol, digoxin, digitoxin, quinidine); I (FPIA)
Amersham International plc, White Lion Road Amersham HP7 9LL, U.K.	C (digoxin); D; F (biotin/streptavidin)
Artek Systems Corporation, 32 Humberstone Road, Cambridge CB4 1JF, U.K.	I; J
Bellco Biotechnology, P O Box B, 340 Edrudo Road, Vineland, New Jersey 08360, USA	G; H
Bioanalysis Ltd., P O Box 88, University Place, Cardiff CFl 1SA, U.K.	B (digoxin, vinblastine/vincristine)
Bio-Rad Labs, 2200 Wright Avenue, Richmond, California 94804, USA	C (digoxin); D; E; I; J
Boehringer Mannheim GmbH, D-6800 Mannheim 31, FRG	C (digitoxigenin, digitoxin); E (visible and fluorescent); G; H; I; J
Dynatech Labs Inc., 900 Slaters Lane, Alexandria, Virginia 22314, USA	A; G; H; I (visible and fluorescent); J (see Fig. 11)
Eli Lilly Labs, Indianapolis, Indiana 46206, USA	B (vinblastine/vincristine);
Flows Labs Ltd., P O Box 17, Irvine KA12 8NB, U.K.	A; G; H; I (visible and fluorescent); J

Food Research Institute, Colney Lane, Norwich NR4 7UA, U. K.	B (aflatoxin B_1, ochratoxin A, quassin, quinidine, quinine, solanine, sterigmatocystin)
GIBCO Europe Ltd., P O Box 35, Paisley PA3 4EF, U. K.	A
Gilson France SA, 72 rue Gambetta, 95400 Villiers le Bel, France	G
Guildhay Antisera, Department of Biochemistry, University of Surrey, Guildford GU2 5XH, U. K.	B (digoxin, morphine, vinblastine/vincristine)
Ilacon Ltd., Gilbert House, Tonbridge TN9 1DT, U. K.	H; I
Kontron AG, Analytical Division, Bernerstrasse Sud 169, CH-8048 Zurich, Switzerland	H; I; J (see Fig. 11)
Miles Labs Inc., P O Box 272, Kankakee, Illinois 60901, USA	D
Northumbria Biologicals Ltd., South Nelson Industrial Estate, Cramlington NE23 9HL, U. K.	A; D; G; H; I (portable)
A/S Nunc, Postbox 280, Kamstrup, DK-4000 Roskilde, Denmark	A; H; I (see Fig. 11)
Pharmacia Biotechnology International AB, S-751 82 Uppsala, Sweden	C (digoxin)

Polysciences Inc., Paul Valley Industrial Park, Warrington, Pennsylvania 18976, USA	E (fluorescent)
Roche Products Ltd., 318 High Street, Dunstable LU6 1BG, U.K.	C (cannabinoids, cocaine, morphine)
Sera-Lab Ltd., Crawley Down RH10 4FF, U.K.	F (biotin/avidin)
Sigma Chemical Co, P O Box 14508, St Louis, Missouri 63178, USA	D; E
Steranti Research Ltd., 141 London Road, St Albans AL1 1TA, U.K.	C (digoxin, digitoxin)
Sterilin, 43–45 Broad Street, Teddington TW11 8QZ, U.K.	A
Syva Corp. Palo Alto, California 94304, USA	C (EMIT or EIA: digoxin, quinidine); I (EMIT)

References

Abraham GE (1974) Radioimmunoassay of steroids in biological material. Acta Endocrinol Suppl 183:7–42

Adler FL, Liu C-T (1971) Detection of morphine by hemagglutination-inhibition. J Immunol 106:1684–1685

Amersham (1984) The biotin-streptavidin system. Amersham International, Amersham, pp 16

Anderson PH, Wells GAH, Jackman R, Morgan MRA (1984) Ochratoxicosis and ochratoxin A residues in adult pig's kidneys – a pilot survey. In: Moss MO, Frank M (eds) Proc 5th meet mycotoxins in animal and human health. Surrey University Press, Guildford, pp 23–29

Arens H, Stockigt J, Weiler EW, Zenk MH (1978) Radioimmunoassays for the determination of the indole alkaloids ajmalicine and serpentine in plants. Planta Med 34:37–46

Atzorn R, Weiler EW, Zenk MH (1981) Formation and distribution of sennosides in *Cassia augustifolia*, as determined by a sensitive and specific radioimmunoassay. Planta Med 41:1–14

Boudene C, Duprey F, Bohuon C (1975) Radioimmunoassay of colchicine. Biochem J 151:413–415
Butler VP, Chen JP (1967) Digoxin-specific antibodies. Proc Natl Acad Sci USA 57:71–78
Campbell GS, Mageau RP, Schwab B, Johnston RW (1984) Detection and quantitation of chloramphenicol by competative enzyme-linked immunoassay. Antimicrob Agents Chemother 25:205–211
Carlier Y, Bout D, Capron A (1979) Automation of enzyme-linked immunoassay (ELISA). J Immunol Methods 31:237–246
Corrie JET (1983) 125-Iodinated tracers for steroid radioimmunoassay: the problem of bridge recognition. In: Hunter WM, Corrie JET (eds) Immunoassays for clinical chemistry. Churchill Livingstone, Edinburgh, pp 353–357
Dandliker WB, Saussure VA de (1970) Fluorescence polarization in immunoassay. Immunochemistry 7:799–828
Ekins RP (1980) More sensitive immunoassays. Nature 284:14–15
Ekins RP (1981) Towards immunoassays of greater sensitivity, specificity and speed: an overview. In: Albertini A, Ekins RP (eds) Monoclonal antibodies and developments in immunoassay. Elsevier/North-Holland, Amsterdam, pp 3–21
Erlanger BF, Borek OF, Beiser SM, Leiberman S (1959) Steroid-protein conjugates II. Preparation and characterization of conjugates of bovine serum albumin with progesterone, deoxycorticosterone and estrone. J Biol Chem 234:1090–1094
Fatori D, Hunter WM (1980) Radioimmunoassay for paraquat. Clin Chim Acta 100:81–90
Fernley HN, Walker PG (1965) Kinetic behaviour of calf-intestinal alkaline phosphatase with 4-methylumbelliferyl phosphate. Biochem J 97:95–103
Fong K-LL, Ho DHW, Carter CJK, Brown NS, Benjamin RS, Freireich EJ, Bodey GP (1980) Radioimmunoassay for the detection of bruceantin. Anal Biochem 195:281–286
Galfre G, Milstein C (1981) Preparation of monoclonal antibodies: strategies and procedures. In: Langone JJ, Vunakis H van (eds) Immunochemical techniques, part B. Academic Press, London, pp 3–46 (Methods Enzymol, vol 73)
Godicke W, Godicke I (1973) The fluorimetric determination of uric acid by the used of the uricase-peroxidase system and 3,5-diacetyl-1,4-dihydrolutidine as secondary substrate. Clin Chim Acta 44:159–163
Greenwood FC, Hunter WM, Glover JS (1963) The preparation of ^{131}I-labelled human growth hormone of high specific radioactivity. Biochem J 89:114–123
Hacker MP, Dank JR, Ershler WB (1984) Vinblastine pharmokinetics measured by a sensitive enzyme-linked immunosorbent assay. Cancer Res 44:478–481
Hamburger RN (1966) Chloramphenicol-specific antibody. Science 152:203–205
Harris CC, Yolken RH, Krokan H, Hsu IC (1979) Ultrasensitive enzymatic radioimmunoassay: application to detection of cholera toxin and rotavirus. Proc Natl Acad Sci USA 76:5336–5339
Harvey MH, McMillan M, Morgan MRA, Chan HW-S (1985) Solanidine is present in sera of healthy individuals and in amounts dependent on their dietary potato consumption. Human Toxicol 4:187–194
Herbert WJ (1978) Laboratory animal techniques for immunology. In: Weir DM (ed) Handbook of experimental immunology, 3rd edn. Blackwell, Oxford, pp A.4.1.-A.4.29
Hunter WM, Corrie JET (eds) (1983) Immunoassays for clinical chemistry, 2nd edn. Churchill Livingstone, Edinburgh, pp 701
Jourdan PS, Mansell RL, Weiler EW (1982) Radioimmunoassay for the *Citrus* bitter principle, naringin, and related flavonoid-7-O-neohesperidosides. Planta Med 44:82–86
Jourdan PS, Weiler EW, Mansell RL (1983) Radioimmunosassay for naringin and related flavone-7-O-neohesperidosides using a tritiated tracer. J Agric Food Chem 31:1249–1255
Jourdan PS, Mansell RL, Oliver DG, Weiler EW (1984) Competative solid phase enzyme-linked immunoassay for the quantitation of limonin in *Citrus*. Anal Biochem 138:19–24

Kang AS, Morgan MRA, Chan HW-S (1984) Production of rabbit anti-sterigmatocystin antisera and its characterisation and use in immunoassay. In: Moss MO, Frank M (eds) Proc the 5th meet mycotoxins in animal and human health. Surrey University Press, Guildford, pp 80–83
Kutney JP, Choi LSL, Worth BR (1980) Radioimmunoassay determination of vindoline. Phytochemistry (Oxf) 19:2083–2087
Landsteiner K (1945) The specificity of serological reactions. Harvard University Press, Cambridge
Langone JJ, Gjika HB, Vunakis H van (1973) Nicotine and its metabolites. Radioimmunoassays for nicotine and cotinine. Biochemistry 12:5025–5030
Langone JJ, D'Onofrio MR, Vunakis H van (1979) Radioimmunoassays for the *Vinca* alkaloids, vinblastine and vincristine. Anal Biochem 95:214–221
Leaback DH, Walker PG (1961) Studies on glucosaminidase IV. The fluorimetric assay of N-acetyl-β-glucosaminidase. Biochem J 78:151–156
Lehtonen O-P, Eerola E (1982) The effect of different antibody affinities on ELISA absorbance and titer. J Immunol Methods 54:233–240
Levitt T (1977) Radioimmunoassay for paraquat. Lancet 1977(2):358
Manganey P, Andriamialisoa RZ, Langlois N, Langlois Y, Potier P (1979) Preparation of vinblastine, vincristine, and leurosidine, antitumor alkaloids from *Catharanthus* spp. (Apocynaceae). J Am Chem Soc 101:2243–2245
Mansell RL, Weiler EW (1980) Radioimmunoassay for the determination of limonin in *Citrus*. Phytochemistry (Oxf) 19:1403–1407
Marrero J (1984) Applications of immunoassay to therapeutic drug monitoring. J Med Technol 1:47–50
Matthew JA, Morgan MRA, McNerney R, Chan HW-S, Coxon DT (1983) Determination of solanidine in human plasma by radioimmunoassay. Food Chem Toxicol 21:637–640
Meisner H, Meisner P (1981) Ochratoxin A, an in vivo inhibitor of renal phosphoenolpyruvate carboxylase. Arch Biochem Biophys 208:146–153
Morgan MRA, McNerney R, Chan HW-S (1983a) Enzyme-linked immunosorbent assay of ochratoxin A in barley. J Assoc Off Anal Chem 66:1481–1484
Morgan MRA, McNerney R, Matthew JA, Coxon DT, Chan HW-S (1983b) An enzyme-linked immunosorbent assay for total glycoalkaloids in potato tubers. J Sci Food Agric 34:593–598
Morgan MRA, Coxon DT, Bramham S, Chan HW-S, Gelder WMJ van, Allison MJ (1985a) Determination of the glycoalkaloid content of potato tubers by three methods including enzyme-linked immunosorbent assay. J Sci Food Agric 36:282–288
Morgan MRA, McNerney R, Coxon DT, Chan HW-S (1985b) Comparison of the analysis of total glycoalkaloids by immunoassays and conventional procedures. In: Morris BA, Clifford MN (eds) Immunoassays in food analysis. Elsevier Applied Science, London, pp 187–195
Morgan MRA, Bramham S, Webb AJ, Robins RJ, Rhodes MJC (1985c) Specific immunoassays for quinine and quinidine: comparison of radioimmunoassay and enzyme-linked immunosorbent assay procedures. Planta Med (3):237–241
Morgan MRA, Kang AS, Chan HW-S (1986a) Production of antisera against sterigmatocystin hemiacetal and its use in an enzyme-linked immunosorbent assay for sterigmatocystin in barley. J Sci Food Agric (in press)
Morgan MRA, Kang AS, Chan HW-S (1986b) Aflatoxin determination in peanut butter by enzyme-linked immunosorbent assay. J Sci Food Agric (in press)
Morgan MRA, Kang AS, Bramham S, Chan HW-S, Butcher G, Galfre G (1986c) Monoclonal antibodies against aflatoxin B_1 – production, characterisation and use in an enzyme-linked immunosorbent assay (in preparation)
Nickel S, Staba EJ (1977) RIA-test of *Digitalis* plants and tissue cultures. In: Barz W, Reinhard E, Zenk MH (eds) Plant tissue culture and its bio-technological application. Springer, Berlin Heidelberg New York, pp 278–284
Niewola Z, Walsh ST, Davies GE (1983), Enzyme linked immunosorbent assay (ELISA) for paraquat. Int J Immunopharmacol 5:211–218

Oliver GC, Parker BM, Brasfield DL, Parker CW (1968) The measurement of digitoxin in human serum by radioimmunoassay. J Clin Invest 47:1035–1042
Owellen RJ, Donigan DW (1972) [^{3}H]-Vincristine. Preparation and preliminary pharmacology. J Med Chem 15:894–898
Owellen RJ, Hartke CA, Hains FO (1977) Pharmacokinetics and metabolism of vinblastine in humans. Cancer Res 37:2597–2602
Owellen RJ, Blair M, Tosh A van, Hains FC (1981) Determination of tissue concentrations of *Vinca* alkaloids by radioimmunoassay. Cancer Treat Rep 65:469–475
Pape BE (1981) Enzyme immunoassay and two fluorimetric methods compared for the determination of quinidine in serum. Ther Drug Monit 3:357–363
Quattrone AJ, Joseph R, Jones T, Sarmir S, Selvaggi A, Maloney K, Lockwood T (1984) A rate nephelometric inhibition immunoassay for quinidine and its active metabolites. J Med Technol 1:381–385
Robins RJ, Morgan MRA, Rhodes MJC, Furze JM (1984a) An enzyme-linked immunosorbent assay for quassin and closely related metabolites. Anal Biochem 136:145–156
Robins RJ, Morgan MRA, Rhodes MJC, Furze JM (1984b) Determination of quassin in picogram quantities by an enzyme-linked immunosorbent assay. Phytochemistry (Oxf) 23:1119–1123
Robins RJ, Webb AJ, Rhodes MJC, Payne J, Morgan MRA (1984c) Radioimmunoassay for the quantitive determination of quinine in cultured plant cells. Planta Med 50:235–238
Robins RJ, Morgan MRA, Rhodes MJC, Furze JM (1985) Cross-reactions in immunoassays for small molecules: use of specific and non specific antisera. In: Morris BA, Clifford MN (eds) Immunoassays in food analysis. Elsevier Applied Science, London, pp 197–211
Rook GAW, Cameron CH (1981) An inexpensive portable battery-operated photometer for the reading of ELISA tests in microtitration plates. J Immunol Methods 40:109–114
Rowley GL, Rubenstein KE, Huisjen J, Ullman EF (1975) Mechanism by which antibodies inhibit hapten-malate dehydrogenase conjugates. J Biol Chem 250:3759–3766
Sethi VS, Burton SS, Jackson DV (1980) A sensitive radioimmunoassay for vincristine and vinblastine. Cancer Chemother Pharmacol 4:183–187
Simpson JSA, Campbell AK, Ryall MET, Woodhead JS (1979) A stable chemiluminescent-labelled antibody for immunological assays. Nature 279:646–647
Smith DS, Al-Hakiem MHH, Landon J (1981) A review of fluoroimmunoassay and immunofluorometric assay. Ann Clin Biochem 18:253–274
Smith TW, Butler VP, Haber E (1970) Characterization of antibodies of high affinity and specificity for the *Digitalis* glycoside digoxin. Biochemistry 9:331–337
Spector S (1971) Quantitative determination of morphine in serum by radioimmunoassay. J Pharmacol Exp Ther 178:253–258
Spector S, Parker CW (1970) Morphine: radioimmunoassay. Science 168:1347–1348
Stuart KL, Kutney JP, Honda T, Lewis NG, Worth BR (1978) The biosynthesis of vindoline using cell free extracts from mature *Catharanthus roseus* plants. Heterocycles 9:647–652
Tanahashi T, Nagakura N, Inouye H, Zenk MH (1948) Radioimmunoassay for the determination of loganin and the biotransformation of loganin to secologanin by plant cell cultures. Phytochemistry (Oxf) 23:1917–1922
Teal JD, Clough JM, Marks V (1977) Radioimmunoassay of vinblastine and vincristine. Br J Clin Pharmacol 4:169–172
Treimer JF, Zenk MH (1978) Enzymic synthesis of corynanthe-type alkaloids in cell cultures of *Catharanthus roseus*: quantitation by radioimmunoassay. Phytochemistry (Oxf) 17:227–231
Vallejo RP, Ercegovich CD (1979) Analysis of potato for glycoalkaloid content by radioimmunoassay (RIA). In: Trace organic analysis: a new frontier in analytical chemistry, Natl Bureau Standards (USA), spec publ 519, pp 333–340 (Proc 9th materials res symp)

Voller A (1978) The enzyme-linked immunosorbent assay (ELISA). Ric Clin Lab 8:289–298

Voller A, Bidwell DE, Bartlett A (1979) The enzyme-linked immunosorbent assay (ELISA). Dynatech Europe, Guernsey

Vunakis H van, Wasserman E, Levine L (1972) Specificities of antibodies to morphine. J Pharmacol Exp Ther 180:514–521

Weeks I, Woodhead JS (1984) Chemiluminescence immunoassay. J Clin Immunoassay 7:82–89

Weeks I, Beheshti I, McCapra F, Campbell AK, Woodhead JS (1983) Acridinium esters as high-specific-activity labels in immunoassay. Clin Chem 29:1474–1479

Weiler EW (1977) Radioimmuno-screening methods for secondary plant products. In: Barz W, Reinhard E, Zenk MH (eds) Plant tissue culture and its bio-technological application. Springer, Berlin Heidelberg New York, pp 266–284

Weiler EW (1980) Radioimmunoassays for the differential and direct analysis of free and conjugated abscisic acid in plant extracts. Planta (Berl) 148:262–272

Weiler EW, Mansell RL (1980) Radioimmunoassay of limonin using a tritiated tracer. J Agric Food Chem 28:543–545

Weiler EW, Westekemper P (1979) Rapid selection of strains of *Digitalis lanata* Ehrh with high digoxin content. Planta Med 35:316–322

Weiler EW, Zenk MH (1976) Radioimmunoassay for the determination of digoxin and related compounds in *Digitalis lanata*. Phytochemistry (Oxf) 15:1537–1545

Weiler EW, Zenk MH (1979) Autoradiographic immunoassay (ARIA): a rapid technique for the semiquantitative mass screening of haptens. Anal Biochem 92:147–155

Weiler EW, Kruger H, Zenk MH (1980) Radioimmunoassay for the determination of the steroidal alkaloid solasodine and related compounds in living plants and herbarium specimens. Planta Med 39:112–124

Weiler EW, Stockigt J, Zenk MH (1981) Radioimmunoassay for the quantitative determination of scopolamine. Phytochemistry (Oxf) 20:2009–2016

Werner G, Mohammad N (1966) Synthetischer Einbau von Tritium in (−)-cocain, (+)-pseudococain und (−)-scopolamin. Liebigs Ann Chem 694:157–161

Westekemper P, Wieczorek U, Gueritte F, Langlois N, Langlois Y, Potier P, Zenk MH (1980) Radioimmunoassay for the determination of the indole alkaloid vindoline in *Catharanthus*. Planta Med 39:24–37

Wieczorek U, Nagakura N, Zenk MH (1984) Radioimmunoassays for opium alkaloids. Phytochem Soc Symp "The chemistry and biology of isoquinoline alkaloids", London, Abstract 36

Zenk MH, El-Shagi H, Arens H, Stockigt J, Weiler EW, Deus B (1977) Formation of indole alkaloids serpentine and ajmalicine in cell suspension cultures of *Catharanthus roseus*. In: Barz W, Reinhard E, Zenk MH (eds) Plant tissue culture and its bio-technological application. Springer, Berlin Heidelberg New York, pp 27–43

Radioimmunoassay and Western Blot Analysis of Acyl Carrier Protein Isoforms in Plants [1]

T. M. Kuo and J. B. Ohlrogge

1 Introduction

Acyl carrier protein (ACP) is the central cofactor protein to which acyl chains are covalently linked as thioesters during de novo fatty acid synthesis and subsequent elongation and desaturation reactions in plants (Stumpf 1980). In yeast and animals, ACP is an integral part of the multifunctional polypeptide, fatty acid synthase, but in bacteria and plants, ACP is a separate, small polypeptide (ca. 9 kDa MW) which is acidic and heat stable (Majerus and Vagelos 1967; Stumpf 1980; Wakil et al. 1983). Despite this basic organizational difference, the structure of ACP has been highly conserved during evolution as evidenced by greater than 25% homology among the primary sequences of an animal ACP domain, bacterial and plant ACPs (Kuo and Ohlrogge 1984a), and by the interchangeability of bacterial and plant ACPs in several fatty acid biosynthetic reactions (Simoni et al. 1967).

ACP occurs at high concentration in bacterial cells, but it is of low abundance (less than 0.1% of the total soluble protein) in plants. The difficulty of measuring low levels of plant ACP in small samples and the need for assaying ACP antigen in the study of the control mechanisms of ACP synthesis in developing seeds have prompted us to raise specific anti-ACP antibodies for use as critical, sensitive immunoprobes for analyzing plant ACPs. The following is a discussion of the development of a radioimmunoassay (RIA) and certain immunological characterizations of plant ACPs.

2 Production of ACP Antibody

2.1 Purification of Spinach ACP Isoforms

Because ACPs are difficult to purify from oilseeds, and structures of plant ACPs are highly conserved, we have developed a large-scale purification procedure using spinach leaves as an alternate source of ACP, to be used for structural analysis and for raising antibodies in rabbits. The initial steps of purification, including heating, ammonium sulfate fractionation, acid precipitation, and ion-exchange

[1] *Abbreviations:* ACP: acyl carrier protein; RIA: radioimmunoassay; MES: 2-(N-morpholino)ethanesulfonic acid; HPLC: high-performance liquid chromatography; DTT: dithiothreitol; OA: Ovalbumin; BSA: bovine serum albumin; PBS: phosphate-buffered saline; PEG: polyethylene glycol; DAF: days after flowering.

chromatography, are modified from the method of Simoni et al. (1967). In a typical preparation, 50 kg of frozen fresh spinach with stems partially removed is homogenized in a Cowles Dissolver (type 7VT, The Cowles Co., Cayuga, New York, USA) with 25 liters of cold 0.1 M K-phosphate, pH 7.5, containing 10 mM $Na_2S_2O_5$, 10 mM β-mercaptoethanol, 0.1% Triton X-100, and 1 mM EDTA. After debris is removed with a Sharples centrifuge, the homogenate is heated to 85 °C for 10 min in a 30-gal heating tank and then cooled. The clear supernatant obtained after centrifugation is adjusted to 70% $(NH_4)_2SO_4$ saturation, centrifuged, and then acid-precipitated at pH 4.0. The acid precipitate, recovered from a small Sharples centrifuge, is suspended in 50 mM K-phosphate, pH 7.5. A second acid precipitation is carried out in the presence of 80% $(NH_4)_2SO_4$, mainly to reduce volume of the solution. The pellets are resuspended in 10 mM 2-(N-morpholino)ethanesulfonic acid (Mes), pH 6.1 (pH adjusted with KOH), so that the conductivity of the solution is below that of 0.1 M NaCl. Ion-exchange chromatography is carried out with a DE52 (Whatman) column (2.5 × 14 cm) equilibrated with 10 mM Mes, pH 6.1, containing 1 mM β-mercaptoethanol and a linear gradient of 0.1 to 0.5 M NaCl in the same buffer. Major ACP-containing fractions, eluted between 0.2–0.3 M NaCl are identified by an enzymatic assay method as described below (Sect. 3.1). ACP is acid-precipitated from the pooled solution containing 1 M NaCl (as a general practice, we adjust to this salt level to facilitate precipitation of ACP from the acid solution). The resulting ACP preparation is about 50% homogeneous, as judged by thin-layer isoelectric focusing on pH 3 to 6 Servalyt Precotes (Serva).

The final purification is carried out by two cycles of high-performance liquid chromatography (HPLC) with a Synchropak AX-300 column (250 × 10 mm; SynChrom, Inc.). All HPLC runs are carried out at 20 °–23 °C with a flow rate of 3 ml min^{-1}. ACP samples are treated with 5 mM dithiothreitol (DTT) before HPLC. In the first cycle, using a linear gradient of 0.2 to 0.9 M NaCl in 50 mM K-phosphate at pH 6.8 for 25 min, the partially purified spinach ACP reveals two peaks containing ACP activitiy. These peaks, designated as I and II, are eluted at 0.5–0.6 M NaCl. ACP-I from a number of HPLC runs is pooled and acid-precipitated in the presence of 1 M NaCl. The acid-precipitated ACP is dissolved in a minimal volume of 0.5 M Tris, pH 8.0 (final pH 6.5–7.0), and diluted with 50 mM K-phosphate, pH 6.8, for the second cycle of HPLC, which employs a linear gradient of 0.2–0.6 M NaCl in the phosphate buffer. Absorbance peaks at 225 nm are assayed enzymatically for ACP. ACP-II is later recycled on a new Synchropak Q-300 column (250 × 4.1 mm) with a linear gradient of 0–0.3 M $MgCl_2$ in 0.02 M Tris, pH 7.5 at a flow rate of 1 ml min^{-1}. Other experimental conditions are the same as for the HPLC of ACP-I. Synchropak Q-300, which is a strong anion exchange HPLC support that became available in 1984, is an excellent HPLC column for the separation of spinach ACP isoforms.

2.2 Immunization Procedure

Spinach ACP is a weak antigen. Purified spinach ACP-I is therefore polymerized to ovalbumin (OA) to increase immune response in rabbits based on the procedure described by Reichlin (1980). Polymerization is carried out in 0.1 M Na-phosphate, pH 7.5, containing spinach ACP and OA (0.15 mM each) in the presence of glutaraldehyde (25 mM). Two young New Zealand white male rabbits (4–5 lb) are injected subcutaneously at multiple sites with a sonicated emulsion containing approx. 0.4 mg of ACP equivalent of ACP-OA polymer in 1.0 ml of Freund's complete adjuvant. Two weekly booster injections are given in the same manner after 4 weeks, using incomplete adjuvant. The titer of antiserum collected from the rabbit ear vein 3 weeks after booster injections is weak, as monitored by its ability to inhibit ACP acylation in the acyl-ACP synthetase enzymatic assay. Two additional booster injections are then given at 2-week intervals with incomplete adjuvant, using an emulsion containing an equal molar ratio of ACP (0.4 mg) and bovine serum albumin (BSA) or cytochrome *c* (horse heart). Serum is collected 2 weeks after the final injection and stored at −15 °C. The antibody titer of the rabbit injected with the ACP and cytochrome *c* mixture shows a rapid decay, whereas the one injected with the ACP and BSA mixtures increases at least twofold in titer. The latter antiserum is used directly for RIA with appropriate dilutions.

The antiserum possesses the ability to completely inhibit ACP acylation and has an approx. twofold higher titer than that previously prepared against spinach ACP (Ohlrogge et al. 1979). Preimmune serum contains no inhibitory activity against ACP acylation. Because of the immunization scheme, the antiserum also possesses cross-reactivities against OA and BSA. Since nonfat dry milk or gelatin can substitute for BSA as a blocking protein in the immunoblot experiments, ACP can be directly polymerized to BSA to enhance immune response in rabbits. Despite our efforts to increase titer through cross-linking, 1.0 μl of antiserum is able to bind only ~10 μg of spinach ACP. Other animals might be used to increase antibody titer. Species differences can affect the quality of a specific antiserum. For instance, Chafouleas et al. (1982) found that in the production of monospecific antibodies to native calmodulin, sheep antiserum is satisfactory, but goat antiserum is inappropriate for use in the development of a RIA for calmodulin.

3 Radioimmunoassay

3.1 Preparation of Radiolabeled ACP

The most commonly used isotopes to label specific proteins in a RIA are ^{125}I and ^{3}H. The commercial preparations of ^{125}I (~2000 Ci mmol^{-1}) have about 70-fold higher specific activity than ^{3}H (e.g., [^{3}H]$NaBH_4$, ~30 Ci mmol^{-1}), but ^{3}H is easier to handle, less expensive, and more stable. Proteins can be labeled with ^{125}I by reaction with the tyrosine residues by the chloramine-T procedure (Hunter and Greenwood 1962) or by reacting free amino groups with an iodinated acylating agent (Bolton and Hunter 1973). Alternatively, proteins can be chemically labeled with ^{3}H by a reductive methylation method (Rice and Means 1971). All chemical methods require high purity of antigen to ensure the specificity of the RIA.

Spinach ACP is devoid of tyrosine, but can be chemically labeled with ^{125}I by a modification of the procedure described by Bolton and Hunter (1973) (Ohlrogge et al. 1979). However, the most convenient way of radiolabeling spinach ACP is carried out by a specific enzymatic reaction using acyl-ACP synthetase isolated from *E. coli* and [^{3}H]palmitic acid (Kuo and Ohlrogge 1984b; Ohlrogge and Kuo 1984). The important feature of this approach is that the enzyme catalyzes acylation only to ACP in a crude plant extract (Kuo and Ohlrogge 1984c), and the acylation does not alter antigenicity of ACP. Therefore, the specificity of the enzymatic-labeling method circumvents the strict requirement of pure antigens needed in a chemical labeling reaction.

E. coli acyl-ACP synthetase can be purified as described by Rock and Cronan (1979) except that the hydroxyapatite step is replaced by Sephadex G-25 gel filtration to remove KSCN. In the enzyme purification, all procedures are performed at 4 °C. Frozen *E. coli* B cells (125 g; Grain-Processing Co.) are thawed and suspended in 50 mM Tris, pH 8.0 (Buffer A) containing 1 mM β-mercaptoethanol in a final volume of 250 ml and 3 mg of DNase II (Sigma) is added. The suspension is homogenized with a Polytron homogenizer and passed through a French pressure cell at 16000 p.s.i. The homogenate is centrifuged at $15000 \times g$ for 20 min to recover supernatant which is adjusted to contain 10 mM Mg^{2+} by the addition of 1 M $MgCl_2$. The resulting membrane pellets recovered from centrifuging the supernatant at $160000 \times g$ for 45 min are resuspended in 50 ml Buffer A and washed with 50 ml Buffer A containing 0.01 M NaCl. The salt-washed membrane pellets are recovered by centrifugation at $160000 \times g$ for 45 min. The membrane pellets are homogenized thoroughly in 50 ml Buffer A with a Polytron homogenizer. A solution of 50 ml Buffer A containing 4% Triton X-100, 20 mM $MgCl_2$ is added and the suspension is stirred gently for 30 min to solubilize acyl-ACP synthetase from membrane pellets. The Triton X-100 supernatant is recovered after centrifugation at $160000 \times g$ for 45 min. An aliquot of 0.1 M ATP in Buffer A is added to supernatant to a final concentration of 5 mM ATP. In the presence of 5 mM ATP, the enzyme preparation can be stored up to 1 week at 4 °C before being applied to a 10-ml blue-Sepharose CL-6B (Sigma) column equilibrated with Buffer A containing 2% Triton X-100 (Buffer B). After sample application, the column is washed with Buffer B overnight (flow rate ca. 12 ml h^{-1}) and then with three bed volumes of Buffer B containing 0.6 M NaCl. Acyl ACP-synthetase is eluted with Buffer B containing 0.5 M KSCN. Three fractions (3×2.5 ml) containing major enzymatic activity to acylate ACP are identified by the enzymatic assay as described in Sect. 3.1. The enzyme preparation is then passed through a Sephadex G-25 (fine size, Pharmacia) column (2.4×14.5 cm) equilibrated with Buffer B to remove KSCN. Fractions containing the enzyme activity are pooled and the solution is adjusted to contain 10 mM ATP and 10 mM Mg^{2+} by the addition of appropriate volume of 0.1 M ATP and 1 M $MgCl_2$. The enzyme preparation is active for about 1 yr at 4 °C. We have also recently observed that enzyme prepared as described above is stable to freezing at −70 °C.

In a reaction mixture containing Tris (0.1 M, pH 8.0), ATP (5 mM), $MgCl_2$ (10 mM), LiCl (0.4 M), Triton X-100 (2%), DTT (2 mM), [^{3}H]palmitic acid (3 µM; 23.5 Ci $mmol^{-1}$, New England Nuclear) and acyl-ACP synthetase (2 milliunits) in a final volume of 50 µl, plant ACPs (1 µg) are completely acylated with palmitic acid in 1 h at 37 °C. The reaction has been used routinely in our laboratory to assay plant ACPs (Kuo and Ohlrogge 1984c; Ohlrogge and Kuo 1984). After incubation, 40 µl of the reaction mixture is transferred to a 1 × 3 cm piece of Whatman 3 mm filter paper and allowed to air dry. The filter papers are then washed with four changes of chloroform:methanol:acetic acid (3:6:1, v/v/v) to remove unreacted palmitic acid. The papers are placed in scintillation vials containing 0.8 ml of 1 M hyamine hydroxide and heated at 65 °C for 15 min to hydrolyze the [^{3}H]palmitoyl ACP formed during the reaction. The amount of ACP can be estimated from the specific radioactivity incorporated into ACP assuming a MW of ca. 9000 for plant ACPs.

In the preparation of radiolabeled [^{3}H]palmitoyl ACP, the enzymatic reaction mixture described above is scaled up to 0.5 ml, and the [^{3}H]-palmitate concentration is increased to 7.5 µM (57 µCi/0.5 ml). Purified or partially purified plant ACP zymatically for functionally active ACP using *E. coli* acyl-ACP synthetase (Sect. The reaction mixture is then diluted to 5.0 ml with 20 mM Mes, pH 6.1 and applied to a 0.8 ml column of DE52. After washing with 5 ml Mes buffer and 3 ml 80% isopropanol/Mes buffer to remove unreacted palmitic acid and Triton X-100, 23 µCi of [^{3}H]palmitoyl ACP is eluted with 0.5 M NaCl/Mes buffer. Octyl-Sepharose chromatography can be used to remove any non acylated ACP (ACP-SH) (Rock and Garwin 1979). The eluent obtained from DE52 chromatography is applied to a 1 ml Octyl-Sepharose (Pharmacia) column equilibrated with 0.5 M NaCl/Mes buffer. The column is washed with 5 ml 10% isopropanol/Mes buffer to remove ACP-SH. [^{3}H]Palmitoyl ACP is then eluted with 50% isopropanol/Mes buffer.

3.2 Radioimmunoassay Procedure

The principle of RIA for ACPs is based on the competition between the test antigen and [^{3}H]palmitoyl ACP for ACP antibodies. The test antigen is first incubated with antiserum in the absence of [^{3}H]palmitoyl ACP which is added later to compete for the limited number of antibody binding sites (nonequilibrium incubation). Bound and free [^{3}H]palmitoyl ACPs are then separated by use of polyethylene glycol (PEG) (Creighton et al. 1973), and the free [^{3}H]palmitoyl ACP in the supernatant fluid is measured. The method has the advantage of an increased sensitivity (Hawker 1973) and simplicity. A solution of 0.01 M K-phosphate, pH 6.8, containing 0.15 M NaCl (PBS) and 1 mg ml^{-1} BSA (Fraction V, Sigma) is used in all dilutions to avoid loss of proteins at low concentrations. The appropriate dilution of antiserum for use in RIA is determined in preliminary experiments as that amount needed to bind approx. 80% of total [^{3}H]palmitoyl ACP. This amount of antibody is chosen to yield a conveniently low background radioactivity without sacrificing accuracy. This situation may change when other techniques are used for the incubation of ACP and antiserum and the separation of bound and free [^{3}H]palmitoyl ACP. The conventional working antibody titer is the dilution of antiserum which binds 50% of the total labeled antigen to allow free reversible competition between labeled and unlabeled antigen (Hawker 1973).

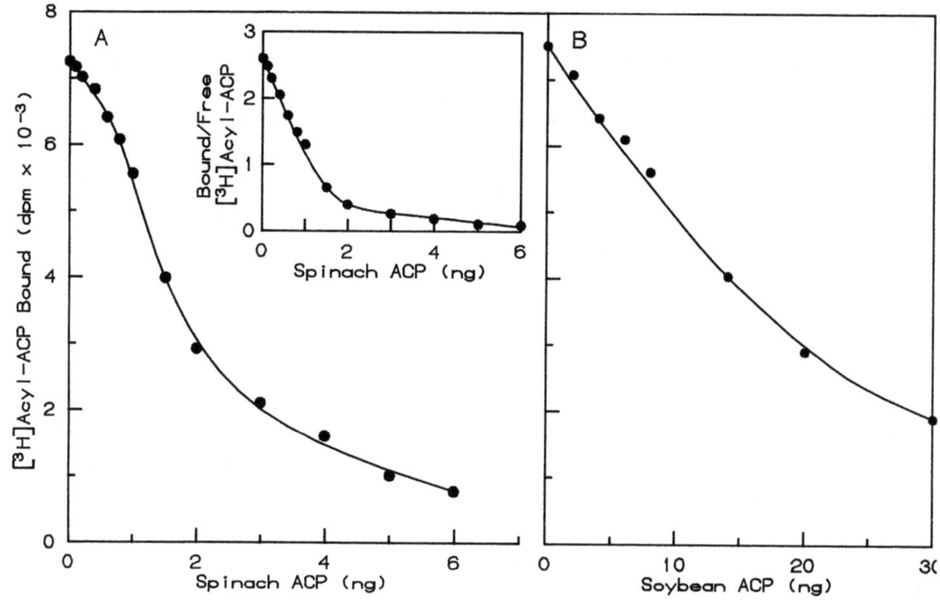

Fig. 1 A. A standard curve for the RIA of spinach ACP. Antiserum (0.2 μl) is first incubated with unlabeled spinach ACP-I and then with spinach [^{3}H]palmitoyl ACP-I. After the reaction, the immunoglobulin fraction is precipitated with PEG. Detailed procedures are described in Sect. 3.2. [^{3}H]palmitoyl ACP-I bound is calculated from the radioactivity in the supernatant (unbound). The *insert* shows the data replotted as the ratio of the bound and free [^{3}H]palmitoyl ACP-I against the amount of spinach ACP-I, Each *point* represents an average of two analyses. **B** A standard curve for the RIA of soybean ACP. Experimental procedures are the same as described above except that the reaction mixture contains 1 μl of antiserum and soybean [^{3}H]palmitoyl ACP is used as the radiolabeled antigen. Each *point* is an average of three separate analyses

Spinach ACP antiserum (20 μl) diluted 20- to 100-fold is allowed to react with 0–45 μl of plant tissue extracts or purified ACP in a final volume of 100 μl (adjusted with PBS/BSA) for 2 h at 4 °C. [^{3}H]palmitoyl ACP-I (8–10×10^3 dpm) in 20 μl of PBS/BSA buffer and 4 μl of normal rabbit serum are added and allowed to react for an additional 1 h at 4 °C before mixing with 0.2 ml of 20% (w/v) PEG 6000 in PBS. After 15 min, the immunoglobulin complex is centrifuged and the radioactivity in 0.15 ml aliquots of the supernatant is determined. Under these assay conditions hydrolysis of [^{3}H]palmitoyl ACP is less than 5%. Using this procedure, a standard RIA curve is established for spinach ACP (Fig. 1 A) in which the highly purified spinach ACP-I competes with spinach [^{3}H]palmitoyl ACP-I for antibody binding sites. If the ratio of the bound and free radioactivity, as separated by PEG precipitation, is plotted against the arithmetic dose of spinach ACP-I, a typical curvilinear function (Benade and Ihle 1982) is obtained. Figure 1 B depicts a standard curve for soybean ACP in a RIA using soybean [^{3}H]palmitoyl ACP as the radiolabeled antigen. Soybean ACP is partially purified (<10% purity) from immature seeds (Kuo and Ohlrogge 1984b) and acylated with [^{3}H]palmitic acid using *E. coli* acyl-ACP synthetase (Rock and Cronan 1979) according to the procedure as described in Sect. 3.1. Immature soybean seeds are homogenized in the K-phosphate buffer with a Waring Blender and a Brinkman Polytron homogenizer (Sect. 4.2.1). Subsequent treatments using 2-propanol and DE52 ion-exchange column chromatography are carried out in the same manner as for the isolation of *E. coli* ACP described by Rock and Cronan (1980). An equal volume of cold 2-propanol is added slowly and mixed with the soybean crude extract. The mixture is allowed to stand with intermittent stirring for 1 h at 4 °C and then centrifuged at 6000 × g for 15 min. The supernatant is adjusted to pH 6.1 with acetic acid and mixed with preswollen DE52 at a ratio of 1 g per 10 g of immature seeds. The suspension is stirred overnight at 4 °C. The DE52 is then collected on a filter funnel and washed with 5 vol of 10 mM Mes, pH 6.1 and packed into a column. The column is further washed with three bed volumes of Mes buffer. ACP is eluted with 1 M NaCl in Mes buffer and acid-precipitated from the salt solution as described in Sect. 2.1. Because of the enzyme specificity, only partial purification of soybean ACP is required to prepare the radiolabeled antigen. Although the standard curve for soybean ACP has about one-tenth the sensitivity of that for spinach ACP (Fig. 1), the procedure is sensitive enough to assay ACP in crude extracts from just one or two immature soybean seeds (0.1–0.4 g).

3.3 Applications

3.3.1 ACP Levels During Soybean Seed Development

During soybean seed development, the in vivo rate of fatty acid synthesis increases sharply between 20 and 50 days after flowering (DAF) and decreases to zero by about 70 DAF (Fig. 2A; Ohlrogge and Kuo 1984). Since ACP participates in each step of fatty acid biosynthesis, we have questioned whether the levels of ACP during seed development may reflect the developmental regulation of fatty acid synthesis. The levels of ACP in developing soybeans were examined enzymatically for functionally active ACP using *E. coli* acyl-ACP synthetase (Sect. 3.1) and immunologically for ACP antigen using RIA (Sect. 3.2). As shown in Fig. 2, the levels of active ACP and ACP antigen correlate closely at all stages of seed development, and both also increase in parallel with the increase in fatty acid synthesis through 50 DAF. These results indicate that de novo synthesis of ACP rather than a post-translational activation mechanism may control the levels of ACP in developing soybeans and that the levels of fatty acid biosynthetic proteins may regulate fatty acid synthesis during early stages of seed development (20–50 DAF).

3.3.2 Immuno-Cross-Reactivity Among ACPs

Because the structure of ACP is highly conserved, antibodies against spinach ACP partially cross-react with ACP from several species (Ohlrogge et al. 1979). The extent of this cross-reactivity was examined by RIA in which extracts of corn,

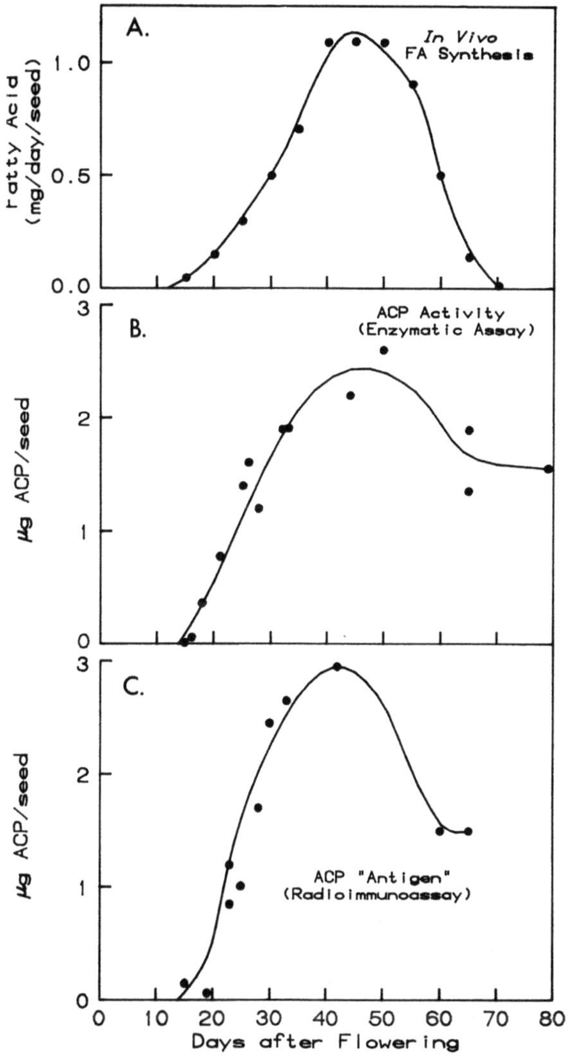

Fig. 2. A–C. Level of acyl carrier proteins during soybean seed development. **A** Rate of fatty acid synthesis per day per seed; **B** ACP levels per seed measured enzymically (Sect. 3.1); **C** ACP levels per seed measured by RIA (Sect. 3.2.)

avocado, safflower, soybean, and *E. coli* were allowed to compete with the spinach [^{3}H]palmitoyl ACP-I for the antibody binding sites. Figure 3 shows that all ACP preparations have the ability to compete with the spinach [^{3}H]palmitoyl ACP-I for antibody binding. The ability to compete with the spinach [^{3}H]palmitoyl ACP-I is approx. 40% for safflower and decreases in order for soybean, avocado, corn ACP preparations to approx. 10% for *E. coli.* These results are in close agreement with those reported for sunflower, soybean, and *E. coli* ACP preparations competing with spinach [^{125}I] ACP for spinach ACP antibodies (Ohlrogge et al. 1979). Thus, the use of [^{3}H]palmitoyl ACP can faithfully substitute for [^{125}I] ACP in a RIA.

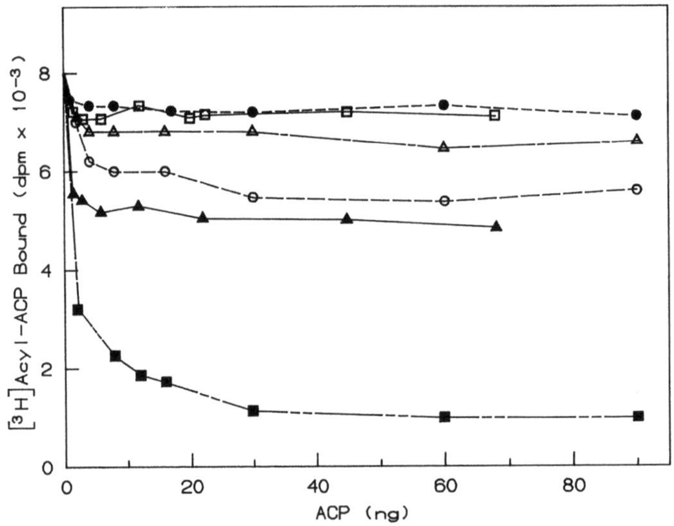

Fig. 3. Comparison of immuno-cross-reactivity among ACPs by RIA using spinach [^{3}H]palmitoyl ACP-I as the radiolabeled antigen. Antiserum (0.2 μl) is first incubated with the indicated amounts of unlabeled ACP preparations and then with spinach [^{3}H]palmitoyl ACP-I. Other procedures are the same as described in Fig. 1. (■) Spinach; (▲) safflower; (○) soybean; (△) avocado; (□) corn; (●) *E. coli*. Each *point* represents an average of at least two separate experiments

4 Immunological Analysis of ACP Isoforms

4.1 Immunorelationship of Spinach ACP Isoforms

Spinach leaves contain two ACP isoforms (ACP-I and ACP-II) separable by HPLC (Sect. 2.1.). They have similar N-terminal primary sequences and amino acid compositions, and they are most probably products of two different ACP genes (Ohlrogge and Kuo 1985). In order to examine their immunorelatedness, the purified spinach ACP isoforms are first radiolabeled with [^{3}H]palmitic acid by enzymatic acylation of the pantethein prosthetic group with *E. coli* acyl-ACP synthetase (Sect. 3.1) and subjected to a precipitation experiment with an increasing amount of antiserum raised against ACP-I. Figure 4 A shows that both radiolabeled isoforms are effectively bound by the antiserum. In this experiment, each [^{3}H]palmitoyl ACP isoform is incubated with different amounts of antiseum in 200 μl of 1 mg ml^{-1} BSA in PBS for 1 h at 4 °C. Antibody-bound [^{3}H]palmitoyl ACP is precipitated from solution by addition of 25 μl IgGsorb (The Enzyme Center, Inc., Boston, Massachusetts, USA). After 15 min, the suspension is centrifuged at 10000 × g for 2 min and aliquots of the supernatant are mixed with scintillation fluid to determine the amount of unbound [^{3}H]palmitoyl ACP. A more direct competitive binding analysis of ACP-I or ACP-II with [^{3}H] palmitoyl ACP-I is also carried out in which ACP-I or ACP-II is first incubated for 1 h at 4 °C with 0.2 μl antiserum, followed by an additional 1 h incubation with [^{3}H]pal-

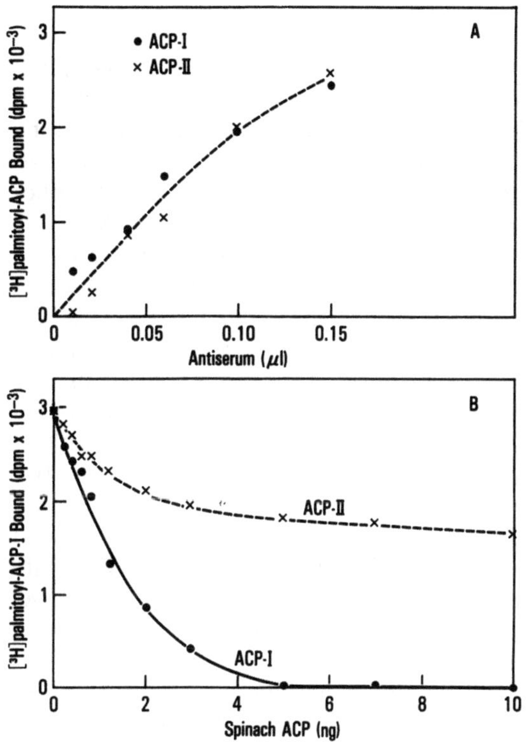

Fig. 4. A, B. Immunological cross-reactivity of spinach ACP-I and ACP-II. **A** Binding of [^{3}H]palmitoyl ACP-I (●) and [^{3}H]palmitoyl ACP-II (×) by antibodies raised against ACP-I. **B** Competition between ACP-I (●) or ACP-II (×) with [^{3}H]palmitoyl ACP-I for binding to antibodies raised against ACP-I

mitoyl ACP-I. Other experimental conditions are the same as described above. As shown in Fig. 4B, ACP-I is able to displace all of the [^{3}H]palmitoyl ACP-I from the antibody binding sites, while ACP-II displaces only approx. 40%. Thus, ACP-I and ACP-II have some epitopes in common, but ACP-II lacks certain ACP-I antigenic determinants.

4.2 Western Blot Analysis of ACP Isoforms in Plant Tissue

4.2.1 Preparation of Plant Tissue Extracts

Spinach leaves (*Spinacea oleracea,* hybrid 424) are harvested from light- (12 h light/12 h dark cycle, 22 °C) and dark-grown plants, 21–28 days old. Immature spinach seeds are harvested while the tissues are still soft and green. Young leaf and immature seed of castor oil plants (*Ricinus communius* L. var. Baker 296) grown in the greenhouse are harvested 25–35 DAF. Spinach chloroplasts are isolated by the method of Nakatani and Barber (1977). Plant tissues are homogenized with a Polytron homogenizer in 10 ml g^{-1} fresh weight, of 0.05 M K-phosphate pH 7.0, containing 1% Triton X-100, 0.1% β-mercaptoethanol, 1 mM sodium metabisulfite, and 15 μM leupeptin. The extracts are centrifuged at 10000 × g for 10 min and the supernatant is passed through a 1 ml column of DE52. After washing the column with 3 ml extraction buffer the ACP is eluted with 0.5 M LiC1 in extraction buffer.

4.2.2 Western Blot Analysis

Crude plant tissue extracts containing 1–4 ng of ACP are subjected to SDS-polyacrylamide gel electrophoresis and immunoblot analysis for the presence of ACP isoforms. Electrophoresis is carried out in the buffer system described by Laemmli (1970) using 1.5-mm-thick slab gels consisting of 3% and 15% acylamide (with 2.7% C) for the stacking and resolving gels, respectively. Electrophoresis is conducted at 30 mA constant current at room temperature. After the tracking dye reaches the end of the resolving gel, the proteins are transferred to nitrocellulose. Transfer to 0.2 μm nitrocellulose is carried out electrophoretically, as described by Towbin et al. (1979), at 1–1.5 A current for 90 min. Colored protein standards, prepared as described by Tzeng (1983), or prestained protein molecular weight standards (BRL, Inc., Gaithersburg, Maryland, USA) are used as markers during gel electrophoresis and for protein transfer to nitrocellulose sheets. After transfer the gel is "blocked" for 45 min with 5% nonfat dry milk in PBS (Johnson et al. 1984) and then incubated overnight in a sealed plastic bag with 10 ml of a 100-fold dilution of anti-spinach ACP antiserum in 1% nonfat dry milk/PBS. The blot is washed four times with 0.025% Tween 20 in 10 mM Tris pH 7.5, 0.9% NaCl for 5 min each. Immobilized antibody is detected with biotin-labeled goat-antirabbit second antibody, avidin and biotinylated horseradish peroxidase, as recommended for the Vectastain ABC procedure (Vector Laboratories, Inc., Burlingame, California, USA).

As shown in Fig. 5A, application of the same quantity of purified ACP-I or ACP-II to Western blot analysis results in similar intensity of bands as detected by the biotin-avidin-peroxidase second antibody system (lanes 3 and 4)., Thus, despite differences in antibody binding to ACP-I and ACP-II detectable when antibody is limiting (Fig. 4B), the sensitivity of detecting ACP-I and ACP-II in the presence of excess antibody, as used in the Western blot, is similar. When crude extracts of spinach leaf tissues are analyzed, two ACP bands are detected which migrate with R_F values corresponding to ACP-I and ACP-II (lane 2). These bands are not detected if the blot is probed with preimmune serum. The same pattern is also observed if fresh tissue is ground in liquid N_2 and then boiled immediately in 1% SDS prior to electrophoresis. Therefore, the two bands we observe are not the result of protease degradation of ACP. Both ACP isoforms are also present in spinach chloroplasts and in the dark-grown spinach leaves as revealed by Western blot analysis (lanes 5 and 6). This result indicates that light is not an absolute requirement for the expression of ACP and supports the conclusion that ACP isoforms are localized in chloroplasts (Høj and Svendsen 1984; Ohlrogge et al. 1979).

In agreement with our recoveries from large-scale ACP isolations (Sect. 2.1), ACP-I is seen to be the predominant ACP species in leaves. In contrast, when extracts of spinach seeds (Fig. 5A, lane 1) or roots (data not shown) are examined only ACP-II is present, with ACP-I barely detectable or totally missing. Similar results are also observed for castor oil plant tissues where endosperm tissues of immature seed contain a single ACP and leaf tissues contain two isoforms (Fig. 5B). Thus, plants contain ACP isoforms that are specifically expressed in different tissues.

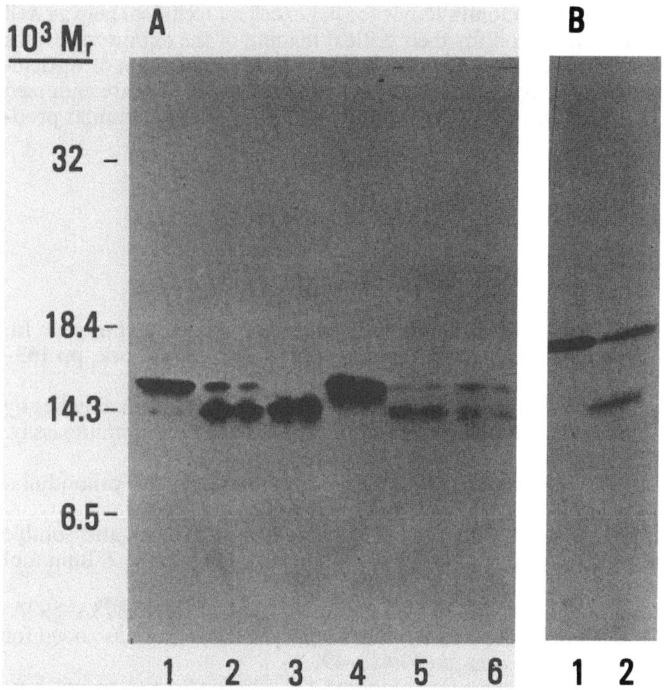

Fig. 5. A, B. Western blot analysis of acyl carrier proteins in spinach and castor bean extracts. Proteins subjected to electrophoresis in 15% SDS gels were transferred to 0.2 μm nitrocellulose membranes and probed with antiserum to spinach ACP-I. **A** *Lane 1*, spinach seed extract; *lane 2*, spinach leaf extract; *lane 3*, purified ACP-I (3 ng); *lane 4*, purified ACP-II (3 ng); *lane 5*, extract of spinach chloroplasts; *lane 6*, extract of dark-grown spinach leaves. **B** *Lane 1*, extract of castor bean seeds; *lane 2*, extract of castor bean leaves

5 Conclusion

We have employed immunological approaches to analyze the expression of ACP in plants. The development of a RIA for plant ACPs takes advantage of enzyme specificity in the preparation of radiolabeled antigens, [^{3}H]palmitoyl ACP and of immuno-cross-reactivity among ACPs of different species. Therefore, the procedure does not require extensive purification of ACP prior to its labeling with ^{3}H and can provide a sensitive assay for ACP from several species.

The use of RIA for ACP antigen and an enzymatic measurement for active ACP has allowed us to conclude that a major mechanism regulating the levels of ACP during soybean seed development is de novo synthesis. Immunocompetition experiments show that spinach ACP isoforms, ACP-I and ACP-II possess both common and isoform specific epitopes. Western blot analyses show that chloroplasts, dark- and light-grown spinach leaves contain both ACP isoforms, and that the expression of ACP isoforms in plants is tissue-specific.

Acknowledgements. We are grateful to Donita Doyle for her excellent technical help as well as Drs. Edward Emken and Walter Wolf for their critical reading of the manuscript. This work is supported by Agricultural Research Service, United States Department of Agriculture. The mention of firm names or trade products does not imply that they are endorsed or recommended by the U.S. Department of Agriculture over other firms or similar products not mentioned.

References

Benade L, Ihle JN (1982) Principles of radioimmunoassays and related techniques. In: Marchalonis JJ, and Warr GW (eds) Antibody as a tool. Wiley, New York, pp 163–187

Bolton AE, Hunter WM (1973) The labelling of proteins to high specific radioactivities by conjugation to a ^{125}I-containing acylating agent-application to the radioimmunoassay. Biochem J 133:529–539

Chafouleas JG, Dedman JR, Means AR (1982) Radioimmunoassay of calmodulin. Methods Enzymol 84:138–147

Creighton WD, Lambert PH, Miescher PA (1973) Detection of antibodies and soluble antigen-antibody complexes by precipitation with polyethylene glycol. J Immunol 111:1219–1227

Hawker CD (1973) Radioimmunoassay and related methods. Anal Chem 45:878A–890A

Høj PB, Svendsen I (1984) Barley chloroplasts contain two acyl carrier proteins coded for by different genes. Carlsberg Res Commun 49:483–492

Hunter WM, Greenwood FC (1962) Preparation of iodine-131 labelled human growth hormone of high specific activity. Nature 194:495–496

Johnson DA, Gautsch JW, Sportsman JH, Elder JH (1984) Improved technique utilizing nonfat dry milk for analysis of proteins and nucleic acids transferred to nitrocellulose. Gene Anal Techn 1:3–8

Kuo TM, Ohlrogge JB (1984a) The primary structure of spinach acyl carrier protein. Arch Biochem Biophys 234:290–296

Kuo TM, Ohlrogge JB (1984b) A novel, general radioimmunoassay for acyl carrier proteins. Anal Biochem 136:497–502

Kuo TM, Ohlrogge JB (1984c) Acylation of plant acyl carrier proteins by acyl-acyl carrier protein synthetase from *Escherichia coli.* Arch Biochem Biophys 230:110–116

Laemmli UK (1970) Cleavage of structural proteins during the assembly of the head of bacteriophage T_4. Nature 227:680–685

Majerus PW, Vagelos PR (1967) Fatty acid biosynthesis and the role of the acyl carrier protein. In: Paoletti R, Kritchevsky D (eds) Advances in lipid research, vol 5. Academic Press, New York, pp 1–33

Nakatani HY, Barber J (1977) An improved method for isolating chloroplasts retaining their outer membranes. Biochim Biophys Acta 461:510–512

Ohlrogge JB, Kuo TM (1984) Control of lipid synthesis during soybean seed development: enzymatic and immunochemical assay of acyl carrier protein. Plant Physiol (Bethesda) 74:622–625

Ohlrogge JB, Kuo TM (1985) Plants have isoforms for acyl carrier protein that are expressed differently in different tissues. J Biol Chem 260:8032–8037

Ohlrogge JB, Kuhn DN, Stumpf PK (1979) Subcellular localization of acyl carrier protein in leaf protoplasts of *Spinacia oleracea.* Proc Natl Acad Sci USA 76:1194–1198

Reichlin M (1980) Use of glutaraldehyde as a coupling agent for proteins and peptides. Methods Enzymol 70:159–165

Rice RH, Means GE (1971) Radioactive labeling of proteins in vitro. J Biol Chem 246:831–832

Rock CO, Cronan JE Jr (1979) Solubilization, purification, and salt activation of acyl-acyl carrier protein synthetase from *Escherichia coli.* J Biol Chem 254:7116–7122

Rock CO, Cronan JE Jr (1980) Improved purification of acyl carrier protein. Anal Biochem 102:362–364
Rock CO, Garwin JL (1979) Preparative enzymatic synthesis and hydrophobic chromatography of acyl-acyl carrier protein. J Biol Chem 254:7123–7128
Simoni RD, Criddle RS, Stumpf PK (1967) Fat metabolism in higher plants. XXXI. Purification and properties of plant and bacterial acyl carrier proteins. J Biol Chem 242:573–581
Stumpf PK (1980) Biosynthesis of saturated and unsaturated fatty acids. In: Stumpf PK (ed) The Biochemistry of plants, vol 4. Academic Press, New York, pp 177–204
Towbin H, Staehelin T, Gordon J (1979) Electrophoretic transfer of proteins from polyacrylamide gels to nitrocellulose sheets: procedure and some applications. Proc Natl Acad Sci USA 76:4350–4354
Tzeng M-C (1983) A sensitive, rapid method for monitoring sodium dodecyl sulfate-polyacrylamide gel electrophoresis by chromophoric labeling. Anal Biochem 128:412–414
Wakil SJ, Stoops JK, Joshi VC (1983) Fatty acid synthesis and its regulation. Annu Rev Biochem 52:537–579

Immunofluorescent Labelling of Enzymes

G. Schmid and H. Grisebach

1 Introduction

Besides knowledge of kinetic and regulatory behaviour of enzymes, the determination of their intra- and intercellular localization is an important prerequisite for understanding cellular metabolism. One of the most sensitive and specific techniques for reaching this goal is the in situ localization of enzymes by immunological methods. In principle the specificity of the antigen(enzyme)-antibody reaction is combined with a labelling technique which allows detection of the enzyme-antibody complex in a tissue section either by light or electron microscopy. For immunoelectron microscopy (Hoyer and Bucana 1982) ferritin, peroxidase, and gold are the most widely used markers for direct and indirect labelling. For example, nitrate reductase was localized in the cell wall-plasmalemma region and in the tonoplast membranes of *Neurospora crassa* cells by incubation with antiserum against nitrate reductase followed by an incubation with ferritin-labelled goat antirabbit IgG (indirect method) (Roldàn et al. 1982). Gold-labelled protein A (a cell wall protein from *Staphylococcus aureus* with a high affinity for the Fc fragment of IgG) was used to localize an extracellular protease of *Nectria galligena* in infected apple tissue (Rey and Noble 1984).

Recently labelling with gold coated with antibodies has been used in immunophotoelectron microscopy (Birrell et al. 1985) which can be considered as the electron optical analogue of immunofluorescent microscopy utilizing emitted electrons from the gold particles instead of emitted light from a fluorescent dye. The excitation source is also an UV lamp, but the optical microscope is replaced by an electron emission-type microscope. This method was applied to visualize microtubule networks with a resolution of 10–20 nm and could be very useful for localization of enzymes with high resolution.

Immunofluorescence analysis, i.e. labelling of antibodies with fluorescent groups and the detection by fluorescence microscopy, was introduced by Coons et al. in 1941. This method, which has been widely used in human and animal biochemistry, has also been successfully applied for localization of enzymes in higher plants and will be described in this chapter in some detail; it has previously been covered in chapters of several books, for example by Sternberger (1979) and DeLuca (1982).

2 General Methodology

2.1 Staining Sequences

The enzyme which is to be localized functions as an antigen in a tissue section. There are several staining sequences for localization of the enzyme-antibody complex in tissues by fluorescence.

Direct Method. The enzyme reacts with a specific antibody which is labelled with a fluorescent dye.

Indirect Method. In the first step the enzyme reacts with an unlabelled specific antibody. In the second step the bound antibody binds a fluorochrome-labelled anti-IgG antibody. With the following methods one can localize two enzymes in the same tissue.

Sequential Technique. In the first step enzyme A binds fluorochrome-labelled antibody and the fluorescent pattern is recorded photographically. The fluorescence is then bleached by prolonged exposure to ultraviolet light. In the second step enzyme B reacts with fluorochrome-labelled antibody.

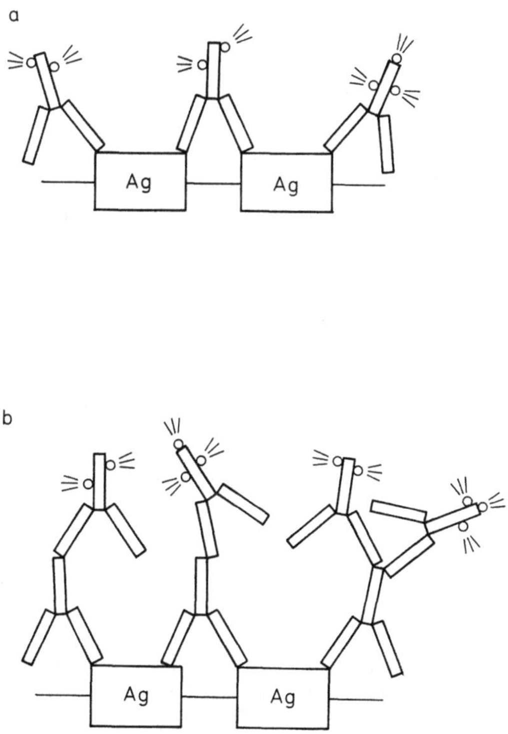

Fig. 1. Direct **a** and indirect **b** labelling of antigen (*Ag*). Redrawn from Hoyer and Bucana (1982)

Double-Staining Method. Enzyme A binds a specific antibody labelled with a green fluorescent dye (e.g.) fluorescein isothiocyanate), while enzyme B reacts with an antibody labelled with an orange-red fluorescing dye (e.g. tetramethylrhodamine isothiocyanate).

In Fig. 1 the direct and the indirect method are shown schematically. Since the indirect method permits the binding of more fluorescent antibody, it is more sensitive than the direct method.

2.2 Antibody Preparation

A prerequisite for obtaining a specific antibody against the desired enzyme is the isolation of the pure enzyme in milligram quantities. The methodology for raising antibodies against plant antigens has been described by van der Veken et al. (1962). Some details are mentioned in the individual examples described in Sect. 3. The antigenic properties of enzymes can vary considerably.

The antiserum which is obtained from the rabbit contains the specific antibody in a maximum concentration of only 3% of total protein. To increase the sensitivity of the immunofluorescent detection it is therefore advantageous to enrich the γ-globulin fraction of the antiserum. This can be achieved, for example, by $(NH_4)_2SO_4$-fractionation, chromatography on DEAE-cellulose or gel filtration (Warr 1982). Affinity chromatography by immunoadsorption onto a column with immobilized protein-A (protein-A Sepharose, Pharmacia) is also often used (Goding 1978; Warr 1982). In species in which a substantial portion of IgG binds to protein-A, this method is recommended for use directly with serum. Protein-A reacts with all IgG antibodies in the serum of rabbits (Goding 1978).

The specificity of the antibody is of great importance. This is usually tested by immunoelectrophoresis (Williams 1971) or Ouchterlony double immunodiffusion (Munoz 1971). The antibody titre can be determined by immunoprecipitation of the enzyme with increasing antibody concentrations and determination of remaining enzyme activity in the supernatant. The specificity tests have to be interpreted with caution, since they are less sensitive than the immunofluorescence method.

For the direct staining method the purified antibody has to be conjugated with a fluorochrome (Coons et al. 1941; Coons 1958). Fluorochromes which are frequently used are listed in Fig. 2 together with their excitation and emission maxima. In chlorophyll-free tissue like root tissue use of tetramethylrhodamine isothiocyanate(TRITC) avoids the problem of fluorescein isothiocyanate-like autofluorescence in plant tissues with green (546 nm) light (Hapner and Hapner 1978). For coupling of these dyes to protein the ratio of the reaction partners is important. The method described below is from DeLuca (1982).

The IgG enriched by protein-A Sepharose chromatography from 10 ml rabbit serum (which should be about 50–60 mg in approximately 10 ml), is dialysed overnight at 4 °C against carbonate-bicarbonate buffer (17.3 g $NaHCO_3$, 8.6 g Na_2CO_3 per litre of destilled H_2O, pH 9.3).

A concentration of protein for fluorochrome conjugation that is convenient and approximately optimal is 5 mg ml^{-1}. The dialysed IgG fraction is placed in a beaker with a small magnetic stirrer at room temperature. About 25 µg of FITC per milligram IgG or

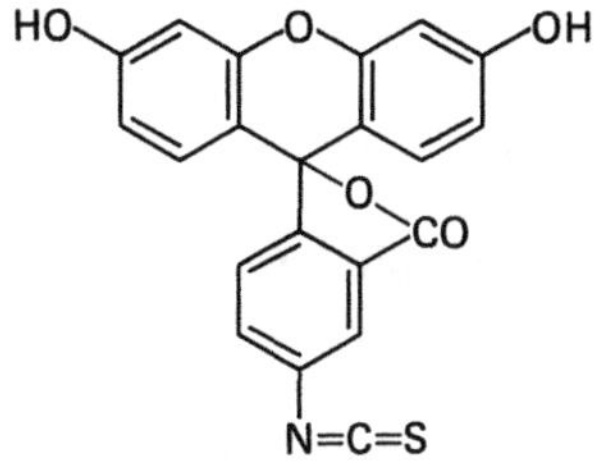

I

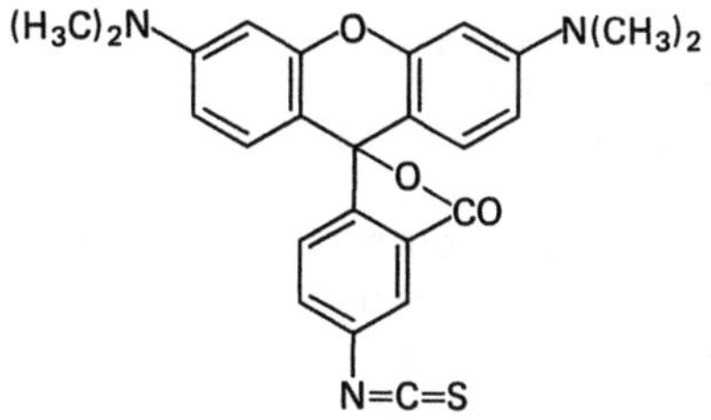

II

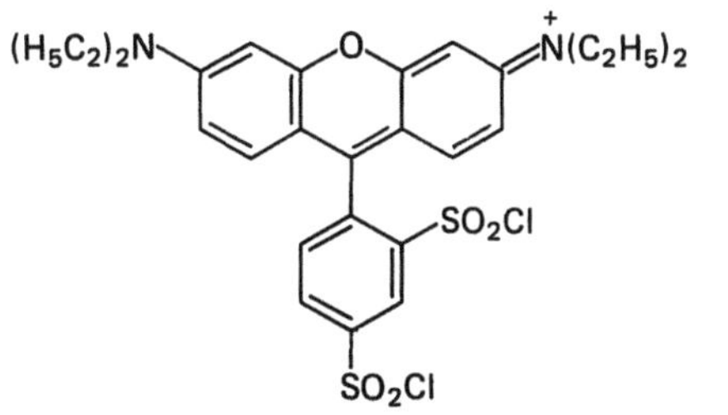

III

Fig. 2. Structural formulae of some commonly used fluorochromes.
I Fluorescein isothiocyanate ($\lambda_{max}=490$ nm, $\lambda_{max}(em)=517$ nm);
II Tetramethylrhodamine isothiocyanate ($\lambda_{max}=510{,}545$ nm, $\lambda_{max}(em)=620$ nm);
III Lissamine rhodamine B200 sulphonyl chloride ($\lambda_{max}=560$ nm, $\lambda_{max}(em)=590$ nm)

10 μg of crystalline TRITC dissolved at a concentration of 1 mg ml^{-1} in dimethylsulphoxide is added dropwise with slow stirring. The solution is stirred for 2 h at room temperature, with the beaker covered with aluminum foil to keep out light.

After coupling, the solution is passed over a 3 × 30 cm Biogel P-6 column (Bio-Rad Laboratories, Richmond, California, USA) equilibrated with 0.01 M phosphate buffer at a flow rate of 1.5 ml min^{-1}. The first coloured band off the column is collected for DEAE-Sephadex chromatography. Unbound fluorochrome should still be left near the top of the column at the time the conjugated protein is eluted. TRITC conjugates often have an additional band of intermediate mobility of undetermined nature which can contaminate the first protein band if the column is too short. If this contamination occurs, longer columns can be run, or the material can be rechromatographed.

A 20 ml plastic syringe filled with DEAE-Sephadex A 50 (Pharmacia) equilibrated with 0.01 M potassium phosphate, pH 8, is used for ion exchange chromatography. The column must be pre-equilibrated with 0.5–1 M phosphate buffer (pH 8) for 1–2 days, then re-equilibrated with 0.01 M phosphate (pH 8) for 1 day before use. The column should then be washed with several column volumes of 0.01 M potassium phosphate to ensure equilibration before adding the conjugate. After the conjugate has run into the column, it is washed with 0.01 M phosphate. The conjugated protein will bind to the column. The conjugate is then eluted using a linear gradient of 0 to 1 M NaCl in 0.01 M phosphate (pH 8). About 200 ml salt gradient should suffice. Fractions of 5 ml should be collected.

FITC conjugates are characterized by reading the OD_{280} and OD_{496} for each fraction. The concentration of bound FITC can be caluclated from the formula below (Goldman 1968):

$$[\text{bound FITC}] = \frac{[\text{free FITC}] \times OD_{496} \text{ of conjugate}}{OD_{490} \text{ of FITC standard} \times 0.75}.$$

Fractions of TRITC conjugates are characterized by the OD_{550}/OD_{280} ratio. An F/P ratio (fluorochromes per protein molecule) of 2–3 for FITC conjugates and an OD_{550}/OD_{280} ratio of about 0.5 for TRITC conjugates should give good results in immunofluorescence, and DEAE-column fractions which give similar values can be pooled. However, before pooling, various fractions should be tested to determine which conjugation ratio is optimal for the investigator's own system. Conjugates can then be stored frozen at −60 °C or kept in the refrigerator with sodium azide added to a final concentration of 10^{-2} M. For preparations of such conjugates see also: McKinney et al. (1964); Goldman (1968); Nairn (1976).

For the indirect method a fluorochrome-labelled anti-rabbit IgG antibody raised in goat is usually used. Such antibodies are commercially available in labelled form. The high affinity of biotin to avidin is applied in a new detection system (Amersham Buchler, Braunschweig, FRG, Research product information No. 938). The first antibody reacts with a second antibody to which several biotin molecules are attached via a spacer arm. Fluorescein-labelled streptavidin then binds to the second antibody. Streptavidin is not a glycoprotein and does therefore not bind to lectines. Besides FITC six other detection systems are available for streptavidin.

2.3 Specimen Preparation

The choice of a plant tissue which contains the enzyme to be localized in high concentration is advantageous. Since activity of enzymes can change drastically during plant development, it is important to determine at what age the respective enzyme is present in highest quantity. During specimen preparation the in vivo localization of the enzyme and its antigenic properties must be preserved. "Soluble" enzymes require special precautions to prevent their loss and redistribution during tissue preparation.

Whether or not fixation is necessary should be determined in each individual case by comparing results with unfixed and fixed preparations. For fixation formaldehyde (freshly prepared from paraformaldehyde), glutardialdehyde, dimethylsuberimidate and ethylacetimidate have been used (Tokuyasu and Singer 1976). Fixation can weaken or destroy the antigenic properties of a protein.

After fixation with aldehydes, free aldehyde groups must be blocked since they could react with amino groups of proteins and therefore cause unspecific binding. Blocking of reactive aldehyde groups can be done by treatment with 0.1–0.5 mol l^{-1} NH_4Cl in PBS, or 0.1 mol l^{-1} glycine or lysine in PBS for 30–60 min, or sodium borohydride (0.5 mg l^{-1}) for 10 min (Roth 1983).

One of the most common methods applied is the use of frozen sections. Several methods are available for preparation of such sections. Tissue blocks (approx. 1 mm^3) are soaked in 1.2 mol l^{-1} sucrose in 0.05 mol l^{-1} Tris/HCl buffer (pH 7.2) for 24 h at 4 °C. Subsequently, they are permeated with 10% (w/v) gelatin in the Tris/HCl buffer containing 1.2 mol l^{-1} sucrose for 30 min at 37 °C. Chilled tissue blocks are then frozen in melting nitrogen and stored in liquid nitrogen (Vernooy-Gerritsen et al. 1984). Soft tissue like hypocotyl tissue can also be embedded in small pieces of liver and the embedded segments dipped into liquid propane (Schmid et al. 1982). Another method frequently used is embedment in tissue tek OCT (Bayer Diagnostik, München, FRG; Ames Co., Indiana, USA) at −30 °C (Marcinowski et al. 1979; Burmeister and Hösel 1981). Embedment in Lowicryl K4M or HM20 which is used in immunoelectron microscopy could also be useful (Roth et al. 1978; Roth et al. 1984).

The embedded specimens are then sectioned (0.1–40 μm) in a freeze microtome at −20 ° to −30 °C. The sections are mounted on microscope slides with an adhesive (egg albumin-glycerol). After washing with buffered saline (PBS) (0.01 mol l^{-1} $NaH_2PO_4 \cdot H_2O$, and 0.14 mol l^{-1} NaCl adjusted with NaOH to pH 7.2 with addition of 0.002% NaN_3) immunofluorescence detection by the staining techniques mentioned above is carried out. Since plant tissue can contain a high concentration of lectines which unspecifically bind IgG as a glycoprotein, Clarke et al. (1975) have suggested incubating the sections in a 2.5% salicin solution in PBS for 5 min to block the lectins.

2.4 Fluorescence Microscopy

It is beyond the scope of this chapter to describe details of this technique. They can be found in articles by Haitinger (1959), Wick et al. (1980), and DeLuca (1982), and in technical bulletins from the manufacturers of such microscopes. For example, "Fluorescence microscopy; instruments, methods, applications" from Leitz; H. M. Holz "Worthwhile facts about fluorescence microscopy" from Zeiss.

The development of epifluorescence systems represented a major breakthrough in fluorescence microscopy. This illumination system is now mostly used. It gives a lower fluorescence background than transmission illumination, avoids absorption or scattering of excitating light, and is compatible with simultaneous use of other forms of transmission illumination, such as phase or interference contrast. It should be remembered that the intensity of the final fluorescent image varies inversely with the square of the magnification.

Of great importance is the choice of a suitable filter combination. The excitation filter usually blocks out all light from the high pressure mercury lamp that has a wavelength below 400 nm (which could excite tissue autofluorescence), whereas it transmits light useful for FITC and TRITC excitation. In fluorescent microscopy the emitted light should be seen against a dark background. To this end, the barrier (or suppression) filter (respectively a dichcroic mirror) eliminates the excitating light so that only the emitted light is seen. Further secondary filters can be used if necessary, for example to eliminate chlorophyll autofluorescence with the short-wave pass filter KP 560 (Zeiss). Figure 3 shows schematically the construction of an epifluorescence microscope and Fig. 4 demonstrates transmission spectra with a filter combination for FITC fluorescence.

Photographic documentation is essential since fluorescence fades on storage and under UV irradiation. The use of high-speed daylight film is recommended. If necessary the rate of photobleaching can be reduced by a factor of about 10 by application of a drop of glycerol containing 0.25 ml l^{-1} n-propyl gallate to the mounted sample (Giloh and Sedat 1982).

The resolution limitation of fluorescence microscopy is about 200 nm. If a higher resolution is desired this can be obtained by immunophotoelectron microscopy and transmission or scanning electron microscopy combined with electron-dense markers (see Introduction).

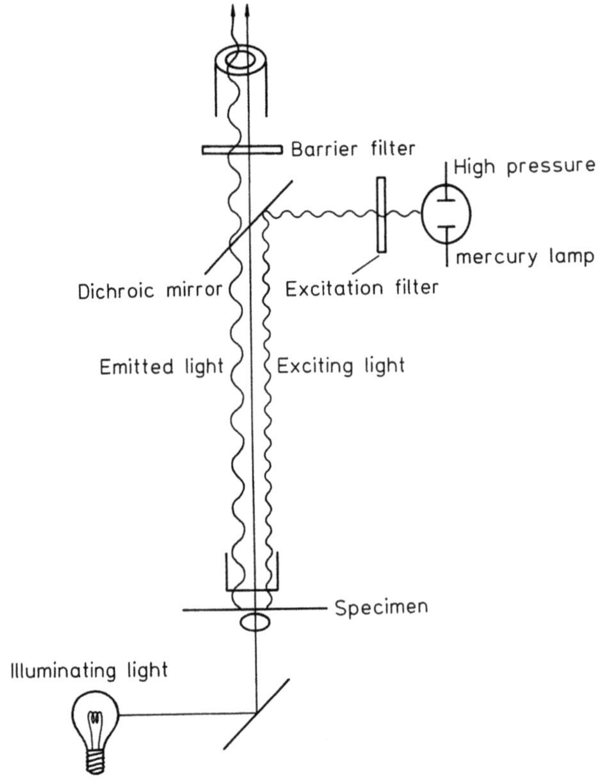

Fig. 3. An epifluorescent microscope with vertical illumination. Redrawn from DeLuca (1982)

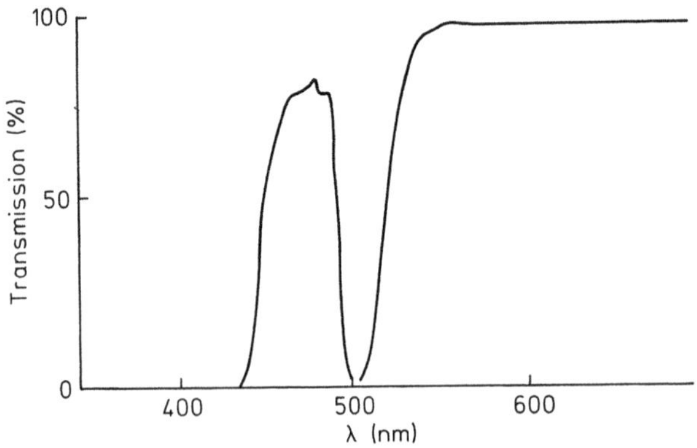

Fig. 4. Transmission spectra of a filter combination for FITC fluorescence. *Left*, excitation filter BP 450-490; *right*, barrier filter LP 520

2.5 Controls

To eliminate the possibility of unspecific staining the following controls should be carried out in each experiment:
a) Incubation with serum from an unimmunized rabbit (normal serum).
b) Incubation with only fluorochrome-conjugated anti-IgG.
c) Incubation in absence of antisera (autofluorescent control).

3 Examples for Localization of Enzymes in Plant Tissue by Immunofluorescent Labelling

In recent years a number of enzymes have been localized in plant tissue by the immunofluorescent method. Examples are listed in Table 1.

Three examples will be described in more detail in the following section.

3.1 In Situ Immunofluorescent Labelling of Ribulosebisphosphate Carboxylase in Leaves of C_3 and C_4 Plants

The chloroplastic enzyme ribulosebisphosphate carboxylase (EC 4.1.1.39) (RuP_2Case) was localized by in situ immunofluorescent labelling in hand-cut leaf blade sections of 40 C_3 and C_4 species and other plant species by Hattersley et al. (1977). Since the antiserum raised against RuP_2Case from one species is known to cross-react with enzyme from a wide variety of other species, antisera raised to wheat and spinach enzyme could be used for these studies.

Table 1. Examples for localization of enzymes in plant tissue by immunofluorescent labelling

Enzyme	Plant material	Ref.
Acid ribonuclease	Morning glory flower tissue	Baumgartner and Matile (1976)
Urease	Cotyledons of jack bean	Murray and Knox (1977)
Ribulosebisphosphate carboxylase	Leaves of C_3 and C_4 plants	Hattersley et al. (1977)
Vicilin peptidohydrolase	Cotyledons of mung bean	Baumgartner et al. (1978)
β-Glucosidase for coniferin	Spruce seedlings	Marcinowsky et al. (1979)
UDP-glucose: coniferyl alcohol glucosyltransferase	Spruce seedlings	Schmid et al. (1982)
Isoenzymes of malate dehydrogenase	Water melon cotyledons	Sautter and Hock (1982)
Extracellular protease of *Nectria galligena*	Infected apple tissue	Rey and Noble (1984)
Phenylalanine ammonia-lyase and chalcone synthase	Anthers of garden tulip	Kehrel and Wiermann (1985)

Table 2. Immunotitration of spinach RuP_2 Case and maize PEP carboxylase with anti-wheat RuP_2 Case serum. (From Hattersley et al. 1977)

Enzyme	Ratio antiserum/ enzyme	Relative activity (% of control)
RuP_2 Case	4	92
	14	80
	36	56
PEP carboxylase	37	120
	74	104
	368	142

RuP_2Case was purified from wheat leaves by $(NH_4)_2SO_4$ fractionation and DEAE and Sepharose 6B chromatography. The enzyme was homogeneous as judged by disc-electrophoresis.

Three-month-old female rabbits were injected in the vicinity of axillary and popliteal lymph nodes with 0.5 ml (10 mg RuP_2Case) wheat enzyme emulsified with 0.5 ml Freund's complete adjuvant. After two booster injections rabbits were bled from the marginal ear vein and serum was obtained in the usual way. Normal and pre-immunization control sera were obtained from untreated rabbits.

The specificity of the antiserum was analyzed by Ouchterlony double diffusion analysis against RuP_2Case from wheat, spinach and *Euglena*. Normal and pre-immunizations sera did not react with wheat RuP_2Case. The results of an immunotitration are shown in Table 2. Spinach RuP_2Case decreased progressively as anti-wheat RuP_2Case concentration was increased. In contrast, the activity of PEP carboxylase from maize was not suppressed, even when the assay contained 368 times more antiserum protein than PEP carboxylase protein.

For in situ immunofluorescent labelling the indirect labelling method was used. Segments of fresh blade material from young leaves were immersed in 70% ethanol for 2 h. They were then held in elder pith ("Holundermark") and 10–15 μm thick transsections were cut with a razor blade. These transsections were rinsed in phosphate saline (PBS), transferred to a glass well containing antiserum and incubated for 1 h. After a three time rinsing with PBS they were incubated for 1 h in the dark with fluorescein isothiocyanate-labelled sheep anti-rabbit immunoglobulin (Wellcome Reagents Ltd, Beckenham, England) which was diluted 1:4 with PBS. The sections were then rinsed three times with PBS and mounted in an aqueous glycerol solution (50%) containing 1% thymol. Slides were kept in the dark.

Microscopic analysis was performed with a Zeiss photomicroscope III set up for epifluorescence with an HBO 200 W/4 mercury vapour lamp, two KP 500 excitation filters, and FI 500 reflector and a B53 barrier Filter. For the penetration of chloroplasts by antibodies ethanol treatment seemed essential. Kodak high-speed Ektachrome reversal films (ASA 125) were used for photographic documentation as they allowed better distinction between specific fluorescence and autofluorescence than black and white films. A normal serum control and an autofluorescent control were run parallel for each test.

C_3 plants and one plant of the CAM type showed specific FITC fluorescence in all leaf chlorenchymatous cells, indicating the presence of RuP_2Case in all leaf chloroplasts. By contrast, in the C_4 leaf anatomy RuP_2Case was shown to be localized almost exclusively in the bundle sheath cells (photosynthetic carbon reduction cells = PCR cells), a finding which is in accordance with other localization studies.

Immunofluorescent labelling of RuP_2Case in species with non-classical C_4 leaf anatomy provided direct evidence of a cellular compartmentation of this enzyme. The interpretation of labelling studies together with a consideration of anatomical and physiological features leads to a better understanding of the photosynthetic metabolism in different plants. For further details see Hattersley et al. (1977).

3.2 In Situ Immunofluorescent Labelling of a β-Glucosidase for Coniferin and of UDP-Glucose: Coniferyl Alcohol Glucosyltransferase

Cinnamyl alcohol glucosides [e.g. coniferin (coniferyl alcohol β-D-glucoside)] have mainly been found in the sap of conifers and other gymnosperms (Sarkanen and Ludwig 1971). Experiments with labelling coniferin have proved that this compound can act as lignin precursor (Brown 1966). The turnover of coniferin in pine seedlings was demonstrated by pulse-labelling experiments with $^{14}CO_2$ (Marcinowski and Grisebach 1977). Two different enzymes in spruce are responsible for cleavage (β-glucosidase) (Marcinowski and Grisebach 1978) and synthesis (glucosyltransferase) (Schmid and Grisebach 1982) of coniferin. In investigations on the role of coniferin in lignification the localization of these two enzymes in the cell in spruce was achieved by indirect immunofluorescent labelling.

3.2.1 Localization of β-Glucosidase

A cell wall fraction from hypocotyls and roots of a 12-day-old spruce seedling contains glucosidase activity towards coniferin. During seedling development highest activity of bound β-glucosidase for coniferin is present in about 12-day-old seedlings (Marcinowski et al. 1979). Part of the glucosidase activity could be solubilized by treatment of the cell wall fraction with 0.6 mol l^{-1} NaCl in buffer. The solubilized enzyme was purified by a four-step procedure (Marcinowski and Grisebach 1978). Two glucosidases with different activity ratios coniferin/4-nitrophenylglucoside were separated on a Sepharose 6B column. The enzyme with the higher activity towards coniferin was further purified by a second Sepharose 6B step and shown to be homogeneous by ultracentrifugation (Mr 58 570), disc electrophoresis and isoelectric focussing. For preparation of the antiserum 400 μg of pure glucosidase were dissolved in 1 ml physiological NaCl and emulsified with 1 ml Freund's complete adjuvant. Antisera for glucosidase were raised in rabbits by two inoculations with 200 μg of the antigen hypodermally and subcutaneously into multiple sites 6 weeks apart. Antisera were collected 2 weeks after the last in-

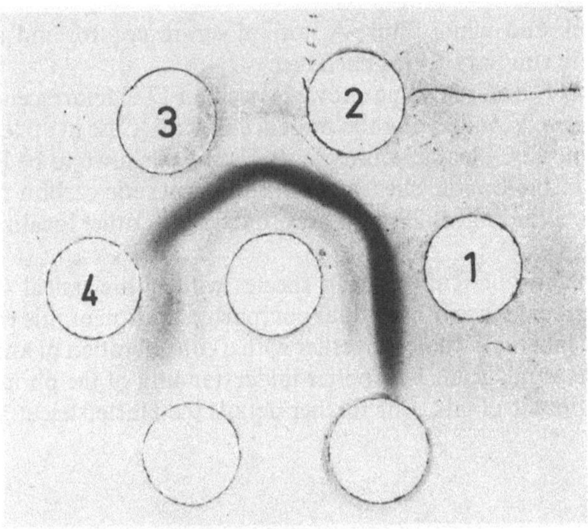

Fig. 5. Ouchterlony double diffusion analysis of crude coniferin glucosidase with rabbit antiserum. The *middle well* contained 20 μl antiglucosidase antiserum and the *outer wells 1–4* contained respectively 20, 15, 10, and 5 μl extract from spruce cell walls with 0.1 mol l^{-1} McIlvaine buffer, pH 5.0, containing 0.6 mol l^{-1}NaCl. 20 μl extract corresponded to 3 μg glucosidase 1. (From Marcinoswky et al. 1979)

jection. The specificity of this anti-glucosidase antiserum was analyzed by Ouchterlony double immunodiffusion and by immunoprecipitation of glucosidase (Marcinowski et al. 1979). The result of the Ouchterlony analysis is shown in Fig. 5. One major precipitin line and a second very weak band were observed when antisera were reacted with crude enzyme extracts of washed cell walls or with pure glucosidase. No precipitin bands were observed with the anti-glucosidase serum or with β-glucosidase from sweet almonds (emulsin). For preparation of microtome sections segments of hypocotyls (5 mm long) from 10- to 19-day-old seedlings taken 5 mm below the cotyledons were embedded at −30 °C in tissue tek OCT. For assay tissue was cut at 10 μm in a cryostat (−18 °C) and subsequently kept at 4 °C until use. All operations with the mounted sections were carried out at 4 °C in the dark. The slides were rinsed for 10 min in buffered saline (10 mmol l^{-1} potassium phosphate in 140 mmol l^{-1} NaCl, pH 7.2, PBS) and subsequently reacted with anti-glucosidase antiserum (diluted 1:10 in PBS) and kept for 25 min in a moist chamber at room temperature. The sections were then rinsed for 10 min with PBS, covered with fluorescein isothiocyanate-conjugated goat anti-rabbit immunoglobulin (diluted 1:30 with PBS) and incubated for 20 min. The sections were subsequently washed 2 × 5 min in PBS and finally embedded in glycerol (9 parts glycerol, 1 part PBS) under a cover slip.

The following controls were run for each experiment

a) Incubation with serum from unimmunized rabbits (normal serum).
b) Sections exposed to fluorescein isothiocyanate conjugated anti-rabbit IgG only.
c) Incubation in the absence of antisera (autofluorescent control).

Mounted sections were analyzed immediately after staining with a Zeiss microscope equipped with a high pressure mercury lamp (HBO 200) and epi-illumination system III S. Excitor filter BP 450/490, barrier filter LP 520 (see Fig. 4). Sections were kept in the dark at 4 °C and were photographed within 24 h after preparations using Ilford HP5 135 or Kodak Ektachrome 160 ASA films.

Fluorescence obtained with anti-glucosidase antiserum was considered to be specific, when all of the controls were shown to be negative. Chloroplast autofluorescence could be eliminated by the filter combination used. The specific labelling of glucosidase is much more clearly visible under the microscope in colour than in black and white photographs. Colour photographs showed a specific fluorescence at the inner layer of the secondary wall. This fluorescence was present in the walls of all cells, but seemed to be stronger in the epidermal layer and in the vascular bundles. The specific fluorescence is demonstrated in the black and white photograph shown in Fig. 6 together with the control, which is completely devoid of fluorescence.

A similar study on localization of β-glucosidases in *Cicer arietinum* seedlings was carried out by Burmeister and Hösel (1981). These authors purified the γ-globulin by a modified method of Hudson and Hay (1976) on a coniferin glucosidase-linked Sepharose column.

3.2.2 Localization of Glucosyltransferase

UDP-glucose: coniferyl alcohol glucosyltransferase was isolated from cambial sap of spruce (*Picea abies*) as a soluble enzyme. An apparently homogeneous enzyme was obtained by a seven-step procedure including dye-ligand chromatography (Schmid and Grisebach 1982). The transferase showed only a single band after dodecyl sulfate slab-gel electrophoresis. During development of spruce seedlings enzyme activity rises from a low level to a maximum at about day 10 and then rapidly declines (Schmid et al. 1982).

Antiserum was obtained by absorbing 400 µg of pure glucosyltransferase on Alu Gel S, incorporating the mixture into complete Freund's adjuvant, and injecting the emulsion into a rabbit. A total of 1 ml of the emulsion was administered at multiple sites, both into the hind foot pads and hypodermally in 0.05 ml doses. A booster injection with 600 µg enzyme was given subcutaneously and intramuscularly 6 weeks later. Fourteen, 21 and 28 days after the second challenge injection antisera were obtained and analyzed by Ouchterlony double diffusion and immunoprecipitation. The antiserum which was collected 28 days after the second challenge was the most active. This antiserum was enriched in the IgG fraction by ammonium sulfate precipitation in the following way. The antiserum was diluted 1:1 with PBS. To this solution was added dropwise in an ice bath an equal volume of $(NH_4)_2SO_4$ solution (pH 7.0, 3.2 mol l^{-1}). The reaction mixture was kept under slow stirring for 12 h at 4 °C. The precipitate was collected by centrifugation and washed two times with 1.6 mol l^{-1} ammonium sulfate. The sediment was dissolved in PBS and dialysed against this buffer. The immunoelectrophoretic appearance of this fraction developed with anti-rabbit serum is shown in Fig. 7. The IgG fraction was quite pure and contained additional proteins only in minor amounts.

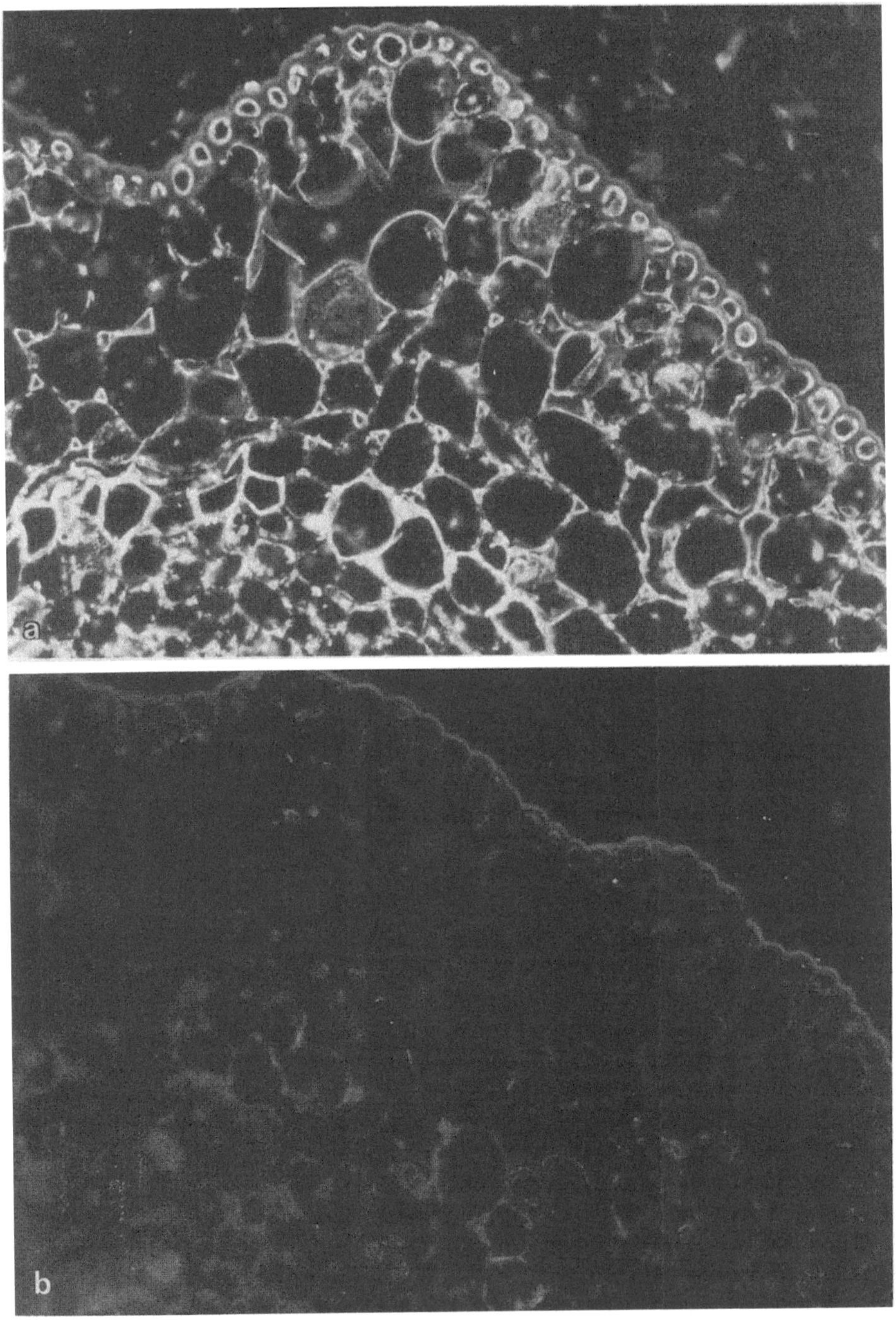

Fig. 6. a Immunofluorescent labelling of glucosidase in a transverse microtome section of 19-day-old hypocotyl showing specific fluorescence at the inner layer of the secondary walls; ×320. **b** Control with serum of unimmunized rabbits. (From Marcinowski et al. 1979)

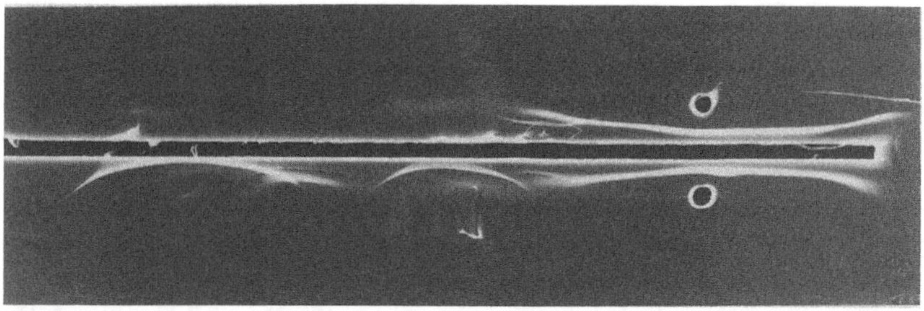

Fig. 7. Immunoelectrophoretic analysis of antiglucosyltransferase IgG. *Upper well* IgG antibody fraction; *lower well* whole rabbit serum; guinea pig anti-whole rabbit serum in the *slot*. (From Schmid et al. 1982)

Double immunodiffusion showed a strong precipitin line and a second weaker band when the IgG antibody fraction was tested with the isolated enzyme. However, when the concentrated crude extract of 10-day-old spruce seedlings was tested against the IgG fraction, only one precipitin line was formed. This proved that the antiserum was monospecific for the glucosyltransferase in the seedlings and that the IgG fraction could therefore be used for localization studies.

Since activity of the transferase was highest 10 days after germination, 10-day-old spruce hypocotyls were used for the in situ immunofluorescent labelling of the glucosyltransferase.

Segments of hypocotyls (3–5 mm long) taken 5–10 mm below the cotyledons were embedded in liver which was fixed to a piece of cork. Some of the segments were treated before embedment with 2.5% glutardialdehyde and washed several times with PBS. The embedded segments were frozen with liquid propane and kept in liquid nitrogen.

For the assay, sections (10–14 μm) were cut with a freeze microtome at −20 °C and mounted on slides which had been wetted with egg albumin-glycerol, after which they were rinsed 3 × 3 min with PBS. The sections were subsequently treated with anti-glucosyltransferase antiserum (IgG fraction diluted 1:100 with PBS, 300 μg protein ml^{-1}) and incubated for 25 min in a moist chamber at 4 °C. They were then rinsed 3 × 3 min with PBS, covered with fluorescein isothiocyanate conjugated goat anti-rabbit IgG (diluted 1:16 with PBS), and incubated in the dark for 25 min at 4 °C. The sections were then washed 3 × 3 min in PBS and finally embedded in glycerol (90%) under a cover slip.

The specificity of anti-glucosyltransferase antibody binding was evaluated by the following controls:

a) Treatment of sections with serum from unimmunized rabbits (normal serum).
b) Exposure of untreated sections to fluorescein isothiocyanate conjugated anti-rabbit IgG only.
c) Incubation of sections in the absence of both anti-glucosyltransferase and FITC-conjugated anti-rabbit IgG (autofluorescent control).

Mounted sections were analyzed immediately after staining with a Zeiss Universal microscope equipped with epifluorescent illumination and the following filter combination:

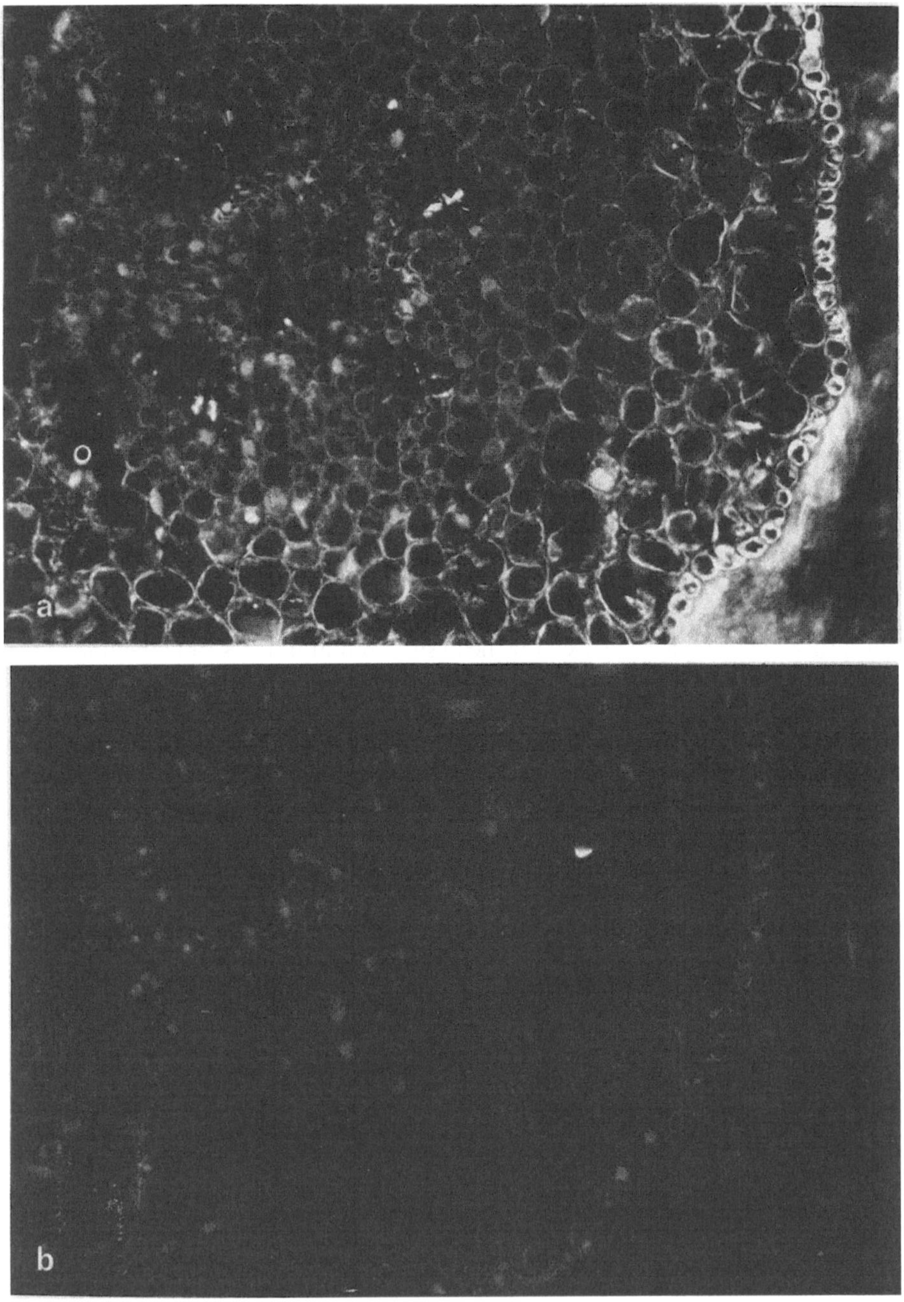

Fig. 8. a Immunofluorescent labelling of glucosyl transferase in transverse microtome section of 10-day-old spruce hypocotyl. Fluorescence is seen predominantly in the epidermal layer and vascular bundles; ×54. **b** Control with normal serum. (From Schmid et al. 1982)

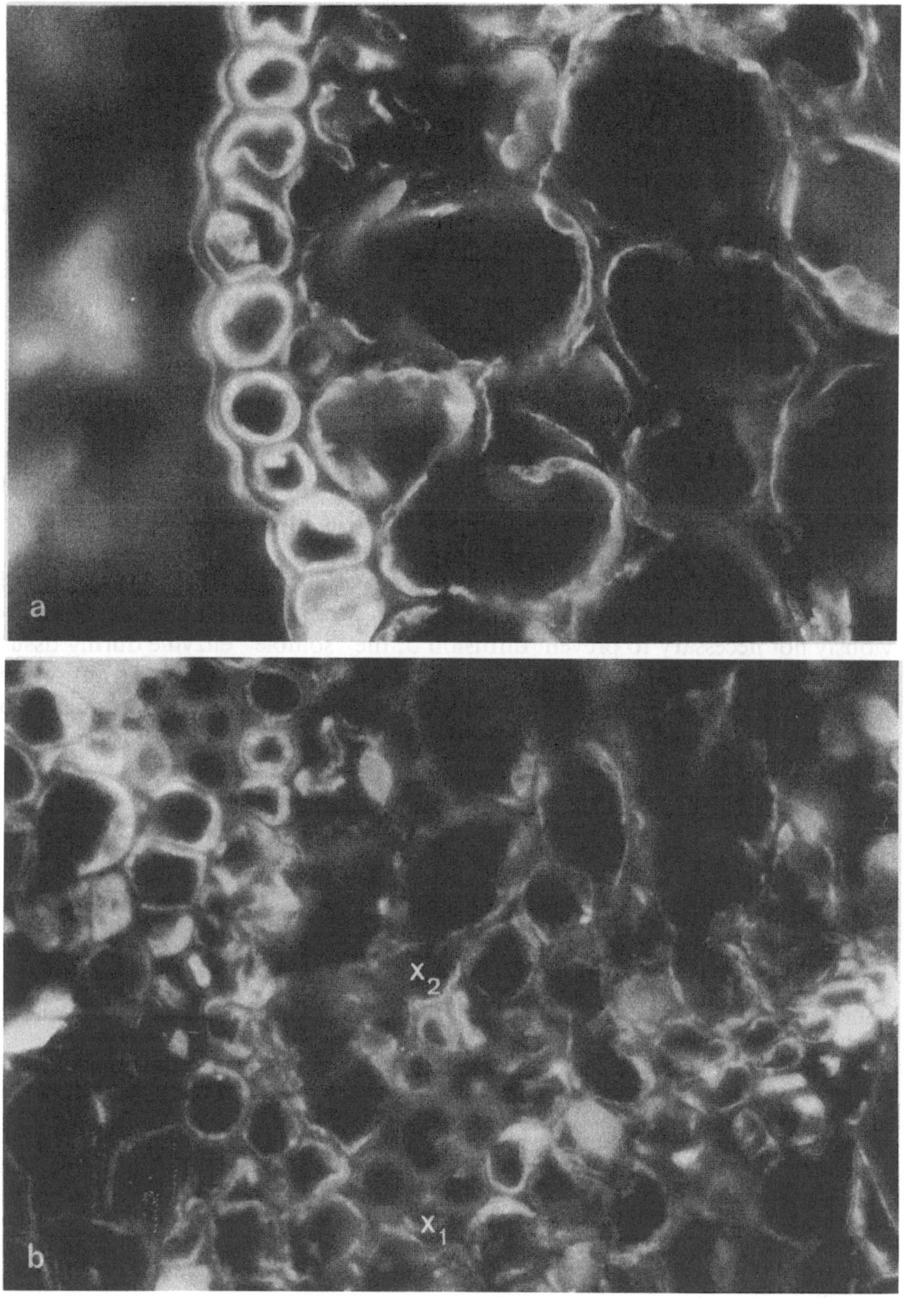

Fig. 9. a Section of Fig. 8 a with epidermal layer; **b** section with vascular bundles; X_1 outer xylem cell; X_2 inner xylem cell; ×216. (From Schmid et al. 1982)

Excitation filter BG 12, broad band filter (450–490 nm) with LP450 and KP 490, barrier filter LP520 and short-wave pass filter KP 560 to eliminate chlorophyll fluorescense. Photographs were taken with Ilford HP5/400 ASA and Kodak Ektachrom 160 ASA films. Fluorescence obtained with anti-glucosyltransferase IgG fraction was specific, since all of the controls were shown to be negative. The specific labelling of the glucosyltransferase and the normal serum control are shown in Fig. 8. The most intense fluorescence is seen in the epidermal layer and in the vascular bundles.

A higher magnification of this region (Fig. 9) demonstrated that the fluorescence is located in the parietal cytoplasmic layer and also covers the chloroplasts. These observations were confirmed by colour photographs taken without elimination of chlorophyll fluorescence. In the epidermis and subepidermis, a green-yellow fluorescence adjacent to a strong red fluorescence of the cortex parenchyma and again a yellow-green fluorescence in the vascular bundles were visible. With higher magnification, chloroplasts in the cortex parenchyma were seen as distinct red spots, whereas they appeared as yellow spots near the epidermis and vascular bundles because of the superposition of the fluorescence of the glucosyltransferase in the cytoplasm to that of chlorophyll.

No differences were found between untreated sections and sections which had been treated with glutardialdehyde. Fixation with glutardialdehyde was obviously not necessary to prevent diffusion of the "soluble" enzyme during tissue preparation. Since the tissue was in a frozen state during sectioning and antiserum was applied when the surface of the sections had begun to thaw, the enzyme could be fixed in its in vivo localization. Whether this method is applicable for localization of all soluble enzymes or whether fixation with glutardialdehyde or another reagent (see Sect. 2.3) is necessary should be investigated in each individual case.

References

Baumgartner B, Matile P (1976) Immunocytochemical localization of acid ribonuclease in morning glory flower tissue. Biochem Physiol Pflanz (BPP) 170:279–285

Baumgartner B, Tokuyasu KT, Chrispeels MJ (1978) Localization of vicilin peptidohydrolase in the cotyledons of mung bean seedlings by immunofluorescence microscopy. J Cell Biol 79:10–19

Birrell GB, Habliston DL, Nadakavukaren KK, Griffith OH (1985) Immunoelectron microscopy: the electron optical analog of immunofluorescence microscopy. Proc Natl Acad Sci USA 82:109–113

Brown SA (1966) Lignins. Annu Rev Plant Physiol 17:223–244

Burmeister G, Hösel W (1981) Immunohistochemical localization of β-glucosidases in lignin and isoflavone metabolism in *Cicer arietinum* L. seedlings. Planta (Berl) 152:578–586

Clarke AE, Knox RB, Hermyn MA (1975) Localization of lectines in legume cotyledons. J Cell Sci 19:157–167

Coons AH (1958) Fluorescent antibody methods. In: Danielli JF (ed) General cytochemical methods, vol 1. Academic Press, New York, pp 399–422

Coons AH, Creech HJ, Jones RN (1941) Immunological properties of an antibody containing a fluorescent group. Proc Soc Expt Biol Med 47:200–202

DeLuca D (1982) Immunofluorescence analysis. In: Marchalonis JJ, Warr GW (eds) Antibody as a tool. Wiley, Chichester, pp 189–231
Giloh H, Sedat JW (1982) Fluorescence microscopy: reduced photobleaching of rhodamine and fluorescein protein conjugates by n-propyl gallate. Science 217:1252–1255
Goding JW (1978) Use of staphylococcal protein A as an immunological reagent. J Immunol Methods 20:241–253
Goldman M (1968) Fluorescent antibody methods. Academic Press, New York
Haitinger M (1959) Fluoreszenzmikroskopie. Akadem Verlagsges, Leipzig, 2. Aufl
Hapner SJ, Hapner KD (1978) Rhodamine immunofluorescence applied to plant tissue. J Histochem Cytochem 26:478–482
Hattersley PW, Watson L, Osmond CB (1977) In situ immunofluorescent labeling of ribulose-1,5-bisphosphate carboxylase in leaves of C_3 and C_4 plants. Aust J Plant Physiol 4:523–539
Hoyer LC, Bucana C (1982) Principles of immunoelectron microscopy. In: Marchalonis JJ, Warr GW (eds) Antibody as a tool. Wiley, Chichester, pp 233–271
Hudson L, Hay FW (1976) In: Practical immunology. Blackwell, Oxford, p 192
Kehrel B, Wiermann R (1985) Immunochemical localization of phenylalanine ammonia-lyase and chalcone synthase in anthers. Planta (Berl) 163:183–190
Marcinowski S, Grisebach H (1977) Turnover of coniferin in spruce seedlings. Phytochemistry (Oxf) 16:1665–1667
Marcinowski S, Grisebach H (1978) Enzymology of lignification. Cell-wall-bound β-glucosidase for coniferin from spruce (*Picea abies*) seedlings. Eur J Biochem 87:37–44
Marcinowski S, Falk H, Hammer DK, Hoyer B, Grisebach H (1979) Appearence and localization of a β-glucosidase hydrolyzing coniferin in spruce (*Picea abies*) seedlings. Planta (Berl) 144:161–165
McKinney RM, Spillane JT, Pearce GW (1964) Factors affecting the rate of reaction of fluorescein isothiocyanate with serum protein. J Immunol 93:232–242
Munoz J (1971) Double diffusion in plates. In: Williams CA, Chase MW (eds) Methods in immunology and immunochemistry, vol 3. Academic Press, New York, pp 146–160
Murray DR, Knox RB (1977) Immunofluorescent localization of urease in the cotyledons of jack bean, *Canavalia ensiformis*. J Cell Sci 26:9–18
Nairn RC (1976) Fluorescent protein tracing, 4th edn. Livingstone, Edinburgh
Rey MEC, Noble JP (1984) Subcellular localization by immunocytochemistry of the extracellular protease produced by *Nectria galligena* Bres. in infected apple tissue. Physiol Plant Pathol 25:323–336
Roldán JM, Verbelen JP, Butler WL, Tokuyasu K (1982) Intracellular localization of nitrate reductase in *Neurospora crassa*. Plant Physiol (Bethesda) 70:872–874
Roth J (1983) The colloidal gold marker system for light and electron microscopic cytochemistry. In: Bullock GR, Petrusz (eds) Techniques in immunocytochemistry, vol 2. Academic Press, London, pp 217–284
Roth J, Bendayan M, Orci L (1978) Ultrastructural localization of intracellular antigens by the use of protein A – gold complex. J Histochem Cytochem 26:1074–1081
Roth J, Carlemalm E, Lucocq JM, Villinger (1984) EMBO practical course "Immunoelectron microscopy". Biocenter, University of Basle, Switzerland
Sarkanen KV, Ludwig CH (1971) Lignins. Wiley Interscience, New York, pp 28–29 and 109–111
Sautter C, Hock B (1982) Fluorescence immunohistochemical localization of malate dehydrogenase isoenzymes in watermelon cotyledons. Plant Physiol (Bethesda) 70:1162–1168
Schmid G, Grisebach H (1982) Enzymic synthesis of lignin precursors. Purification and properties of UDP glucose: coniferyl alcohol glucosyltransferase from cambial sap of spruce (*Picea abies* L.) Eur J Biochem 123:363–370
Schmid G, Hammer DR, Ritterbusch A, Grisebach H (1982) Appearance and immunohistochemical localization of UDP-glucose: coniferyl alcohol glucosyltransferase in spruce (*Picea abies* L. *Karst.*) seedlings. Planta (Berl) 156:207–212
Sternberger LA (1979) Immunocytochemistry, 2nd edn. Wiley, New York

Tokuyasu KT, Singer SJ (1976) Improved procedures for immunoferritin labeling of ultrathin frozen sections. J Cell Biol 71:894–906
van der Veken JA, van Slogeren DHM, van der Want JPH (1962) Immunological methods. In: Linskens HF, Tracey MV (eds) Modern methods of plant analysis, vol 5. Springer, Berlin Heidelberg New York, pp 422–463
Vernooy-Gerritsen M, Leunissen JLM, Veldink GA, Vliegenthart JFG (1984) Intracellular localization of lipoxygenases −1 and −2 in germinating soybean seeds by indirect labeling with protein A-colloidal gold complexes. Plant Physiol 76:1070–1079
Warr GW (1982) Purification of antibodies. In: Marchalonis JJ, Warr GW (eds) Antibody as a tool. Wiley, Chichester, pp 59–96
Wick G, Baudner S, Herzog F (1980) Immunofluoreszenz. Med Verlagsges Marburg/Lahn
Williams CA (1971) Immunoelectrophoretic analysis in agar gels. In: Williams CA, Chase MW (eds) Methods in immunology and immunochemistry, vol 3. Academic Press, New York, pp 237–273

Quantitative Immunochemistry of Plant Phosphoenolpyruvate Carboxylases

J. BRULFERT and J. VIDAL

Since its discovery (Bandurski and Greiner 1953) phosphoenolpyruvate carboxylase (PEPC, EC 4.1.1.31) has attracted increasing interest among plant scientists. The enzyme catalyses the reaction of CO_3H^- and phosphoenolpyruvate to produce oxaloacetate, immediately reduced to form malate; this latter can be oxidatively decarboxylated by NADP malic enzyme, and thus, appears to be a physiological vector for carbon (CO_2) and energy (reducing power). Extensive studies established the ubiquitous presence of PEPC in plants and its functional, regulatory and physico-chemical properties have been described by several groups. PEPC appears to be involved in number of physiological roles, wich were recently extensively reviewed [Physiol Vég 21:5 (1983)]. More particularly PEPC seems to play a fundamental role in adaptation of plant organisms to changes in physiological and environmental parameters; for this reason PEPC can be considered as a good marker for differentiation of physiological processes and for operation of adaptive metabolic pathways. In some cases isoforms involved in specific physiological roles, were described as typical.

The purpose of this article is to discuss the utilization of immunochemical tools for quantitative measurements of plant PEPCs. Methods for the purification of the enzyme required to raise a monospecific immune serum are presented in the first section of the paper. The second and third sections deal with the utilization of immunoprecipitation in gels and immunotitration sensu stricto for PEPC titration. Techniques based on quantitative immunoprecipitation of the enzyme in plant extracts and on immunoaffinity chromatography will be described in the final sections.

1 Purification of PEPC – Preparation of Immune Sera

All plant materials can be used for extraction and purification of PEPC; however some plant species provide high amounts of enzyme, mainly C_4 plants, in which PEPC can reach 10–15% of the total soluble proteins (Uedan and Sugiyama 1976). It has been shown that heterologous antibodies can cross-react with PEPC from any other plant species, but when very high specific affinity towards the enzyme is sought, obtention of homologous immune sera is recommended.

Reviewing the literature on this topic affords the basis of a standard general technique for the extraction and purification of PEPC.

1.1 Extraction of PEPC

Extraction of PEPC can be performed at 0 °–4 °C, in 100–200 mM Tris or phosphate buffer, final pH 7.0 to 8.0 containing 100 mM mercaptoethanol, 5% glycerol, 1 mM $MgCl_2$; glycerol and magnesium were shown to be strong stabilizers of enzyme activity (Manetas 1982). Deaeration of the brei by a stream of nitrogen is usually convenient to avoid polyphenol oxydase activity.

Notes. In order to obtain total recovery of enzyme activity the following points should be carefully checked in each case:
1. Grinding of tissues can be done by usual methods; for analytical purposes a Potter-Elvejhem is the most convenient system allowing quantitative extraction.
2. Because of considerable amounts of malic acid accumulating in leaf tissues of CAM plants, the pH of the extraction medium must be carefully adjusted in order to obtain a final pH between 7.0 and 8.0.
3. For most CAM plants, extraction of PEPC requires addition of polyvinylpyrrolidone (PVP), or polyethylene glycol (PEG, MW 20 000) to protect the enzyme against inactivation by phenolic compounds.
4. PEPC from C_4 plants exhibits a low-temperature dependent inhibition which is reversible upon moderate heating (Hatch and Oliver 1978).
5. Excessive dilution of extracts can result in enzyme deactivation (Robertson and Kerr 1971).

1.2 Purification of PEPC

Purification of PEPC can be achieved utilizing a standard scheme involving the following successive steps:
1. Hydrophobic chromatography on octyl- or phenyl-Sepharose, equilibrated with 30% $(NH_4)_2SO_4$ phosphate or Tris-buffer, pH 7.0 to 8.0; elution performed either by increasing ethylene glycol concentration (0 to 40%) or by decreasing salt concentration (30% to 0); both elution systems can be applied together.
2. Chromatography on hydroxyapatite columns equilibrated with 25 mM Na-phosphate buffer pH 6.8; elution by increasing phosphate concentration (25 to 300 mM).
3. Chromatography on DEAE cellulose columns equilibrated with 25 mM Tris-HCl pH 8.0; elution either by increasing buffer molarity (25 to 300 mM) or by a NaCl gradient (0 to 400 mM).
4. Polyacrylamide gel electrophoresis (PAGE) carried out on homogenous (5% acrylamide) or gradient pore (4–27% acrylamide) gels, according to Laemmli (1970).

Notes. The right succession in the purification steps must be selected in order to minimize handling of the enzyme; for instance, if $(NH_4)_2SO_4$ fractionation (40 to 55% saturation) of the crude extract is utilized, hydrophobic chromatography can be immediately performed without any preliminary treatment. However, PAGE must be the last purification step: PEPC bands, detected on the gels by specific staining (Vidal et al. 1976) are directly used to inject the rabbits; it is therefore necessary to assess the purity of the enzyme by SDS-gel electrophoresis (Laemmli 1970).

All buffers used through the purification procedures can be supplemented with 5% glycerol, 14 mM mercaptoethanol.

When changing the buffer or removing salts from extracts is necessary, precipitation of the enzyme by polyethylene glycol (MW 4000 to 6000) 30% final concentration appears as the most convenient procedure.

Temperature is usually kept close to 4 °C during purification; nevertheless, in the case of PEPC from *Sorghum* leaves, it has been shown that the stability of the enzyme was considerably improved when working at 16 °C provided protective agents are present.

Purified PEPC can be stored for months, precipitated by $(NH_4)_2SO_4$ (70% saturation), at −25 °C.

1.3 Preparation of Antisera

Rabbits are commonly used for preparation of PEPC antisera. The strongest titre of the final serum has been obtained with the following protocol: for the primary injections (0.2 to 1 mg protein per rabbit) acrylamide gel bands containing the pure antigen are homogenized in PBS buffer (Na-phosphate, 25 mM, pH 7.5, NaCl 150 mM) and mixed v/v with complete Freund's adjuvant; both subcutaneous and intramuscular injections are simultaneously done. About 1 month later, booster injections consist in intravenous administration of antigen into the marginal vein of the ear, twice at 1 day intervals. For the primary injection, presence of acrylamide enhances the production of antibodies; in contrast, for booster injections, the antigen must be freed from acrylamide by electroelution or by extruding the gel band through a French press and recovering PEPC by centrifugation (Hirel, unpublished).

Notes. PEPC antisera can be raised using the monomer of the enzyme obtained by SDS treatment (see Sect. 4.1.), this provides a means of preparing new antisera from dissociated immunoprecipitates obtained from previous immunizations (Crétin et al. 1984).

The purity and specificity of the sera can be checked by double diffusion method (Ouchterlony 1958), immunoprecipitation followed by SDS-PAGE (Vidal et al. 1983b) or immunoaffinity chromatography (Vidal et al. 1980).

1.4 Purification of Antisera

Optimal and rapid results are obtained by three main methods.

1.4.1 Fractionation by Ion-Exchange Chromatography

DEAE-Trisacryl columns are equilibrated in 25 mM Tris-HCl, 35 mM NaCl, pH 8.8 and eluted by the same buffer. IgG are eluted in the void volume of the column; other types of plasma proteins are retained. Sera must be first equilibrated in the same buffer through a column of Ultrogel AcA 202 (IBF, Pharmindustrie, France).

1.4.2 Affinity Chromatography Utilizing Protein A Immobilized on a Support

Basis of this technique is the remarkable affinity of protein A towards certain classes of immunoglobulins. Elution of IgG is obtained with 200 mM citric acid-citrate buffer, pH 2.8; recovery of IgG is performed by $(NH_4)_2SO_4$ precipitation (50% saturation) and dissolution of the precipitate in any suitable buffer. Detailed operations of this methodology are described by the manufacturers. It should be kept in mind that some types of IgG are not recognized by protein A.

1.4.3 Affinity Chromatography Utilizing Enzyme Immobilized on an Activated Support

Proteins from partially purified extracts containing PEPC are bound to a glutaraldehyde activated gel as described in Sect. 5; passing antisera through the column results in specific binding of the anti-PEPC IgG. Elution is achieved by acid washing as described for protein A chromatography.

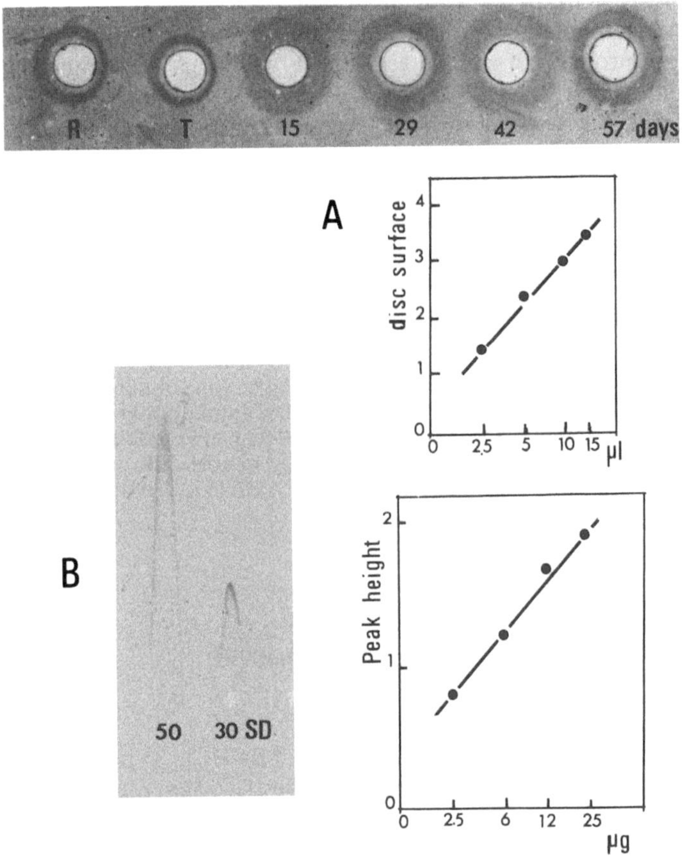

Fig. 1 A,B. Quantitative immunoprecipitation of PEPC in agarose gels. **A** Quantification of PEPC during root nodule formation in soybean by radial immunodiffusion (Vidal et al. 1986); comparison of precipitin lines obtained with aliquots of extracts (corresponding to the same weight of plant material) from roots (*R*), upper parts of taproots (*T*) and after 15, 29, 42, and 57 days of nodule development; quantification of PEPC can be achieved through a calibration curve obtained with successive dilutions of the same extract. **B** Quantification of PEPC during CAM development in *Kalanchoe blossfeldiana* under short days (30 and 50 short days). Immunoelectrophoresis of aliquots of extracts corresponding to the same dry weight of plant material; quantification of PEPC by comparison with a calibration curve obtained with 2.5, 6.2, 12.5, and 25 µg of pure PEPC. See also Hayakawa et al. (1981)

2 Quantitative Immunoprecipitation in Gels

Usual and well-documented methods of diffusion in agarose gels (Mancini et al. 1965) or immunoelectrophoresis (Laurell 1966) are valuable techniques to estimate amounts of PEPC in plant crude extracts. Precipitin lines are visible as circles in the case of radial immunodiffusion and as fuse-rockets in the case of immunoelectrophoresis; comparison of amounts of antigens in extracts is done by measuring circle radius or surface, peak height or surface; calibration curves give absolute or relative values depending on the purity of PEPC used as reference.

Due to oligomeric structure of plant PEPCs, artifactual multiple bands may arise; to avoid these drawbacks, it is necessary to find out suitable conditions of ionic strength, temperature and to prevent association-dissociation phenomena of the enzyme.

Typical examples of these techniques are shown in Fig. 1, A and B.

3 Immunotitration Coupled to Enzyme Activity

3.1 Measurement of PEPC Residual Activity in the Supernatant

Addition to a given volume of plant extract (corresponding to a known PEPC activity), of increasing volumes of anti-PEPC IgG results in progressive formation of immunoprecipitates and concomitant decrease in PEPC activity in the extract. The equivalence zone corresponds to the volume of IgG solution required for complete immunoinhibition of enzyme activity (provided that no inhibitory effect interferes with the titration process).

Classical methodology is as follows:

Incubate the mixtures between a fixed volume of crude or $(NH_4)_2SO_4$ precipitated extract and increasing amounts of antiserum (brought to equal volumes with buffer). Let the mixture stand for a prolonged period of time (usually overnight), at 4 °C. Centrifuge the immunoprecipitates and measure the remaining PEPC activity in the supernatants.

Two controls must be run in parallel: one of them without any antibody; the other with preimmune serum in the same concentration range as in the assays. Ideally, PEPC activity in these controls should stay constant over the experiment.

Notes. Conditions of precipitation, duration, buffers, pH, ionic strength, temperature, enzyme stabilizers, must be carefully determined in each case in order to preserve PEPC from inactivation by any other process except immunoinhibition.

Excess of reducing compounds such as dithiothreitol, mercaptoethanol must be avoided.

Addition of formalin-fixed *Staphylococcus aureus* ghosts (immunoprecipitin, BRL) was shown (I. Krüger and M. Kluge, unpublished) to allow a rapid (3 to 5 min) removal of immunocomplexes from the precipitation medium by centrifugation.

Results are classically expressed by plotting the residual PEPC activity in the supernatant (absolute values, or as % of the control) versus the amounts of IgG in the successive samples. Determination of the equivalence zone or more precisely of the 50% immunoinhibition point can be obtained from the immunotitration curve.

Interpretation is based on a linear relationship between the amount of PEPC and the volume of antiserum needed to reach either the 50% point of precipitation or the equivalence zone. The purpose of titration experiments is to provide a means:

1. To compare amounts of the same enzyme (assessed by Ouchterlony test) in a plant following changes in the physiological situation; Fig. 2, A schematizes the effect of switching situation 1 to situation 2 (increasing PEPC amount) or to situation 3 (decreasing PEPC amount) (see also, Sect. 3.3.1).
2. To evidence activation, or inactivation of the enzyme (Fig. 2, B); two situations can be distinguished: either changes in specific activity of PEPC following post-translational modifications or appearance of a new enzyme form displaying different immunochemical properties. Changes both in enzyme amounts and specific activities can occur simultaneously (see Sect. 3.3.1).
3. To compare the structure of the enzyme extracted from different sources (plant species) on an immunochemical basis, by their affinity towards a given antiserum (see Sect. 3.3.2).

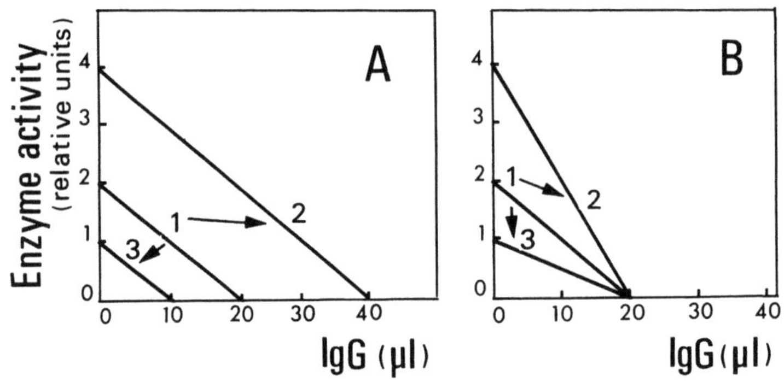

Fig. 2 A, B. Basic schemes for interpretation of immunotitration of PEPC from plant extracts. **A** Change in amount of PEPC protein extracted from plants which were given different physiological situations. Switch from physiological situation *1* [2 enzyme units (EU) totally precipitated by 20 μl IgG] to situation *2* (4 EU totally precipitated by 40 μl IgG): twofold increase in the amount of PEPC protein; switch from situation *1* to situation *3* (1 EU totally precipitated by 10 μl IgG): twofold decrease in the amount of PEPC protein. **B** Change in specific activity of PEPC. Switch from physiological situation *1* (as in **A**) to situation *2* (4 EU totally precipitated by 20 μl IgG): enzyme activation, twofold increase in specific activity of the enzyme; switch from situation *1* to situation *3* (1 EU totally precipitated by 20 μl IgG): enzyme inactivation, twofold decrease in specific activity. A more complicated figure would result from simultaneous changes in protein amount and specific activity of the enzyme

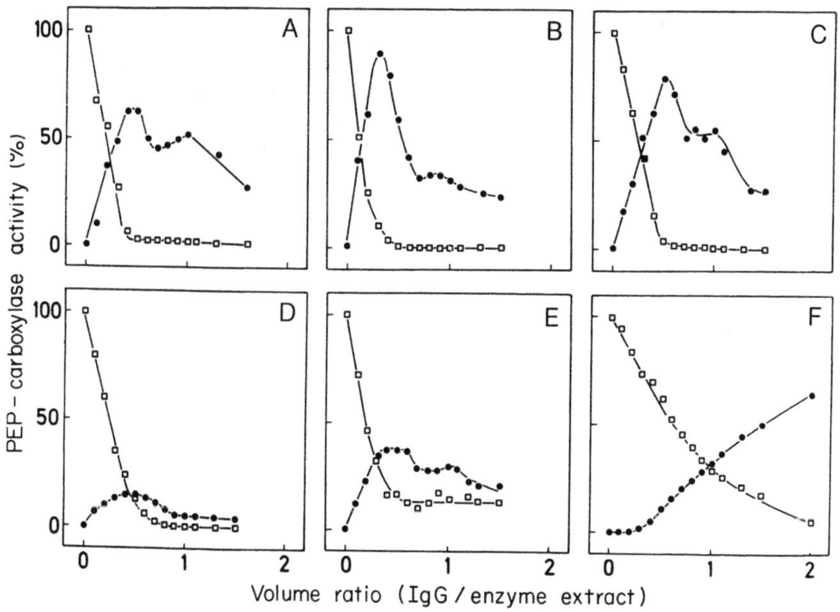

Fig. 3 A–F. Immunotitration of PEPC from various plant species. Equal volumes of extracts were incubated with increasing volumes of IgG; PEPC activity was measured in the supernatant (□–□) and in the resuspended immunoprecipitate (●–●); results are plotted as % of the control without IgG and against IgG/enzyme extract volume ratio. **A** *Sedum coeruleum*; **B** *Sedum compressum*; **C** *Kalanchoe blossfeldiana*; **D** *Triticum vulgare*; **E** *Vicia faba*; **F** *Zea mays*. (From Müller et al. 1982)

3.2 Measurement of PEPC Activity in the Immunoprecipitates

Immunoprecipitated PEPC was generally shown to keep some catalytic activity, when resuspended in Tris-HCl buffer (Fig. 3) (Müller et al. 1982). Advantage has been taken from this property to establish a titration curve complementary to that using the residual PEPC activity in the supernatant. For each assay of a titration curve, PEPC from immunoprecipitates can also be separated from antibodies by HPLC as described in Sect. 4.1. In this case direct amounts of PEPC protein are measurable.

3.3 Applications

Immunotitration of PEPC was used to obtain information on the mechanism(s) controlling the changes in PEPC activity in plant organisms during the course of developmental and physiological processes. This section is devoted to specific examples which could help the experimentator in solving his own problems in other fields.

3.3.1 PEPC During Establishment of CAM and C_4-Type Photosynthesis

Utilization of the immunotitration technique established that the strong increase in PEPC activity observed during the greening process of C_4-plant leaves was due to light-triggered neosynthesis of a specific isoform (Vidal and Gadal 1983; Vidal et al. 1983a, b). Similarly, transferring facultative CAM plants from non-inductive to inductive environmental conditions (photoperiod, drought) results in neosynthesis of an isoform which can be considered as CAM specific (Brulfert et al. 1982b; Brulfert and Queiroz 1982). During full-CAM behaviour of plants, this isoform displays a 24-h oscillation in capacity and in enzymic properties; results

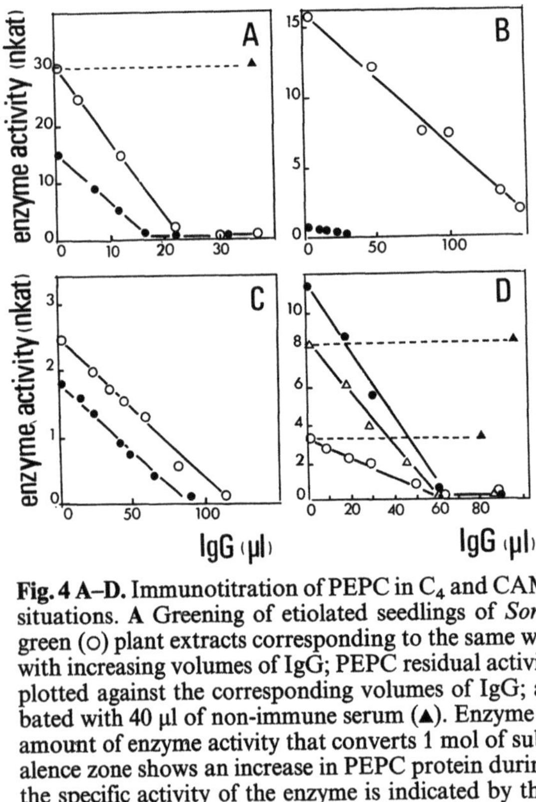

Fig. 4 A–D. Immunotitration of PEPC in C_4 and CAM plants under different physiological situations. **A** Greening of etiolated seedlings of *Sorghum*; PEPC from etiolated (●) and green (○) plant extracts corresponding to the same weight of plant material was incubated with increasing volumes of IgG; PEPC residual activity in each fraction was measured and plotted against the corresponding volumes of IgG; as a control, PEPC extract was incubated with 40 µl of non-immune serum (▲). Enzyme activity was expressed in nkat (katal: amount of enzyme activity that converts 1 mol of substrate s^{-1}). The change in the equivalence zone shows an increase in PEPC protein during greening; a simultaneous change in the specific activity of the enzyme is indicated by the change in the slope of the titration curves. **B** Induction of CAM in *Kalanchoe blossfeldiana* transferred from long-day to short-day conditions; immunotitration of PEPC (corresponding to 10 mg of leaf dry weight) in extracts from long-day (●) and short-day (○) grown plants; experiment and results as in **A. C** Development of CAM in *K. blossfeldiana* given increasing number of short days; immunotitration of PEPC (corresponding to 5 mg of leaf dry weight) in extracts from plants at the 50th and 80th short-day; the change in the equivalence zone and the parallelism of the two curves indicate an increase in the amount of the same PEPC isoform. **D** Mechanism underlaying the diurnal oscillation in PEPC capacity in CAM-displaying *K. blossfeldiana*; immunotitration of PEPC (corresponding to 10 mg of leaf dry weight) in extracts performed at 9.00 h (○), 15.00 h (△), and 23.00 h (●); typical results showing a change in the specific activity of the enzyme without modification in the amount of PEPC protein. ▲: control utilizing a non-immune serum

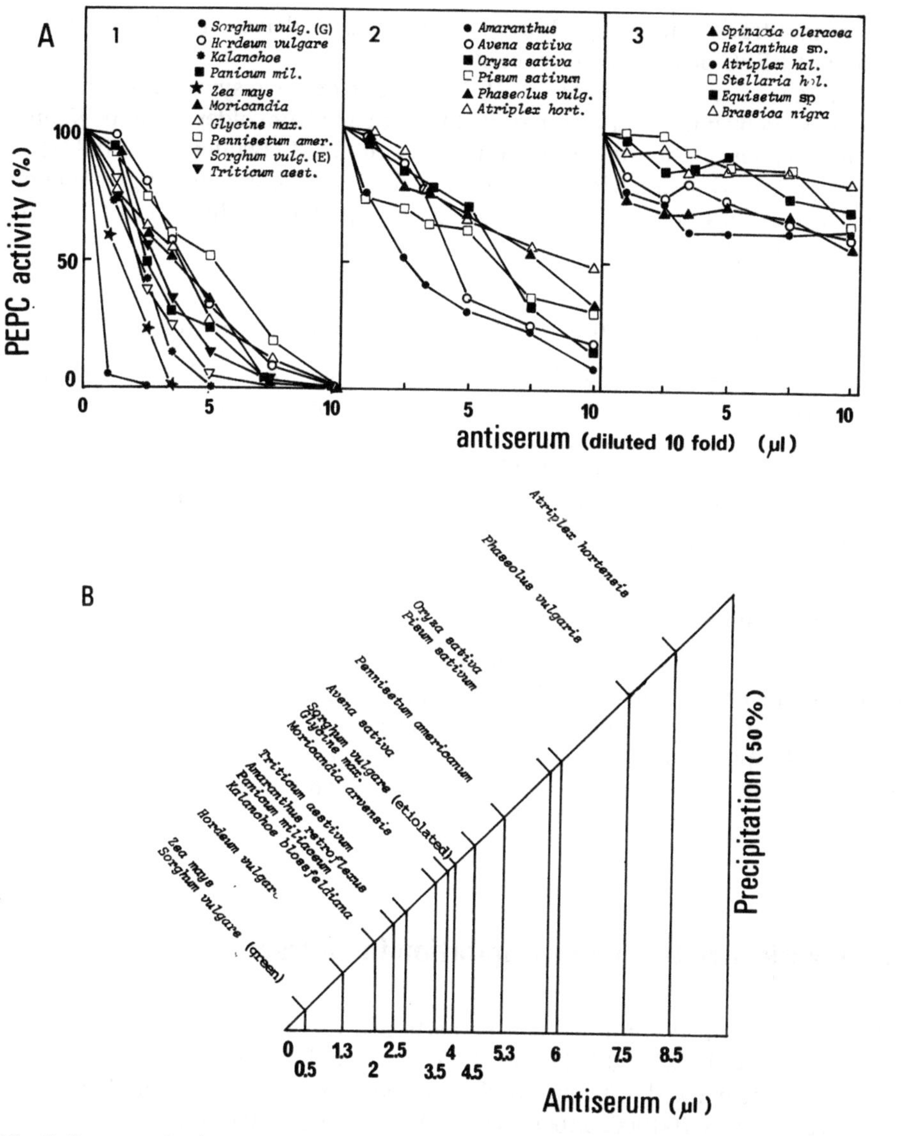

Fig. 5. Comparative immunotitration of PEPCs extracted from different botanical species. **A** PEPCs (0.33 nkat) were incubated by increasing volumes of immune serum 12 h at 4 °C. After centrifugation at 10 000 g for 15 min, the residual activity was measured in the supernatant, calculated as % of the control without immune serum and plotted against the corresponding volumes of immune serum; three main types of responses (*1, 2, 3*) can be distinguished depending on the affinity of PEPC towards the immune serum. **B** Classification of the botanical species as a function of the volume of immune serum used to precipitate 50% of the initial PEPC activity

of immunotitration along the diurnal cycle failed to detect any change in the amount of PEPC protein (Brulfert et al. 1982c).

Ageing is an inducing factor for PEPC through de novo synthesis of enzyme, as established by immunotitration of PEPC extracted from older leaves of *Kalanchoe blossfeldiana* in non-inductive conditions for CAM (Brulfert et al. 1982a).

Figure 4, A–D summarizes experimental results obtained during greening of etiolated seedlings of *Sorghum* and during photoperiodic induction of CAM in *Kalanchoe blossfeldiana*.

3.3.2 Immunotitration of PEPCs in Relation to Plant Photosynthetic Types

Titration of PEPCs extracted from different botanical species by the same immune serum (raised against the C_4 PEPC from *Sorghum* leaves) (Crétin et al. 1983) showed (Fig. 5) that: all C_4 and CAM plant PEPCs were immunochemically related proteins; C_3 plant PEPCs displayed low affinity towards the immune serum; exceptions were found for PEPC from gramineous C_3 plants in which the enzyme is well recognized by the immune serum.

3.3.3 Miscellaneous

Immunotitration of PEPC appeared also as a useful tool for investigations of other specific physiological problems. Enhancement of PEPC activity was shown to be achieved through de novo synthesis of enzyme in wheat seedlings at an early stage of phosphorus deficiency (Miginiac-Maslow et al. 1983), during growth of cell cultures from *Nicotiana tabacum* (Nato 1983) and during nodule formation on soybean roots (Vidal et al. 1986).

4 Quantitative Immunoprecipitation in Extracts

Quantitation of PEPC protein can be perfomed by direct analysis of immunoprecipitates by HPLC and SDS-PAGE. These techniques are allowed to be used because of the marked difference in the molecular weight of PEPC and IgG subunits, respectively 100000, 52000 (heavy chain) and 25000 (light chain).

4.1 HPLC Techniques

Immunoprecipitates are prepared at equivalence between PEPC in extracts and IgG, and collected by centrifugation. Pellets are successively washed three times with 50 mM phosphate or Tris-buffer pH 8.0, 1% Triton X 100, 1% NaCl, and twice with 50 mM phosphate or Tris-buffer pH 8.0.

Precipitates are solubilized in a minimal volume of dissociating buffer (50 mM phosphate or Tris-buffer pH 8.0, 1% SDS, 10% mercaptoethanol) and heated 10 min at 100 °C.

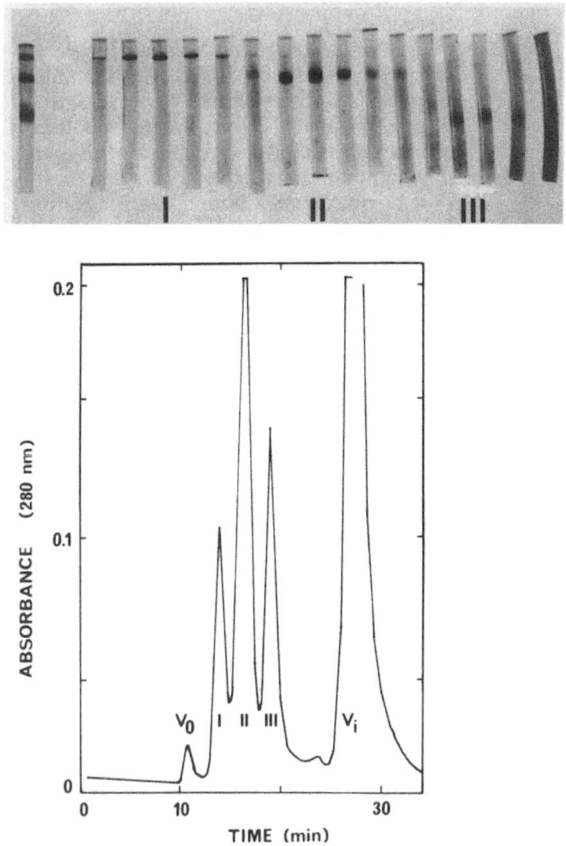

Fig. 6. High performance liquid chromatography of an immunoprecipitate of PEPC from green *Sorghum* leaves (80 μg total proteins; 18 μg PEPC protein). Immunoprecipitates were dissociated at 100 °C for 10 min in 1 ml of 100 mM Na-phosphate buffer, pH 7.0, 1% SDS, 5% mercaptoethanol (v/v). Proteins were filtered through a TSK SW 4000 column (Altex) equilibrated with 10 mM Na-phosphate buffer, pH 7.0, 0.1% SDS at a flow rate of 0.5 ml min^{-1} (Waters SA pump F 6000 A). Proteins were recorded at 280 nm. V_0: void volume; V_i: internal volume; *I*: PEPC subunit; *II* and *III*: heavy and light chains of antibodies, respectively. The purity of the eluted proteins was checked in fractions of the effluent by SDS-polyacrylamide gel electrophoresis (*top* of the figure). (From Vidal et al. 1983a)

Solubilized precipitates are passed through a HPLC gel filtration column equilibrated with 10 mM Na-phosphate buffer, pH 7.0, 0.1% SDS.

A typical experiment with *Sorghum* leaf PEPC is described in Fig. 6. Control by SDS-PAGE of the eluted fractions shows that almost the whole amount of PEPC protein is recovered in the first peak of the elution profile, without any contamination (Fig. 6, peak I). Quantification of PEPC can be obtained in the fractions by any method of protein determination.

Notes. Duration of heating required for complete dissociation of immunoprecipitates is a crucial step and must be tested in each specific case.

Load volume of the column should be as small as possible for the best resolution of the technique.

Flow rates for elution should be set up around 0.5 ml min^{-1}.

4.2 SDS-PAGE Techniques

Preparation and solubilization of immunoprecipitates are as above; SDS-PAGE (10% acrylamide, discs or slabs) are performed according to Laemmli (1970).

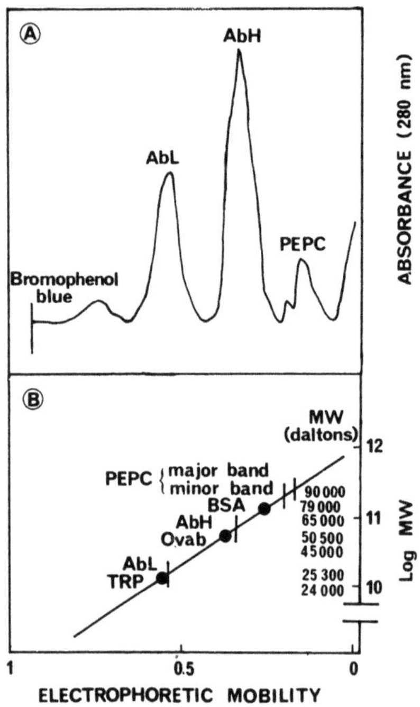

Fig. 7. A, B. SDS-polyacrylamide gel electrophoresis (SDS-PAGE) of an immunoprecipitate of PEPC from green *Sorghum* leaves (165 μg total proteins; 10.8 nkat). **A** Scanning of the gel (10% acrylamide, 0.1% SDS) at 280 nm. **B** Molecular weight determination of the immunoprecipitated proteins; *AbL* and *AbH*: light and heavy chains of antibodies; *BSA*: bovine serum albumin; *Ovab*: Ovalbumin; *TRP*: trypsinogen. (From Vidal et al. 1983a)

Scanning of the protein band and comparison to a calibration assay allow a relative quantification of PEPC.

Figure 7 shows a typical example of this technique applied to *Sorghum* leaf PEPC.

5 Immunoaffinity Chromatography

The method requires immobilization of IgG on a matrix, usually Sepharose (Pharmacia Fine Chemicals, Sweden) or Ultrogel AcA 22 (IBF, Pharmindustrie, France) activated by glutaraldehyde or cyanogen bromide. A convenient protocol for immobilization of IgG on the activated supports is as follows:

1. Mix the activated gel with IgG (5 to 10 mg of protein per ml of gel) in 500 mM phosphate buffer pH 7.6. Let stand 18 h at 4 °C.
2. Wash several times, with the same buffer and estimate the amount of fixed protein.
3. Block the remaining aldehyde groups by incubating the gel with 100 mM ethanolamine or 200 mM Tris-HCl buffer, pH 7.5 to 8.5, 3 h at 4 °C.
4. Before first use, wash the column successively with all buffers needed for the affinity chromatography experiment (see below).

Plant crude or partially purified extracts containing PEPC are passed through the column equilibrated with 50 mM phosphate or Tris-buffer pH 7.0 to 8.0. En-

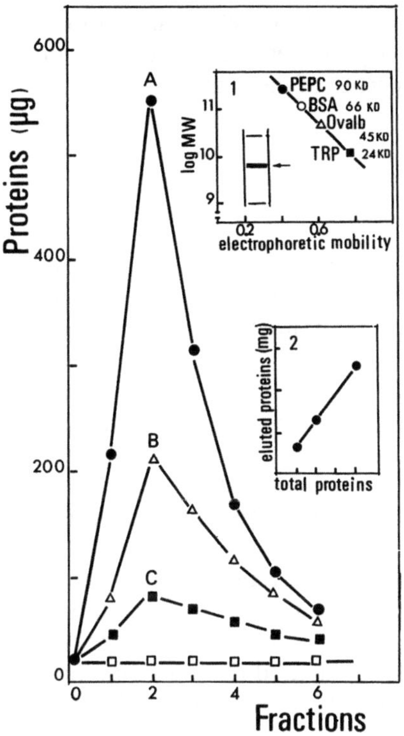

Fig. 8. Quantification of PEPC on an immunoadsorbent column. Semi-purified extracts containing 3.32 mg total proteins, 622 nkat PEPC (*A*), 1.66 mg total proteins, 311 nkat PEPC (*B*) and 0.83 mg total proteins, 155 nkat PEPC (*C*) were passed through the immunoadsorbent; protein were eluted by 100 mM citrate buffer, pH 2.8. *Insert 1*: the purity of the desorbed PEPC was checked by SDS-PAGE and the molecular weight determined as in Fig. 7. *Insert 2*: proportionality between PEPC recovery after acidic elution and the amount of total proteins loaded onto the column

zyme stabilizers can be added to the extracts: Mg, glycerol, substrates, activators; high concentrations of thiol reagents should be avoided.

Columns are rinsed several times with a large excess of equilibrating buffer supplemented with 2 M NaCl and 1% Triton X 100 (in order to wash out any remaining protein), then with 10 mM phosphate or Tris-buffer. Adsorbed PEPC is eluted with citrate buffer pH 2.8, at least 100 mM. Columns can be re-used after washing with the equilibrating buffer.

In a standard assay concerning the *Sorghum* leaf PEPC and in the experimental conditions described in Fig. 8, it was established that all PEPC activity was adsorbed on the column, and that the desorbed protein through acid washing was a highly pure PEPC (Fig. 8, Insert 1). Moreover, Fig. 8, Insert 2, shows that the amount of PEPC protein recovered was proportional to the total amount of protein in the extracts loaded onto the columns.

Notes. Comparison between cyanogen bromide and glutaraldehyde activated gels showed a significant release of IgG by acid elution from the cyanogen bromide activated gel and not from the glutaraldehyde activated gel.

Several types of eluents can be used depending on the affinity constant of antigen-antibody complexes: high salt concentrations (>1 M), low (2.8) or high (up to 11.0) pH values, polarity reducing agents (polyethylene glycol up to 50%), urea, chaotropic ions (thiocyanates). In some cases, coupling low affinity IgG to gels could be more favourable to desorb PEPC by mild procedures, allowing recovery of active enzyme.

Immunoaffinity chromatography provides a fast and efficient method for quantitative recovery of PEPC; moreover, the high degree of purity of the protein would allow studies on physico-chemical properties of the enzyme.

6 Concluding Remarks

Quantitative immunochemistry of PEPC has been developed in the recent years. Application of these techniques allowed to solve important physiological problems in which PEPC was implied; in this respect the contribution of immunology appeared as crucial.

For the study of PEPC by immunotitration no specific method has been developed besides those classically employed for other proteins; surprisingly, radioimmunoassays were not utilized.

Recent developments in immunocytology in plants provided a means to visualize PEPC inside cells and tissues, semiquantitatively in some cases (Perrot et al. 1981; Perrot-Rechenmann et al. 1982).

Increasing sharpness in the problems, demands increasing efficiency in the technical approaches. Evidence of PEPC polymorphism was recently established by the existence of PEPC isozymes probably encoded by multiple gene family (Harpster and Taylor 1986), of post-translational modifications resulting in changes in enzymic properties (Brulfert and Queiroz 1982) and of isosubunits (Vidal et al., unpublished). Investigations on these points require more powerful tools. Raising monoclonal antibodies affording discrimination between isozymes or isoforms, appears as a priority in the immediate future of research in the field.

References

Bandurski RS, Greiner CM (1953) The enzymatic synthesis of oxalacetate from phosphoryl-enolpyruvate and carbon dioxide. J Biol Chem 204:781–786

Brulfert J, Queiroz O (1982) Photoperiodism and Crassulacean acid metabolism III. Different characteristics of the photoperiod-sensitive and non-sensitive isoforms of phosphoenolpyruvate carboxylase and Crassulacean acid metabolism operation. Planta (Berl) 154:339–343

Brulfert J, Guerrier D, Queiroz O (1982a) Photoperiodism and Crassulacean acid metabolism II. Relations between leaf ageing and photoperiod in Crassulacean acid metabolism induction. Planta (Berl) 154:332–338

Brulfert J, Müller D, Kluge M, Queiroz O (1982b) Photoperiodism and Crassulacean acid metablism I. Immunological and kinetic evidences for different patterns of phosphoenolpyruvate carboxylase isoforms in photoperiodically inducible and non-inducible Crassulacean acid metabolism plants. Planta (Berl) 154:326–331

Brulfert J,. Vidal J, Gadal P, Queiroz O (1982c) Daily rhythm of phospoenolpyruvate carboxylase in Crassulacean acid metabolism plants. Immunological evidence for the absence of a rhythm in protein synthesis. Planta (Berl) 156:92–94

Crétin C, Perrot-Rechenmann C, Vidal J, Gadal P, Loubinoux B, Tabach S (1983) Study on plant phosphoenolpyruvate carboxylase: sensitivity to herbicides and immunochemical reactivity. Physiol Veg 21:927–933

Crétin C, Vidal J, Suzuki A, Gadal P (1984) Isolation of plant phosphoenolpyruvate carboxylase by high-performance size-exclusion chromatography. J Chromatogr 315:430–434

Harpster MH, Taylor WC (1986) Differential expression of the maize PEPC gene family. J Biol Chem 261:6132–6136

Hatch MD, Oliver JB (1978) Activation and inactivation of phosphoenolpyruvate carboxylase in leaf extract from C_4 species. Aust J Plant Physiol 5:571–580

Hayakawa S, Matsunaga K, Sugiyama T (1981) Light induction of phosphoenolpyruvate carboxylase in etiolated Maize leaf tissue. Plant Physiol (Bethesda) 67:133–138

Laemmli UK (1970) Cleavage of structural proteins during the assembly of the head of bacteriophage T_4. Nature 227:680–685

Laurell CB (1966) Quantitative estimation of proteins by electrophoresis in agarose gel containing antibodies. Anal Biochem 15:45–52

Mancini G, Carbonara AO, Heremans SJF (1965) Immunochemical quantitation of antigens by single radial immunodiffusion. Immunochemistry 2:235–254

Manetas Y (1982) Changes in properties of phosphoenolpyruvate carboxylase from the CAM plant *Sedum praealtum* DC. upon dark/light transition and their stabilization by glycerol. Photosynth Res 7:321–333

Miginiac-Maslow M, Vidal J, Bismuth E, Hoarau A, Champigny ML (1983) Effets de la carence et la réalimentation en phosphate sur l'équilibre énergétique et l'activité phosphoenolpyruvate carboxylase de jeunes plantes de Blé. Physiol Vég 21:325–335

Müller D, Kluge M, Gröschel-Stewart U (1982) Comparative studies in immunological and molecular properties of phosphoenolpyruvate carboxylase in species of *Sedum* and *Kalanchoe* performing crassulacean acid metabolism (CAM). Plant Cell Environ 5:223–230

Nato A, Vidal J (1983) Phosphoenolpyruvate carboxylase activity in relation to physiological processes during the growth of cell suspension cultures from *Nicotiana tabacum*. Physiol Veg 21:1031–1039

Ouchterlony H (1958) Diffusion in gel methods for immunological analysis. Prog Allergy 5:1–78

Perrot C, Vidal J, Burlet A, Gadal P (1981) On the cellular localization of phosphoenolpyruvate carboxylase. Planta (Berl) 151:226–231

Perrot-Rechenmann C, Vidal J, Brulfert J, Burlet A, Gadal P (1982) A comparative immunocytochemical localization study of phosphoenolpyruvate carboxylase in leaves of higher plants. Planta (Berl) 155:24–30

Robertson A, Kerr HW (1971) Properties of PEPC isolated from Maize leaves. Biochem J 125:34

Uedan K, Sugiyama T (1976) Purification and characterization of phosphoenolpyruvate carboxylase from maize leaves. Plant Physiol (Bethesda) 57:906–910

Vidal J, Gadal P (1983) Influence of light on phosphoenolpyruvate carboxylase in *sorghum* leaves I. Identification and properties of two isoforms. Physiol Plant 57:119–123

Vidal J, Cavalié G, Gadal P (1976) Etude de la phosphoenolpyruvate carboxylase du haricot et du sorgho par électrophorese sur gel de polyacrylamide. Plant Sci Lett 7:265–270

Vidal J, Godbillon G, Gadal P (1980) Recovery of active, highly purified phosphoenolpyruvate carboxylase from specific immunoadsorbent column. FEBS Lett 118:31–34

Vidal J, Crétin C, Gadal P (1983a) The mechanism of photocontrol of phosphoenolpyruvate carboxylase in *sorghum* leaves. Physiol Vég 21:977–986

Vidal J, Godbillon G, Gadal P (1983b) Influence of light on phosphoenolpyruvate carboxylase in *sorghum* leaves II. Immunochemical study. Physiol Plant 57:124–128

Vidal J, Nguyen J, Perrot-Rechenmann C, Gadal P (1986) De novo synthesis of phosphoenolpyruvate carboxylase during development of soybean root nodules. Planta 167:190–195

Immunochemical Methods for Higher Plant Nitrate Reductase

A. Kleinhofs, K. R. Narayanan, D. A. Somers, T. M. Kuo, and R. L. Warner

1 Introduction

Immunochemical methods provide excellent probes for the detection and study of virtually all cells or cell components. As such, these techniques have been and continue to be used for study of macromolecule structure, function, and evolutionary relationships. Immunochemical methods can be conducted with crude or partially purified extracts permitting the study of labile components in or near native condition. This latter aspect of immunochemical methods has been very useful in our studies on higher plant nitrate reductase, a protein that is still difficult to obtain in pure and undegraded form from most plant species.

In this paper we describe the techniques of nitrate reductase purification, antiserum preparation, and the use of antiserum in genetic, physiological and evolutionary studies of nitrate reductase. The techniques described have proven to be useful and reliable in our laboratory. Immunochemical methods, however, are rapidly evolving and being improved. While newer techniques may be available, only techniques we have used or are developing will be discussed in this paper.

2 Nitrate Reductase Purification

The purification of the antigen constitutes the first important step in obtaining monospecific polyclonal antiserum. This has been a problem with higher plant nitrate reductases (NR). We have spent considerable effort in characterizing the physiological conditions of the plant growth and the extraction conditions for obtaining pure NR in undegraded state. The conditions that we have found to be most useful are described in this section.

2.1 Seedling Growth Conditions

Nitrate reductase in plant cells is subjected to a rapid turnover (Zielke and Filner 1971) and it is labile in plant tissue extracts due to tissue complexity, proteolytic degradation, and the instabilitiy of NR molecular structure (Beevers and Hageman 1969; Hewitt 1975). The level of extractable NR also depends on growth temperature, light intensity, and water stress (Hageman and Hucklesby 1971; Nicholas et al. 1976; Tischler et al. 1978; Travis et al. 1969). Optimal seedling growth conditions are necessary to provide both a high level and good stability of NR for purification.

In barley (*Hordeum vulgare*), the in vitro stability of NR decreases as a function of seedling age and growth temperature (Kuo 1979; Brown et al. 1981). We grow barley seedlings at 18 °C in plastic pans containing vermiculite under continuous illumination provided by cool, white, fluorescent tubes and incandescent bulbs. The quantum flux (400–700 nm) at plant height is 300 $\mu E\ m^{-2}\ s^{-1}$. Plants are watered daily for 5 days by subirrigation with deionized water or with a nitrate-free nutrient solution (Warner and Kleinhofs 1974) containing 4 mM $CaSO_4$, 2.5 mM $(NH_4)_2SO_4$, 2.5 mM K_2SO_4, 3 mM $MgSO_4$, 1 mM KH_2PO_4, 15 mg l^{-1} sodium ferric diethylene triamine pentaacetate and 1.5 ml l^{-1} micronutrients (70 µM H_3BO_3, 40 µM $MnCl_2$, 2 µM $CuSO_4$, 5 µM $ZnSO_4$, 0.29 µM $(NH_4)_6Mo_7O_{24}$). To induce NR, the seedlings are watered with a nutrient solution containing 15 mM nitrate 24 h and 12 h before harvest of leaf tissues. The nitrate solution contains 5 mM $Ca(NO_3)_2$ and 5 mM KNO_3 replacing $CaSO_4$, $(NH_4)_2SO_4$, and K_2SO_4 in the nitrate-free solution.

2.2 Extraction

The primary leaves are excised above the coleoptiles. The excised seedling leaves are frozen immediately in liquid N_2 and ground to a powder in a Waring blender containing liquid N_2. The Waring blender is equipped with a vent hole to prevent pressure build-up within the container. After evaporating the liquid N_2, the powder is stirred gently for 30 min in ice-cold extraction buffer (0.25 M Tris-HCl, 1 mM EDTA, 1 µM Na_2MoO_4, 5 µM FAD, 3 mM DTT, 10 µM antipain, 250 µM PMSF, 1 µg ml^{-1} pepstatin, 1.5% casein w/v, pH 8.2) at a ratio of 1 g seedling fresh weight to 4 ml buffer. The brei is filtered through cheesecloth and centrifuged at 12000 × g for 30 min. The supernatant is decanted, slowly adjusted to 50% saturation with solid $(NH_4)_2SO_4$ and the precipitate collected by centrifugation. The pellets can be used immediately for further purification or stored frozen at −20 °C.

The extraction buffer components were optimized by studying the in vitro stability of NR in crude extracts (Kuo et al. 1982b). Although the exogenous protein and the protease inhibitors did not contribute significantly to the enzyme stability as measured by activity decay in crude extracts, we have not been able to obtain good NR purification when these components were left out of the extraction buffer. The use of antipain and exogenous protein was particularly important. Bovine serum albumin (BSA) could be substituted for casein in most cases, but for purification, casein was preferred since high BSA concentrations interferred with the binding of NR to the Blue A Sepharose affinity gel. The use of plant material induced with nitrate for 24 h was much superior to the use of seedlings grown continuously on nitrate. Although the latter had higher NR activity, the NR was partially degraded even in our extraction buffer (Fig. 1).

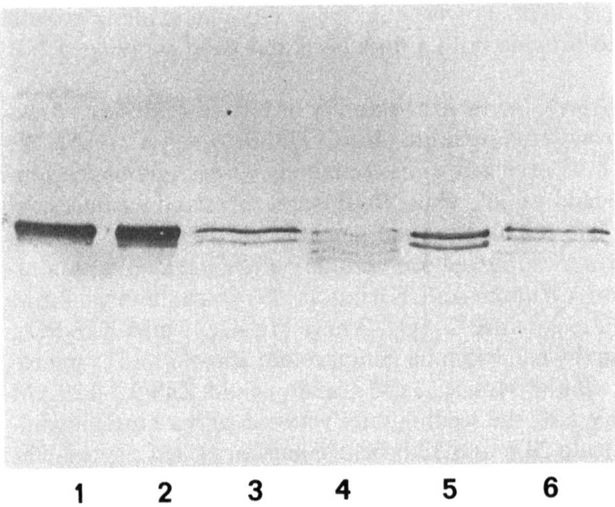

Fig. 1. Effects of growth conditions and extraction buffer on the apparent subunit nature of nitrate reductase. Wild-type barley (cv. Steptoe) seedlings were grown on nitrate-containing nutrient solution for 6 days or in deionized water for 5 days and induced with nitrate-containing nutrient solution 24 h and 12 h prior to extraction. NR was extracted, subjected to SDS-PAGE, and analyzed by Western blot techniques as described in text. *1*, *2* Nitrate-induced seedlings, complete extraction buffer; *3* seedlings grown continuously on nitrate, complete extraction buffer; *4* seedlings grown continuously on nitrate, extraction buffer without protease inhibitors and casein; *5* seedlings grown continuously on nitrate, extraction buffer without casein; *6* seedlings grown continuously on nitrate, extraction buffer without protease inhibitors

2.3 Affinity Chromatography

The $(NH_4)_2SO_4$ pellets are dissolved in a volume of column buffer (25 mM Tris-HCl, 1 mM EDTA, 1 μM Na_2MoO_4, 1 μM FAD, 1 mM DTT, 10 μM antipain, 250 μM PMSF, 1 μg/ml pepstatin, pH 8.2) equal to 10% of the original extraction buffer. The solution is clarified by centrifugation, the supernatant decanted through two layers of Miracloth and desalted on a G-25 Sephadex column equilibrated with the column buffer minus antipain, PMSF, and pepstatin. About twice the applied volume is collected beginning with the sample front and loaded immediately on a Blue A Sepharose (Amicon) column (3.8×22.5 cm) preequilibrated with column buffer containing BSA (10 μg ml^{-1}). The loaded Blue A Sepharose column is washed overnight with wash buffer (25 mM Tris-HCl, 1 mM EDTA, 10 μM Na_2MoO_4, 10 μM FAD, 1 mM DTT, 1 μg ml^{-1} BSA, pH 8.2). The NR is eluted using a linear 0–50 μM NADH gradient in elution buffer (10 mM Tris-HCl, 1 mM EDTA, 1 μM Na_2MoO_4, 1 μM FAD, 1 mM DTT, pH 8.2). The fractions are collected in tubes containing sufficient 1 M KNO_3 to give a final concentration of 10 mM. All steps in NR purification are carried out at 4 °C. Fractions containing the NR activity are dialyzed against chilled glycerol for 1 h and stored at −20 °C. Alternatively, the fractions are concentrated using ultrafiltration, frozen in small pellets with liquid N_2, and stored in liquid N_2. Results of a typical purification are presented in Table 1.

Table 1. Affinity purification of nitrate reductase from barley seedlings

Step	Total activity (μmol min^{-1})	Total protein (mg)	Specific activity (μmol min^{-1} mg^{-1})	Purification (Fold)	Yield (%)
1. Homogenate	20.7	2592[a]	0.008	1	100
2. 0–50% $(NH_4)_2SO_4$	13.8	810	0.02	3	70
3. Blue A Sepharose	8.2	1.1	7.5	932	39

[a] Includes the exogenous protein contained in the extraction buffer.

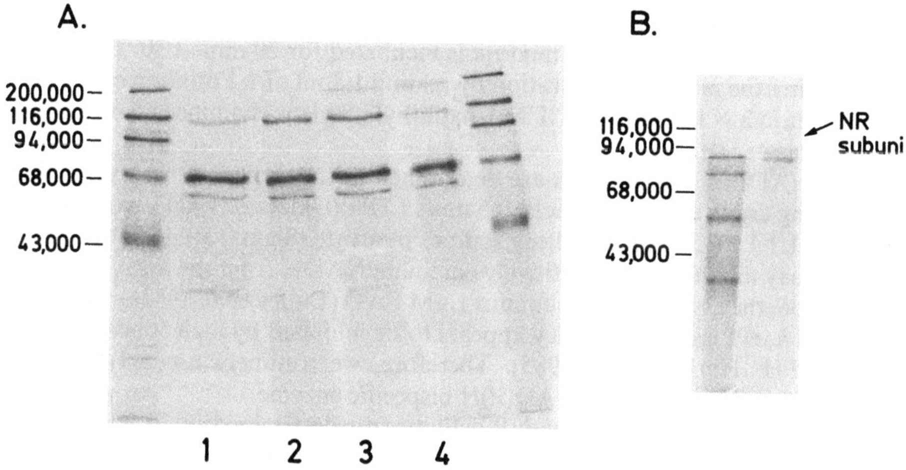

Fig. 2. SDS-PAGE analysis of fractions from a blue A-Sepharose affinity column (**A**) and NR purified by preparative gel electrophoresis (**B**). Nitrate reductase from barley was extracted and purified on a blue A-Sepharose column as described in text. Aliquots of the nitrate reductase peak were lyophilized and analyzed by SDS-PAGE. Gels were stained with Coomassie Brilliant Blue R-250. *1* Start of the NR peak; *2* and *3* NR activity peak; *4* tail of NR peak. Fractions *2* and *3* were pooled, concentrated, and the NR purified by native-PAGE as described in the text. Part of the gel slice containing the electrophoretically purified NR was analyzed by SDS-PAGE. The molecular weight markers are: myosin, 200 000; β-galactosidase, 116 000; phosphorylase b, 94 000; bovine serum albumin, 68 000; ovalbumin, 43 000

The Cibacron Blue F3GA dye linked to various inert polymers has been the most widely used affinity ligand for NR purification. The chromophore binds proteins that possess a "dinucleotide fold" supersecondary structure (Thompson et al. 1975) such as NR (Solomonson 1975). We have used blue dextran (Kuo et al. 1980) and Blue A Sepharose (Kuo et al. 1982a; Somers et al. 1983a) to purify barley NR. The Blue A Sepharose is preferred for large-scale purification because of its higher binding capacity. The Blue A Sepharose purified barley NR is, however, only partially pure (Fig. 2 A). Pure NR is obtained by preparative gel electrophoresis (Fig. 2 B, Sect. 3.1) or immunological procedures (Sect. 5.3).

2.4 Assays

NR activity in higher plant extracts is usually measured by determining the nitrite concentration in the assay medium. NADH or NADPH (Beevers and Hageman 1969; Dailey et al. 1982b; Jolly et al. 1976) are used as specific electron donors. In addition, NR can accept electrons from $FMNH_2$ or reduced viologens to reduce nitrate. In our laboratory, NADH NR activities are determined as described by Warner and Kleinhofs (1974). The 2.0 ml final assay mixture contains 25 mM potassium phosphate, pH 7.5, 10 mM KNO_3, 0.2 mM NADH, and an appropriate amount of enzyme (usually in 0.1 ml) which is added to initiate the reaction at 30 °C for 15 min. The control reaction mixture contains no electron donor. The reaction is terminated by adding 0.2 ml of 1:1 mixture of 1 M zinc acetate and 0.3 mM phenazine methosulfate to remove the excess NADH from the assay medium (Scholl et al. 1974). This mixture is incubated for 20 min at 30 °C prior to determining the nitrite concentration by adding 1.8 ml of 1:1 mixture of 1% sulfanilamide in 3 N HCl and 0.02% N-1-naphthyl-ethylene-diamine and measuring absorbance at 540 nm.

The NADPH NR activities are determined in the same manner except that when using crude enzyme extracts, 0.5 units LDH (L-lactate:NAD^+ oxidoreductase E.C.1.1.1.27, Sigma) and 1 mg sodium pyruvate (Sigma) are included in the 2.0 ml assay mixture to competitively eliminate NADH from the assay medium. In addition, the assay mixture contains 1 μM FAD (Dailey et al. 1982a). NADPH NR from *nar*1a mutant of barley appears to be inhibited by high concentrations of NADPH (Harker et al. 1985). Therefore, we routinely use only 100 μM NADPH to assay the barley NAD(P)H bispecific enzyme.

The reduced methyl viologen NR activities are determined as the NADH NR except that the 2.0 ml assay mixture contains 0.2 mM methyl viologen in place of NADH and 3.2 mM sodium hydrosulfite ($Na_2S_2O_4$). The enzyme is added to the assay mixture, incubated at 30 °C for 1 min and the reaction initiated with 0.1 ml of sodium hydrosulfite (14 mg ml^{-1} in 25 mM phosphate, pH 8.2, prepared fresh prior to use) (Dailey et al. 1982b). The assay is stopped after 30 min by mixing vigorously (ca. 10 s) to oxidize the methyl viologen. To reduce interference from substances derived from sodium hydrosulfite, 0.2 ml of 1.5% (v/v) formaldehyde in H_2O is added and incubated for 5 min (Senn et al. 1976). Nitrite determinations are made as above. Controls containing all of the reaction components and the enzyme are stopped immediately upon the addition of the sodium hydrosulfite.

$FMNH_2$ NR activity is determined as described by Schrader et al. (1968). The 2.0 ml final assay mixture contains 25 mM potassium phosphate, pH 7.5, 10 mM KNO_3, 0.8 mM FMN, an appropriate amount of enzyme, and 3.2 mM sodium hydrosulfite ($Na_2S_2O_4$) which is added to initiate the reaction at 30 °C. The reaction is stopped after 15 min incubation by mixing vigorously for 10 s. Formaldehyde is used to reduce interference with the nitrite color reaction and controls are prepared as above.

NR also possesses cytochrome c reductase (CR) activity which is measured as described by Wray and Filner (1970). Crude extracts for NADH CR assay are made as for NR assay except that DTT in the extraction buffer is reduced to

1 mM to minimize interference with the CR assay. The 1.0 ml final assay mixture contains 50 mM potassium phosphate, pH 7.5, 1 mg cytochrome c (horse heart), 0.2 mM NADH, and 0.1 ml of an appropriately diluted enzyme which is added to initiate the reaction. The rate of cytochrome c reduction is monitored by absorbance changes at 550 nm. (The millimolar extinction coefficient for the reduced horse heart cytochrome c at 550 nm is 29.5.) The control reaction mixture contains no NADH.

Protein concentrations are measured by either the Lowry et al. (1951) or Bradford (1976) assays. Bradford dye-binding assay solution is commercially available.

3 Production of Nitrate Reductase Antiserum

3.1 Preparation of Nitrate Reductase

The NR obtained from the affinity column is purified to homogeneity by nondenaturating preparative slab gel PAGE (Davis (1964). One ml of affinity purified NR (ca. 0.54 mg protein) is applied to a 130 mm trough in a 1.5 mm slab gel. Both the stacking [2.5% acrylamide, 0.625% N′-N′-methylene-bis-acrylamide (BIS)] and resolving (7% acrylamide, 0.184% BIS) gels are prepared in 378 mM Tris-HCl, pH 8.9. Bromophenol blue is added to the upper buffer reservoir and the electrophoresis started at 10 mA per slab gel. The current is increased to 15 mA per slab after the tracker dye enters the resolving gel and is maintained at 15 mA until the tracker dye reaches the bottom of the gel (10 cm). All steps of the purification are carried out at 4 °C.

NR is identified by staining slab gels for reduced methyl viologen NR activity (Lund and DeMoss 1976). Reduced methyl viologen (purple) donates electrons to NR driving the reduction of nitrate to nitrite, becoming oxidized (colorless). This provides a nondestructive assay for NR activity in gels. Slab gels are incubated in 50–100 ml of 10 mM methyl viologen in 50 mM K phosphate, pH 7.5, for 10–15 min at room temperature on a slow reciprocal shaker platform. Methyl viologen is reduced by adding 2 mg sodium hydrosulfite per ml of reaction solution. The gel and solution are gently shaken for 10 min prior to addition of 0.2 ml 1 M KNO_3 per ml reaction mixture. The gel is continuously shaken in the solution. A clear band in the purple reduced methyl viologen background is observed after 5–15 min incubation. The clear band is excised from the gel. Methyl viologen is eluted from gel slices by washing in distilled water for 1 h. A portion of the gel slice is subjected to SDS-PAGE to determine purity of the NR (Fig. 2 B). The gel is frozen in liquid N_2, lyophilized, and stored at −20 °C until used.

3.2 Immunization

Gel slices containing pure NR protein are used to immunize rabbits for antiserum production. The gel slices are pulverized using a mortar and pestle. Thirty mg of the powder are emulsified in 1.5 ml of 1:1 mixture of distilled water and Freund's adjuvant using a Sorvall Omnimixer at maximum speed for 15 min at room temperature. French Lop rabbits are injected interdermally at multiple sites four times at 10 day intervals using NR purified from two slab gels (ca. 200 µg protein) per rabbit. The first injection is made with Freund's complete adjuvant and subsequent injections with Freund's incomplete adjuvant. Preimmune control serum is obtained from the rabbit before immunization.

3.3 Serum Collection and Isolation

Titer and monospecificity are determined on a small antiserum sample obtained by bleeding from an ear vein. Large-scale antiserum collection is carried out by terminating the rabbit. We found that repeated booster injections led to antiserum production that was no longer monospecific for NR presumably due to contaminating antigens that were introduced at low levels during the booster injections resulting in the eventual elicitation of non-NR antibodies. Rabbits are anesthetized by intramuscular injection of Innovar Vet (Pitman Moore Co.). Blood is isolated by cardiac puncture using a 18 ga (gauge) needle fitted to a 50 ml disposable syringe. Up to 150 ml of blood can be collected per rabbit prior to termination. Blood is allowed to clot in a glass beaker for 1–2 h at room temperature and 5 h at 8 °C. Serum is isolated by centrifugation at 2000 × g for 15 min and the supernatant stored at −20 °C. No further treatment of the antiserum was required for immunological studies.

4 Characterization of Nitrate Reductase Antiserum

4.1 Antiserum Titration

Antiserum titer was determined by the levels of antiserum required to inhibit NR-associated activities in crude extracts of wild-type barley grown with or without nitrate. Dilutions of antiserum and preimmune control serum in 100 µl of 50 mM K-phosphate buffer, pH 7.5, are added to 1 ml of wild-type barley extract in 1.5 ml Eppendorf microcentrifuge tubes. The samples are incubated for 1.5 h at 4 °C, centrifuged at 12000 × g for 5 min and supernatants assayed for NADH NR, $FMNH_2$ NR, and NADH CR activities. Antiserum produced against purified barley NR inhibited the in vitro NADH NR activity of extracts from nitrate-induced wild-type seedlings (Fig. 3). The associated $FMNH_2$ NR and NADH CR activities of the extract were also inhibited by the antiserum.

Inhibition was not dependent on centrifugation or precipitation indicating that the antibodies directly bind and inactivate NR-associated activities. Preimmune control serum had no effect on NR-associated activities (Fig. 3). The $FMNH_2$ NR activity of the wild-type enzyme was inhibited with slightly less antiserum than the NADH NR of the same extract. About 20% of the NADH CR

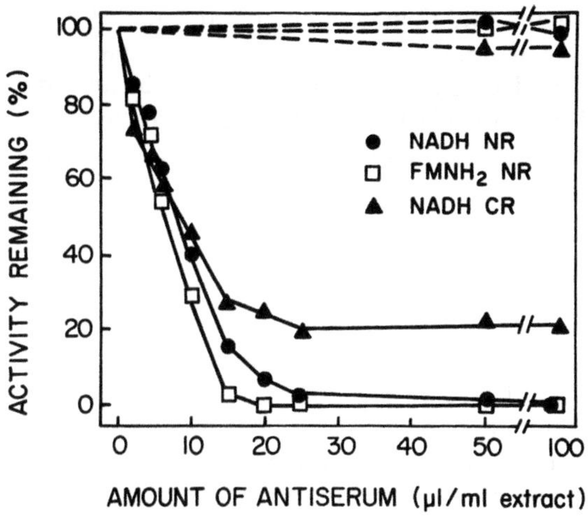

Fig. 3. Inhibition of nitrate reductase and associated activities by preimmune control (---) and anti NR (——) sera. Extracts were prepared from 6-day-old nitrate-induced barley (cv. Steptoe) seedlings. One g of seedlings was ground in a mortar and pestle in 6 ml extraction buffer containing 0.25 M Tris-HCl, 1 mM EDTA, 3 mM DTT, 5 µM FAD, 1 µM Na_2MoO_4, 2 µM antipain, 1 µg ml^{-1} pepstatin, 1 mM PMSF, 3% (w/v) BSA, pH 8.2. The brei was centrifuged 30000 × g 20 min and the supernatant used as the enzyme source. Reaction with serum was as described in text. Initial enzyme activities were *NADH NR*, 0.092 units ml^{-1}; *FMNH₂ NR*, 0.082 units ml^{-1}; and *NADH CR*, 0.641 units ml^{-1}

activity was not inhibited even by 25 µl antiserum per ml of extract. This NADH CR activity is probably not related to NR and appears to be microsomal in origin. The NADH CR in the cytosolic fraction (after centrifugation at 120000 × g of wild-type barley was completely inhibited by NR antiserum (Narayanan et al. 1983).

4.2 NR Antiserum Monospecificity

4.2.1 Immunodiffusion

The most common immunodiffusion test for qualitative analysis of antigen-antibody reactions is the Ouchterlony double diffusion technique (Ouchterlony 1968) in which both antigen and antibody are free to diffuse in a semi-solid agar medium to form precipitin lines. The technique allows analyses for the titer of antiserum, the purity of antigen used to raise antiserum, the immuno-cross-reactivity of NRs from different plant species (Nakagawa et al. 1984; Smarrelli and Campbell 1981) and from NR-deficient mutants (Kuo 1979). In general, a 1% Noble agar (Difco) slab is prepared in a 0.1 M sodium barbital, pH 8.5, containing 0.9% (w/v) NaCl and 1:10000 (w/v) thimersol. Antiserum is placed in the center well (2–3 mm diam) which is surrounded by adjoining outer wells of the same size filled with different NR samples. NR and antibody are allowed to develop precipitin bands for 24–48 h. The agar slab is rinsed first with saline solution (0.9%, w/v NaCl) to remove unreacted proteins and then with deionized water. Precipitin bands are stained with 0.5% (w/v) of amido black or Coomassie Brilliant Blue R-250 in a solution of acetic acid:methanol:H_2O (10:45:45) for 15–30 min and then destained with the acetic acid-methanol-H_2O solution. Precipitin bands can also be stained with Crowle's Double Stain solution (Polysciences) and subsequently destained with deionized water.

4.2.2 Crossed Immunoelectrophoresis

This technique is extremely sensitive for determining antiserum monospecificity because proteins are separated by agarose gel electrophoresis before being subjected to a second-dimension immunoelectrophoresis into an antiserum containing gel. If multiple immunoprecipitates are observed, their relationship can be determined by observing the interaction between the overlapping immunoprecipitates. A complete cross of immunoprecipitates indicates no cross-reaction of two proteins and that the serum is polyspecific. Partial identity is indicated by a spur in a fused cross and complete identity by complete fusion. If only related immunoprecipitates are observed from crossed immunoelectrophoresis of crude extracts and antiserum, the antiserum is monospecific (Otterness and Karush 1982).

NR extracts are first separated by agarose gel electrophoresis. Agarose (1% w/v, Bio-Rad) is melted by boiling in Tris-barbital buffer (73 mM Tris, 24.3 mM diethylbarbituric acid, 1 mM Ca lactate, and 2 mM NaN_3, pH 8.6) and cooled to 50 °–55 °C before pouring into gel molds. After 15–20 min cooling at room temperature, the gaskets are pried away from the solidified gel and one glass plate slid off the gel. Sample wells 3 mm in diam are cut across the narrow end of the gel about 10–15 mm from the gel edge. The glass plate and gel are placed on a horizontal electrophoretic unit with sample wells nearest the cathode buffer reservoir. The gel is cooled to 4 °C by a circulating water bath. Electrode wicks, constructed of Whatman 3 mm filter paper are laid on both ends of the gel. The wicks are saturated with Tris-barbital buffer and submerged in the buffer reservoirs. A volume of 9 µl of crude extract or purified NR is applied to the wells. A 2 µl aliquot of 0.1% bromophenol blue is spotted on the gel in line with the wells.

Immediately following sample application, the current is turned on and electrophoresis performed at 5 V cm^{-1} until the bromophenol blue has migrated to the anode. Electrophoresis is terminated and the gel cut into 10 mm slices parallel to the direction of electrophoresis. Meanwhile, a 0.8% w/v agarose gel is prepared as above except NR antiserum is added prior to pouring the gel. A 1% agarose solution in Tris-barbital is used to bond the gel slice containing the separated proteins to the antiserum gel. The same agarose solution is also used to construct an electrode bridge on the side where the gel slice is bonded. Electrode wicks are wetted with Tris-barbital buffer and attached to the gel such that the first dimension gel slice is nearest the cathode. Immunoelectrophoresis is conducted at 1 V cm^{-1} for 20 h at 4 °C.

After immunoelectrophoresis the gels are soaked in several changes of 0.1 M NaCl for 48 h to elute proteins that are not immunoprecipitated. Immunoprecipitates are insoluble and are retained in the gel even during extended washing. NaCl

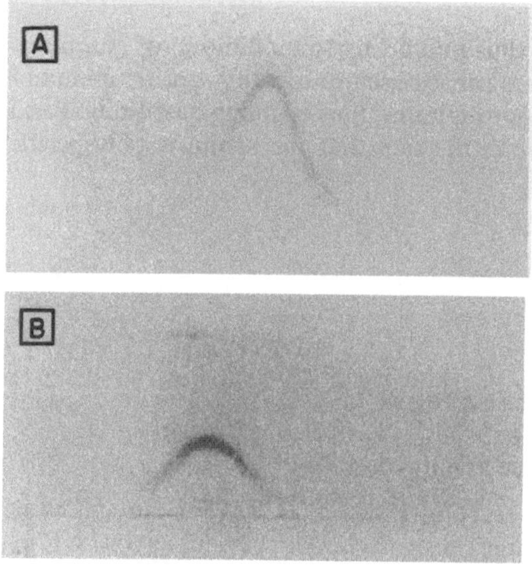

Fig. 4. Crossed immunoelectrophoresis of blue A-Sepharose purified barley NR (**A**) and crude extract from barley seedlings (**B**). Samples were separated on agarose in the first dimension and then subjected to electrophoresis into the antiserum containing gel as described in the text. **A** 0.006 units NR, 0.3% (v/v) antiserum in gel; **B** 0.005 units NR, 0.4% (v/v) antiserum in gel

is eluted from the gel by one wash in distilled water and the gels are dried by blotting with paper towels. Gels are transferred to the hydrophilic side of Geld Bond[R] (FMC) film and dried at 50 °–60 °C. Immunoprecipitates are visualized by staining the dried gels for 5–10 min in 0.1% Coomassie Brilliant Blue R-250 dissolved in 8% (v/v) glacial acetic acid and 25% (v/v) methanol. Destaining is in the same solution without the stain.

Single, rocket-shaped immunoprecipitates resulted from the interaction of the NR antiserum and NR in the affinity gel purified preparation and crude extracts showing that the antiserum was monospecific (Fig. 4). The crossed immunoprecipitates retained NR-associated NADH diaphorase and methyl viologen NR activities when assayed immediately after electrophoresis indicating that the precipitated protein was NR. The gels are stained for diaphorase by incubating for 30 min at room temperature in 100 mM K phosphate buffer, pH 7.5, containing 1 mg NADH and 1 mg p-iodonitrotetrazolium violet (Sigma) per 3 ml of buffer. A red staining band corresponding with the immunoprecipitate in the gel indicates that the antigen has diaphorase activity. Methyl viologen NR staining is carried out as described for the preparative PAGE of purified NR. A clear band of nitrate specific oxidation of reduced methyl viologen indicates the protein is NR.

It seems contradictory that while the antiserum inhibits NR associated activities in the titration experiments, the immunoprecipitates in gels have enzymatic activity. Similar results have been reported by Amy and Garrett (1979) with *Neurospora crassa* NR immunoprecipitates. Apparently the three-dimensional antigen-antibody matrix of the immunoprecipitate within the agarose gel restricts the number of antigen determinants bound per NR molecule. Therefore, a certain subset of enzymatically active sites may be left unbound and free to react with the substrates. Titration experiments, performed in solution, presumably do not restrict antibody-antigen interactions thus permitting full inhibition of enzymatic activities. Enzyme staining techniques can be used to identify specific immunoprecipitates among complex immunoprecipitates, thus enabling quantitation and characterization of the antigen despite the fact that the serum is polyspecific (Owen and Smyth 1977).

5 Applications

5.1 Quantification of Nitrate Reductase Antigen

5.1.1 Protection of Nitrate Reductase Inactivation

The amount of NR-specific cross-reacting material (CRM) in an unknown sample can be estimated by determining the protection it provides against NR inactivation with limiting quantities of antiserum. Aliquots of diluted antiserum are mixed with 0.5 ml of the sample to be assayed and incubated for 1.5 h on ice in a total volume of 0.6 ml. An aliquot (0.5 ml) of NR is added and the mixture in-

Table 2. Nitrate reductase cross-reacting material (NR CRM) in nitrate reductase-deficient barley mutants

Selection no.	Gene designation	NR CRM[a]		
		−CRM	+CRM	Antigenicity[b]
		% of control		
Group I				
Az 12	*nar* 1a	0	11	4
Az 13	*nar* 1b	0	38	64
Az 23	*nar* 1c	0	29	18
Az 29	*nar* 1e	0	8	3
Az 30	*nar* 1f	0	8	4
Az 31	*nar* 1g	0	46	77
Az 33	*nar* 1i	0	28	49
Group II				
Az 28	*nar* 1d	136	128	138
Az 32	*nar* 1h	128	137	125
Group III				
Az 34	*nar* 2a	32	43	46
Xno 18	*nar* 3a	54	67	–
Xno 19	*nar* 3b	51	47	–
Xno 29	*nar* 1j	20	20	–

[a] NR CRM was estimated by rocket immunoelectrophoresis either in the presence of purified NR (+CRM) or with no additives (−CRM).

[b] Antigenicity was determined by protection of nitrate reductase inactivation assays.

cubated for an additional 1.5 h. Controls are prepared in the same manner, but containing no antiserum. The reaction mixture is centrifuged and the supernatant immediately assayed for NADH NR activity. A standard curve is established for each experiment under identical conditions by reacting the NR used in the protection assay directly with different aliquots of antiserum. These experiments were used to compare the NR-specific CRM in extracts of the induced wild-type NR with the CRM in NR-deficient mutants (Table 2) (Kuo et al. 1981). The difference between the quantities of antiserum required for 50% inactivation of a constant amount of NR in the presence and absence of test sample is used as an indication of the antigenicity of the test sample. This value is dependent upon either the amount of antigen of the relative antigenicity of the antigen or both in the assayed extract.

5.1.2 Rocket Immunoelectrophoresis

Rocket immunoelectrophoresis is similar to crossed immunoelectrophoresis except the antigen is not subjected to the first-dimension agarose electrophoresis, but is subjected to electrophoresis directly into an antiserum containing gel. As the antigen enters the antibody gel, an arc-shaped precipitin matrix of antigen-antibody is formed. Initially at antigen excess, the antigen continues to migrate into the gel until all antigen is complexed in an insoluble precipitin matrix which no longer migrates in the electrophoretic gel (Fig. 5). Since the antigen-antibody ratio is a constant for a specific protein, the amount of antigen in the test sample determines the migration distance of the rocket immunoprecipitate at equilibrium, i.e., the height of the rocket.

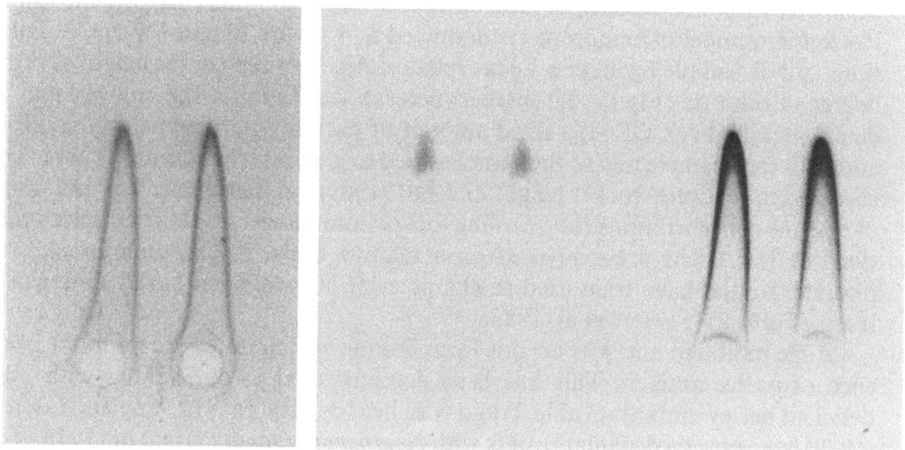

Fig. 5. Rocket immunoelectrophoresis of barley NR. Affinity column purified NR was subjected to electrophoresis into NR antiserum containing gel as described in text. The immunoprecipitates were stained for: protein with Coomassie Brilliant Blue R-250 (*left*); NR activity with nitrite stain (*middle*); and NR associated diaphorase activity with p-iodonitrotetrazolium violet and NADH (*right*). Nitrate was detected by 1:1 mixture of 1% sulfanilamide in 3 N HCl and 0.02% N-1-naphthyl-ethylenediamine. Diaphorase was stained as described in text

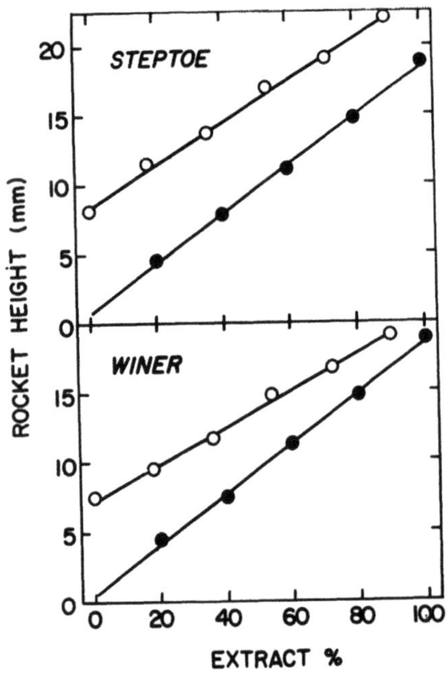

Fig. 6. Relationship of rocket immunoprecipitate height to dilution of barley (cv. Steptoe and Winer) seedling extracts. One g of seedlings was ground in a mortar and pestle in 1 ml extraction buffer containing 0.5 M Tris-HCl, 1 mM EDTA, 3 mM DTT, 25 μM FAD, 5 μM Na_2MoO_4, 10 μM antipain, 5 μg ml^{-1} pepstatin, 5 mM PMSF, 3% BSA, pH 8.6. The brei was centrifuged 30000 × g 20 min and the supernatant used as enzyme source. Electrophoresis was as in text. Rocket heights were measured as precisely as possible to the nearest mm. The extracts were diluted (●) or diluted and mixed with 0.0015 units of purified NR (○)

Agarose gels (0.8% w/v) containing 0.4% (v/v) NR antiserum in Tris-barbital buffer (pH 8.6) are poured into 1.5 × 70 × 100 mm molds. Sample wells 3 mm in diam are cut across the cathode end of the gel and 9 μl of sample are applied to each well while the electrophoretic current is on to minimize sample diffusion. Rocket immunoelectrophoresis is conducted at 4 °C for 20 h at 1 V cm^{-1}. Dilutions of NR sample result in a linear relationship between rocket height and the degree of dilution (Fig. 6). To further increase sensitivity of the analysis and to detect modified NR CRM, a fixed amount of partially purified wild-type NR is added to each sample and to the samples used to generate the standard curve. The relationship between rocket height and NR content is linear over a broad range of enzyme concentrations (Fig. 6). The rocket immunoelectrophoretic determinations of NR CRM are expressed on a relative basis. Rocket immunoelectrophoretic studies have been used to characterize NR-deficient barley mutant extracts (Table 2) (Somers et al. 1983a).

Some modified antigens do not form immunoprecipitates except in the presence of native antigen. This has been demonstrated in our studies with NR-deficient barley mutants (Table 2) and with heat denatured NR. Amy and Garrett (1979) have observed similar results with *Neurospora crassa* mutant *nit*-1 NR. The modified antigen-antibody complexes formed with denatured NRs are most likely soluble and migrate out of the immunoelectrophoretic gels. The added wild-type NR forms an immunoprecipitate which probably facilitates the altered NR CRM to interact with the complex in such a way as to increase the rocket height.

We have also used rocket immunoelectrophoresis to study the evolutionary relationships of NR from nine different crop species (Snapp et al. 1984). Results showed that NR from monocotyledonous species were structurally similar, but not identical to barley NR. The NR from dicotyledonous species did not form rockets with barley NR antiserum in the absence of added barley NR. These results indicated that the dicotyledonous species' NR structure was considerably different from the barley NR. NR activity from all species tested was inactivated by barley NR antiserum indicating that the active site(s) antigenic properties have been conserved.

5.1.3 Other Methods

Rocket immunoelectrophoretic techniques as described above provide sensitivity at the submicrogram level. Other immunoassay methods, such as radioimmunoassay and the enzyme-linked immunosorbent assay detect antigens at picogram or nanogram levels. The higher sensitivity of these immunoassays is acquired from using radioisotopes (e.g., ^{125}I, ^{3}H) or enzyme-linked (e.g., β-galactosidase, peroxidase, phosphotase, elastase) antigen-antibody reactions. The amount of a labeled or enzyme-linked (marked) antigen bound by a fixed level of antibody is inversely proportional to the amount of unmarked antigen present to compete with it. For a detailed description of these immunoassay techniques, readers are referred to other articles (Van Vunakis and Langone 1980).

5.2 Western Blot

The procedure, wherein proteins separated by polyacrylamide gel elctrophoresis are transferred onto nitrocellulose sheets and subsequently detected immunologically, is commonly referred to as Western blot (Renart et al. 1979; Towbin et al. 1979). This procedure has a high resolution and sensitivity and is ideally suited for the investigation of the structure and physiology of proteins in biological systems. The Western blot procedure consists of four steps: PAGE, transfer of proteins to nitrocellulose sheets, treatment with antiserum, and detection of antigen-antibody complex.

5.2.1 Native Polyacrylamide Gel Electrophoresis

Nitrate reductase extracts are separated by discontinuous PAGE (Davis 1964) using 5–30% linear gradient. The Davis separating and stacking gel buffers are supplemented with 1 mM EDTA, 10 µM sodium molybdate, 1 µM FAD, and 1 mM DTT. The tank buffer also contains FAD (0.1 µM), DTT (0.1 mM), and sodium molybdate (1 µM). Electrophoresis is peformed at 4 °C. Details of PAGE are as in Sect. 3.1. After the native PAGE, the gels are electroblotted (Sect. 5.2.3; Towbin et al. 1979) for immunodetection.

5.2.2 Sodium Dodecyl Sulfate-Polyacrylamide Gel Electrophoresis (SDS-PAGE)

NR is denatured with SDS and 2-mercaptoethanol and separated by PAGE (Laemmli 1970). The extracts (crude or affinity purified) are adjusted to 4% (w/v)

SDS and 5% (v/v) 2-mercaptoethanol and placed in boiling water bath for 5 min. SDS-PAGE is carried out in 1.5 mm gels (7.5% acrylamide, 0.1% BIS) at 4 °C. After SDS-PAGE, the gels are electroblotted (Sect. 5.2.3) onto nitrocellulose sheets for immunodetection of NR. If molecular weight markers are included in the gel, the appropriate region in the nitrocellulose sheet (usually either end) is cut and stained with amido black. Amido black is easier to destain from nitrocellulose sheets than Coomassie blue.

5.2.3 Electrophoretic Transfer of Proteins from Polyacrylamide Gels to Nitrocellulose Sheets

After native or SDS-PAGE, the gels are rinsed in electroblot buffer (50 mM Tris, 200 mM glycine, pH 8.3) for 5 min. For SDS gels, the buffer also contains 20% (v/v) methanol (Towbin et al. 1979; Towbin and Gordon 1984). A nitrocellulose sheet is cut to fit and laid on the gel. The gel and nitrocellulose sheets are placed between filter papers and dacron sponges, excluding all air bubbles and electroblotted in a transblotter fitted with cooling coils (Fig. 7). The blotting is carried out at 4 °C for 2 h (native gels) or 4 h (SDS gels) at 0.8 to 1.0 A. The long blotting period is necessary to insure complete transfer of high molecular weight proteins such as NR. While we routinely use electroblotting, other useful procedures include blotting by diffusion (Reinhart and Malamud 1982) and vacuum blotting (Peferoen et al. 1982). For a detailed review of these blotting procedures, readers are referred to Towbin and Gordon (1984).

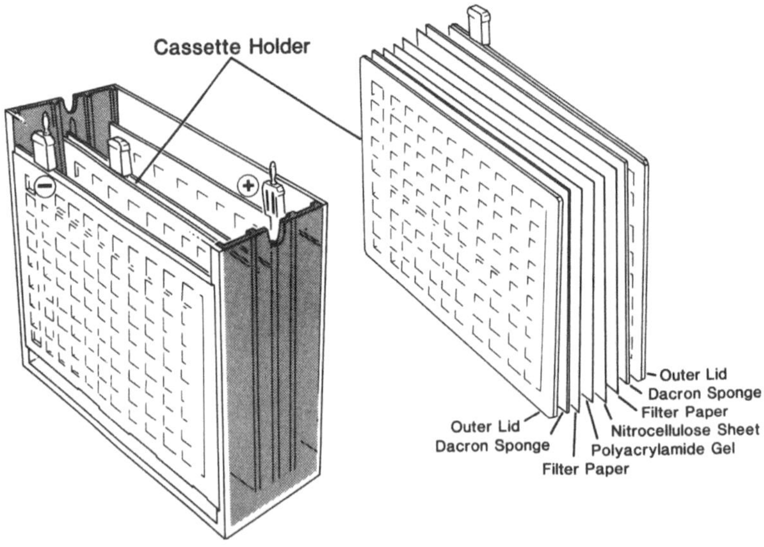

Fig. 7. Assembly of polyacrylamide gel, nitrocellulose sheet, dacron sponges, and filter papers for electrophoretic transfer of proteins from gels to nitrocellulose sheet

5.2.4 Immunodetection of Nitrate Reductase

After the electrophoretic protein transfer, the nitrocellulose sheets are treated with liquid gelatin (Hipure liquid fish gelatin, Norland Products Inc.) and Tween 20 to block the nonspecific protein binding sites (Saravais 1984; Tsang et al. 1983). Liquid gelatin (3%, v/v) and Tween 20 (0.05%, v/v) are dissolved in phosphate saline buffer (PBS), 10 mM K phosphate, 150 mM NaCl, pH 7.3. The minimum blocking time is 1 h at 40 °C. However, we normally block at room temperature for 4–12 h with no appreciable difference in the quality of the blots. After blocking, the nitrocellulose sheets are treated with monospecific antiserum raised against barley NR for 2–4 h. We routinely use antiserum diluted 1:1000 in PBS containing 3% liquid gelatin, 0.05% Tween 20, and 3% (w/v) PEG 4000. The dilution is empirically determined and needs to be adjusted for each batch of antiserum to accommodate variabilities in the titer. The 15 × 10 cm nitrocellulose sheet with 10 ml diluted antiserum is sealed in a plastic bag, excluding air bubbles, and incubated with vigorous shaking for 4–6 h at room temperature. The nitrocellulose sheets are next rinsed with PBS and then extensively washed with PBS containing 0.05% Tween 20 for 30 min with several changes of the buffer. This is critical since most of the nonspecifically-bound antibodies can be washed away in this step.

The antigen-antibody complex can be visualized by different methods. We use two techniques to detect the NR-antibody complex: (1) peroxidase-conjugated second-antibody method, and (2) soluble peroxidase-antiperoxidase (PAP) method. The peroxidase-conjugated, second-antibody method consists of directing a peroxidase conjugated antibody against the NR-antiNR complex. The PAP method involves the use of goat antirabbit serum as a bridge between anti-NR rabbit serum and the peroxidase rabbit antiperoxidase complex. In both cases, the resulting complex is visualized by staining for peroxidase activity.

In the second-antibody method, the NR antiserum-treated nitrocellulose sheets are treated with peroxidase-conjugated antibody mixed in PBS containing 3% liquid gelatin and 0.05% Tween 20. This incubation is carried out in sealed plastic bags in a rotary shaker for 2–4 h at room temperature. The nitrocellulose sheets are then washed with ice-cold PBS (three times 5 min each) and ice-cold 50 mM phosphate buffer, pH 7.3 (three times 5 min each).

The PAP method, originally developed by Sternberger et al. (1970) for immunocytochemical detection of antigens in tissue sections, has been adapted for Western blots (Peferoen et al. 1982). This method is about 100 times more sensitive than the peroxidase-conjugated second-antibody method (Towbin and Gordon 1984). The nitrocellulose sheets are sequentially treated with the appropriate antisera as illustrated (Fig. 8).

We use two different peroxidase staining methods. The preferred method is as follows. For 100 ml of staining buffer, add 60 mg of 4-chloro-1-naphthol dissolved in 20 ml methanol to 80 ml of 50 mM phosphate buffer, pH 7.3. Just prior to use, add 100 µl of 30% hydrogen peroxide. The second method involves the use of o-dianisidine (3,3′-dimethoxybenzidine) which may be carcinogenic and, therefore, must be handled with caution. For 100 ml staining buffer, add 500 µl of 0.5% (w/v) 3,3′-dimethoxybenzidine and 333 µl of 3% hydrogen peroxide to

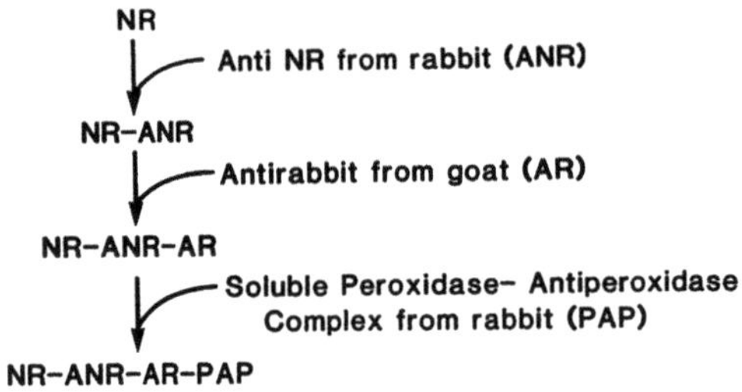

Fig. 8. Steps involved in immunodetection of NR using peroxidase-antiperoxidase method

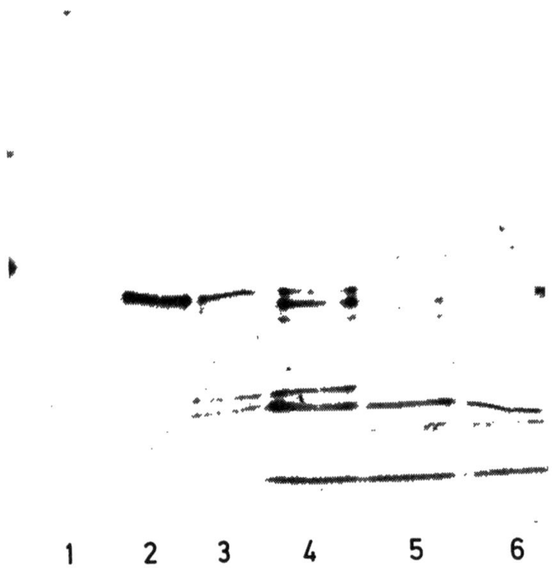

Fig. 9. Immunological detection of tryptic digests of nitrate reductase. Affinity purified NR was incubated with trypsin for 10 min on ice. The reaction was stopped with antipain (10 μM final concentration) and prepared for SDS-PAGE. SDS-PAGE and Western blot analysis (peroxidase-conjugated second-antibody method) were carried out as in text. *1* Blank incubated with trypsin (90 μg ml^{-1}; *2* NR incubated without trypsin; *3* NR incubated with 15 μg ml^{-1} trypsin; *4* NR incubated with 30 μg ml^{-1} trypsin; *5* NR incubated with 60 μg ml^{-1} trypsin; *6* NR incubated with 90 μg ml^{-1} trypsin

100 ml of 50 mM phosphate buffer, pH 7.3. The staining solution is poured over the nitrocellulose sheet and gently shaken. NR bands (actually peroxidase bands) appear within 2–30 min depending on the amount of NR, amount of antibody and specific activity of peroxidase conjugate. When the desired staining intensity is reached, the nitrocellulose sheets are washed in distilled water and dried between filter papers. The sheets are stored in the dark (the bands tend to fade in light) or photographed for permanent records.

We routinely incubate the nitrocellulose sheets in sealed plastic bags. Another technique has been published recently which allows the use of much smaller quantities of antiserum (Douglas and King 1984). In this method a filter paper wetted with antiserum solution (with all the necessary additives) is laid on top of the nitrocellulose sheet making sure no air bubbles are trapped between the filter paper and nitrocellulose sheet. These are then clamped between two glass plates for incubation.

We have used Western blotting techniques to investigate the nature of the NR specific cytochrome c reductase in NR-deficient barley mutants (Narayanan et al.

1983) and the synthesis and degradation of NR in young barley seedlings (Somers et al. 1983b). In the first case, we were able to identify the 4S sedimenting CRM as a dissociated subunit of NR. In the second case, we were able to use crude extracts to demonstrate that the appearance and disappearance of NR activity in barley seedlings in response to nitrate is correlated with presence of NR CRM.

More recently, we have used the Western blot technique to address the question of NR subunit size in different plant species. This question has persisted in NR literature probably due to the lability of NR in most plant extracts (for review, see Kleinhofs et al. 1985). By using Western blot techniques on crude extracts prepared so as to minimize proteolytic degradation, we have been able to demonstrate that higher plant NR from all species we have examined has a high molecular weight subunit ranging from 110000–140000 (Kleinhofs et al. 1985). We have also demonstrated that the Western blot technique is sensitive enough to detect numerous proteolytic degradation products of NR (Fig. 9).

5.3 Immunopurification of Nitrate Reductase

Once a monospecific antibody is obtained, it can be used to purify the antigen. We have used two immunochemical methods (immuneline electrophoresis and antibody column) to purify NR. However, these techniques are only as good as the original antibody since any impurities in the antibody will be reflected in the antigen purified by these methods.

5.3.1 Immuneline Electrophoresis

This method is similar to rocket immunoelectrophoresis described in Sec. 5.1.2. The only difference is that the NR sample is applied as a line instead of in wells. Agarose (2% w/v) is dissolved in Tris-barbital buffer, pH 8.6 at 90 °C. The solution is cooled to 50 °–52 °C, NR antiserum added (1.2% v/v), and the mixture poured as a slab gel (70 × 100 × 1.5 mm). Water is layered on top to give a sharp interface. Low melting point Seaplaque[R] (FMC) agarose (1% w/v) is dissolved in Tris-barbital buffer, pH 8.6, at 90 °C, cooled to 40 °–45 °C, mixed with the NR sample, added to the top of the gel (after pouring off the water layer), and allowed to set. When the gel is set, horizontal electrophoresis is conducted for 18 h (5 mA/ 7.5 cm gel). Since the pH of the separating gel (antibody gel) and buffer is 8.6 (pI of IgG), the antibodies do not migrate in the electric field. NR moves into the antibody-containing gel complexing with the antibody as it migrates through the gel. When the concentrations of NR and NR antibody reach equilibrium, the antigen-antibody complex forms a precipitate which will essentially be a line and will not move in the electric field. All other contaminating proteins will migrate off the gel. If the concentration of NR is too high for the concentration of the NR antibody in the gel, the antigen-antibody complex will not form a precipitate and will migrate through the gel. On the other hand, if the concentration of NR is too low, then the antigen-antibody complex will precipitate at the interface between the antigen-containing gel and the antibody-containing gel resulting in little or no purification of the antigen. The precipitin line (easily visible to the naked eye or with illumination) is excised and subjected to preparative SDS-PAGE. After SDS-PAGE, the gel is treated with 1 M KCl to visualize the protein bands. The NR band is excised from the gel and the NR electroeluted.

5.3.2 Antibody Column

Affi-Gel is coupled with NR antibodies under aqueous coupling conditions (Staehelin et al. 1981; Bio-Rad Bulletin 1983). Affi-Gel 10 slurry (10 ml) is transferred to a Buchner funnel and washed with 50 ml of ice-cold isopropanol followed by 50 ml of ice-cold deionized water. The moist cake is mixed with 4 ml of NR antiserum and the suspension gently agitated at 4 °C for 4–5 h. One ml of 1 M ethanolamine HCl, pH 8, is added to the slurry to block active ester sites. After an additional 60 min, the slurry is poured into the column (1 × 10 cm) and the column is washed with 0.1 M HEPES (N-2-hydroxyethyl-piperazine-N′-2-ethanesulfonic acid) buffer, pH 7.5, containing 1 μM FAD, 1 mM DTT, and 1 μM sodium molybdate for several hours. Meanwhile, affinity-purified NR is dialyzed against 10 mM HEPES buffer, pH 7.5, containing 1 μM FAD, 1 mM DTT, and 1 μM sodium molybdate. Alternatively the buffers can be exchanged by Sephadex G-25 chromatography. One ml of concentrated, affinity column purified NR is loaded onto the antibody column and washed with dialytic buffer for 2–3 h. The NR is eluted with 0.01 M glycine-HCl, pH 2.5. The 0.5 ml fractions are collected in tubes containing 50 μl of 1 M HEPES buffer, containing 10 μM FAD, 10 mM DTT, and 10 μM sodium molybdate, pH 7.5. The fractions are tested for NR homogeneity by SDS-PAGE. This procedure results in a significant loss of NR enzymatic activity (approximately 50%), presumably due to inactivation by the acid elution.

6 Summary

Nitrate reductase was purified from nitrate-induced barley seedlings using affinity chromatography and electrophoretic methods. The electrophoretically pure nitrate reductase was used to prepare monospecific rabbit antiserum. The nitrate reductase antiserum was characterized by titrating against nitrate reductase-associated enzymatic activities, immunodiffusion, and crossed immunoelectrophoresis. The nitrate reductase antiserum has been used to characterize the nitrate reductase antigen in nitrate reductase-deficient barley mutants, in wild-type barley seedlings during nitrate induction and decay, and in different species. Techniques used include protection of nitrate reductase inactivation, rocket immunoelectrophoresis, and Western blots. We have also used immunochemical methods to further purify nitrate reductase using immuneline and antibody column methods.

Acknowledgements. Information Paper. College of Agriculture Research Center, Washington State University, Pullman. Project Nos. 0605 and 0233. The work in our laboratory is supported in part by USDA-CRGO Grant No. 82-CRCR-1-1112 and NSF Grant No. PCM-8119096.

References

Amy NK, Garrett RH (1979) Immunoelectrophoretic determination of nitrate reductase in *Neurospora crassa*. Anal Biochem 95:97-107

Beevers L, Hageman RH (1969) Nitrate reductase in higher plants. Annu Rev Plant Physiol 20:495–522

Bio-Rad Bulletin (1983) Immunoaffinity chromatography, no 1099, pp 1–4

Bradford M (1976) A rapid and sensitive method for the quantitation of microgram quantities of protein utilizing the principle of protein-dye binding. Anal Biochem 72:248–254

Brown J, Small IS, Wray JL (1981) Age-dependent conversion of nitrate reductase to cytochrome c reductase in barley leaf extracts. Phytochemistry (Oxf) 20:389–398

Dailey FA, Kuo TM, Warner RL (1982a) Pyridine nucleotide specificity of barley nitrate reductase. Plant Physiol (Bethesda) 69:1196–1199

Dailey FA, Warner RL, Somers DA, Kleinhofs A (1982b) Characteristics of a nitrate reductase in a barley mutant deficient in NADH nitrate reductase. Plant Physiol (Bethesda) 69:1200–1204

Davis BJ (1964) Disc electrophoresis. Ann NY Acad Sci USA 121:404-427

Douglas GC, King BF (1984) A filter paper sandwich method using small volumes of reagents for the detection of antigen electrophoretically transferred onto nitrocellulose. J Immunol Methods 75:333–338

Hageman RH, Hucklesby DP (1971) Nitrate reductase from higher plants. Methods Enzymol 23:491–503

Harker AR, Narayanan KR, Warner RL, Kleinhofs A (1986) NAD(P)H bispecific nitrate reductase in barley leaves: partial purification and characterization. Phytochemistry (Oxf) 25:1275–1279

Hewitt EJ (1975) Assimilatory nitrate-nitrite reduction. Annu Rev Plant Physiol 26:73–100

Jolly SO, Campbell W, Tolbert NE (1976) NADPH- and NADH-nitrate reductase from soybean leaves. Arch Biochem Biophys 174:431–439

Kleinhofs A, Warner RL, Narayanan KR (1985) Current progress towards an understanding of the genetics and molecular biology of nitrate reductase in higher plants. In: Miflin BJ (ed) Oxford surveys of plant molecular and cell biology. Oxford University Press 2:91–121

Kuo TM (1979) Nitrate reductase in barley: nature and biochemical characterization of mutants. Dissertation. Washington State University, Pullman

Kuo TM, Kleinhofs A, Warner RL (1980) Purification and partial characterization of nitrate reductase from barley leaves. Plant Sci Lett 17:371–381

Kuo TM, Kleinhofs A, Somers DA, Warner RL (1981) Antigenicity of nitrate reductase-deficient mutants in *Hordeum vulgare* L. Mol Gen Genet 181:20–23

Kuo TM, Somers DA, Kleinhofs A, Warner RL (1982a) NADH-nitrate reductase in barley leaves: identification and amino acid composition of subunit protein. Biochim Biophys Acta 708:75–81

Kuo TM, Warner RL, Kleinhofs A (1982b) in vitro stability of nitrate reductase from barley leaves. Phytochemistry (Oxf) 21:531–533

Laemmli UK (1970) Cleavage of structural proteins during the assembly of the head of bacteriophage T4. Nature 277:680–685

Lowry OH, Rosebrough NJ, Farr AL, Randall RJ (1951) Protein measurement with Folin phenol reagent. J Biol Chem 193:265–275

Lund K, DeMoss JA (1976) Association-dissociation behavior and subunit structure of heat-released nitrate reductase from *E. coli*. J Biol Chem 251:2207–2216.

Nakagawa H, Poulle M, Oaks A (1984) Characterization of nitrate reductase from corn leaves (*Zea mays* cv WA64 × W182E). Two molecular forms of the enzyme. Plant Physiol (Bethesda) 75:285–289

Narayanan KR, Somers DA, Kleinhofs A, Warner RL (1983) Nature of cytochrome c reductase in nitrate reductase-deficient mutants of barley. Mol Gen Genet 190:222–226

Nicholas JC, Harper JE, Hageman RH (1976) Nitrate reductase activity in soybeans (*Glycine max* L. Merr.). I. Effects of light and temperature. Plant Physiol (Bethesda) 58:731–735
Otterness I, Karush F (1982) Principles of antibody reactions. In: Marchalonis JJ, Warr GW (eds) Antibody as a tool. Wiley, Chichester, pp 139–161
Ouchterlony O (1968) Handbook of immunodiffusion and immunoelectrophoresis. Ann Arbor Sciences, Ann Arbor, pp 215
Owen P, Smyth CJ (1977) Enzyme analysis by quantitative immunoelectrophoresis. In: Salton MRJ (ed) Immunochemistry of enzymes and their antibodies. Wiley, New York, pp 147–202
Peferoen M, Huybrechts R, DeLoof A (1982) Vacuum-blotting: a new simple and efficient transfer of proteins from sodium dodecyl sulfate-polyacrylamide gels to nitrocellulose. FEBS Lett 145:369–372
Reinhart MP, Malamud D (1982) Protein transfer from isoelectric focusing gels: The native blot. Anal Biochem 123:229–235
Renart J, Reiser J, Stark GR (1979) Transfer of proteins from gels to diazobenzyloxymethyl-paper and detection with antisera: method for studying antibody specificity and antigen structure. Proc Natl Acad Sci USA 76:3116–3120
Saravais CA (1984) Improved blocking of nonspecific antibody binding sites on nitrocellulose membranes. Electrophoresis 5:54–55
Scholl RL, Harper JE, Hageman RH (1974) Improvement of nitrite color development in assays of nitrate reductate by phenazine methosulfate and zinc acetate. Plant Physiol (Bethesda) 53:825–828
Schrader LE, Ritenour GL, Elrich GL, Hageman RH (1968) Some characteristics of nitrate reductase from higher plants. Plant Physiol (Bethesda) 43:930–940
Senn DR, Carr PW, Klatt LN (1976) Minimization of a sodium dithionite-derived interference in nitrate reductase-methyl viologen reactions. Anal Biochem 75:464–471
Smarrelli J, Campbell WH (1981) Immunological approach to structural comparisons of assimilatory nitrate reductase. Plant Physiol (Bethesda) 68:1226–1230
Snapp S, Somers DA, Warner RL, Kleinhofs A (1984) Immunological comparisons of higher plant nitrate reductases. Plant Sci Lett 36:13–18
Solomonson LP (1975) Purification of NADH-nitrate reductase by affinity chromatography. Plant Physiol (Bethesda) 56:853–855
Somers DA, Kuo T, Kleinhofs A, Warner RL (1983a) Nitrate reductase-deficient mutants in barley. Immunoelectrophoretic characterization. Plant Physiol (Bethesda) 71:145–149
Somers DA, Kuo TM, Kleinhofs A, Warner RL, Oaks A (1983b) Synthesis and degradation of barley nitrate reductase. Plant Physiol (Bethesda) 72:949–952
Staehelin T, Hobbs DS, Kung H-F, Pestka S (1981) Purification of recombinant human leukocyte interferon (IFLrA) with monoclonal antibodies. Methods Enzymol 78:505:512
Sternberger LA, Hardy PH Jr, Cuculis JJ, Meyer HG (1970) The unlabeled antibody enzyme method of immunohistochemistry. Preparation and properties of soluble antigen-antibody complex (horseradish peroxidase-antihorseradish peroxidase) and its use in identification of spirochetes. J Histochem Cytochem 18:315–333
Thompson ST, Kathleen HC, Stellwagen E (1975) Blue dextran Sepharose: An affinity column for the dinucleotide fold in proteins. Proc Natl Acad Sci USA 72:669–672
Tischler CR, Purvis AC, Jordan WR (1978) Factors involved in in vitro stabilization of nitrate reductase from cotton (*Gossypium hirsutum* L.) cotyledons. Plant Physiol (Bethesda) 61:714–717
Towbin H, Gordon J (1984) Immunoblotting and dot immunobinding. Current status and outlook. J Immunol Methods 72:313–340
Towbin K, Staehlin T, Gordon J (1979) Electrophoretic transfer of proteins from polyacrylamide gels to nitrocellulose sheets: procedure and some applications. Proc Natl Sci USA 76:4350–4354
Travis RL, Jordan WR, Huffaker RC (1969) Evidence for an inactivating system of nitrate reductase in *Hordeum vulgare* L. during darkness that requires protein synthesis. Plant Physiol (Bethesda) 44:1150–1156

Tsang VCW, Peralta JM, Simons AR (1983) Enzyme-linked immunoelectrotransfer blot techniques (EITB) for studying the specificities of antigens and antibodies separated by gel electrophoresis. Methods Enzymol 92E:377–391
Van Vunakis H, Langone JJ (eds) (1980) Immunochemical techniques, part A. Methods Enzymol 70:201–438
Warner RL, Kleinhofs A (1974) Relationships between nitrate reductase, nitrite reductase, and ribulose diphosphate carboxylase activities in chlorophyll-deficient mutants of barley. Crop Sci 14:654–658
Wray JL, Filner P (1970) Structural and functional relationships of enzyme activities induced by nitrate in barley. Biochem J 199:715–725
Zielke HR, Filner P (1971) Synthesis and turnover of nitrate reductase induced by nitrate in cultured tobacco cells. J Biol Chem 246:1772–1779

Immunological-Cytochemical Localization of Cell Products in Plant Tissue Culture

K. J. WILSON

1 Introduction

One of the most important objectives of those who work with and study plant tissue cultures is to understand the dynamic cell, tissue, and organ differentiation and development that occur in cultured materials. Numerous laboratories, working on crop species especially, often are concerned with testing scores of different culture media and growing and regenerating plants from hundreds of different species and varieties to overcome recalcitrant growth. The immediate goals of such studies are usually twofold. Large-scale studies can exploit the genetic changes, called somaclonal variation, resulting from the process, per se, of culturing tissue (Snowcroft and Larkin 1982), and can select culture lines with superior qualities of interest (stress tolerance, disease resistance, high yield, improved nutritive qualities, etc.). Such studies also seek to identify efficient tissue culture systems to accommodate genetic engineering of important crops, such as rice, corn, and soybean. Of equal importance to such goals, however, is basic research that seeks to understand the developmental process of embryogenesis in vitro, that is, somatic embryo (embryoid) development in culture, as well as to gain knowledge of cell development of specific cell types within embryoids and callus in culture systems. The purpose of the following descriptions and discussions of techniques is to suggest immunocytochemical approaches to developmental studies that may elucidate the process of cultured cell and embryoid development. Such approaches, by virtue of their specificity and sensitivity, will increase our understanding of cell development, in its multiple aspects, in both cultured and zygotic plant tissues, and, such understanding is neccessary, more specifically, to overcome recalcitrant embryogenesis.

2 The Laticifer

The specialized cell type, the laticifer, is interesting because its differentiation and development is well understood in several plants. Its differentiation occurs only during the heart stage of embryogenesis making its differentiation a unique event in the life of the plant (Wilson and Mahlberg 1977), and therefore an event one may more easily target in developmental studies. The laticifer is important also because many valuable industrial products come from latex which is the cytoplasm and/or vacuolar sap of the laticifer. Therefore, an understanding of the growth and development, in tissue culture, of this cell type in particular may prove economically important. At the same time, such an understanding would be a significant contribution to knowledge of cell development in callus in general. Such knowledge is critical in view of our general desire to obtain economically important cell products from plant tissue cultures (Chaleff 1983). Finally, since immunocytochemical methods allow identification of both laticifers and laticiferlike cells in any environment (Wilson et al. 1984), as described below, it should be possible to sort out and proliferate laticifer cells in pure cultures.

2.1 Laticifer Morphology and Differentiation in the Asclepiadaceae

There are two major classes of laticifers, the articulated and the nonarticulated types and two forms of nonarticulated laticifers, branched and unbranched (Esau 1965). Many members of the plant families Asclepiadaceae and Euphorbiaceae possess nonarticulated, branched laticifers. In young heart stage embryos of the genus *Asclepias* a specific number of laticifer initials (16) differentiates between the ground meristem and procambium. Figure 1 shows a late heart stage embryo with elongating laticifer initials. The initials elongate at either end and grow into the embryonal hypocotyl-root axis, cotyledons, and shoot meristem. The cells are coenocytic. At maturity the extremely elongated, branched laticifer system has proliferated throughout the entire embryo (Fig. 2). Upon seed germination the laticifer system proliferates throughout all parts of the plant axis by penetration of laticifer cell apices (intrusive growth) among other cells (Chaveaud 1891; Wilson et al. 1976) (Fig. 3).

The mature laticifer cell type is characterized best by its production of an exudate, latex. The cells of several nonarticulated, branched laticifer systems, including, for instance, those of *Asclepias syriaca* (Wilson and Mahlberg 1980), *A. tuberosa* (Wilson and Frantz 1981), *A. incarnata* (Wilson, unpublished), *Stapelia bella* (Wilson and Ellis 1982), *Euphorbia marginata* (Schulze et al. 1967), *E. characias* (Marty 1968, 1970, 1971), and *Ficus elastica* (Heinrich 1970), possess peripheral cytoplasm and a large central vacuole, the contents of which are represented by latex. Depending on the species, latex contains various important secondary plant products including drugs (opium in the opium poppy that bears articulated laticifers), natural rubber, proteolytic enzymes (Brockbank and Lynn 1979), and steroids (Shukla and Murti 1971). Rubber has been identified in latex of several desert *Asclepias* species (Buehrer and Benson 1945). *A. syriaca*, in particular, contains up to 4% rubber as well as 23% resin (including lipids and complex carbohydrates) (Paul et al. 1943). Several plant species possessing nonarticulated branched laticifers have been investigated as alternative sources of rubber or petroleum-producing compounds (Nielson et al. 1977; Nishimura et al. 1977; Biesboer and Mahlberg 1979a).

As has been reviewed elsewhere (Wilson and Mahlberg 1977, 1980; Wilson et al. 1984) laticifer cells are difficult, although not impossible, to fix well for either light or electron microscopy because they have high osmotic pressure and tend to surge during fixation and embedment procedures. Differential staining is difficult because laticifers share the precursors of their more exotic components, such as rubber, with a host of other cell components in other cells. Also, because

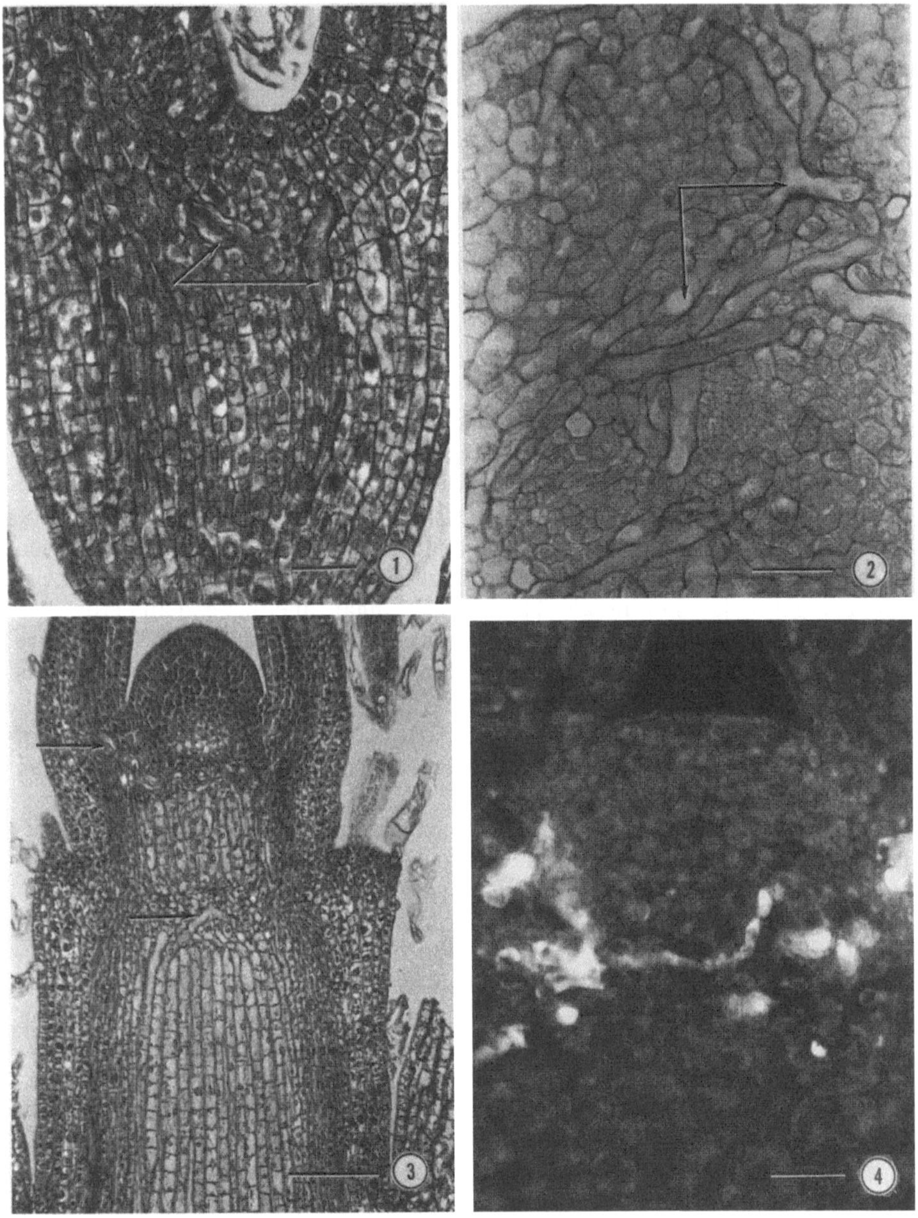

Figs. 1–4. Zygotic material from *Asclepias syriaca*

Fig. 1. Longitudinal section of paraffin-embedded late heart stage embryo. *Arrows* indicate elongating, branching laticifer initials. *Bar* = 20 μm

Fig. 2. Cross-section of paraffin-embedded mature embryo cut at the cotyledonary node. *Arrows* indicate highly branched coenocytic laticifer cells. Bar = 30 μm

Fig. 3. Longitudinal section through a paraffin-embedded shoot apex. *Upper arrow* indicates laticifer entering a leaf primordium. *Lower arrow* indicates plexus of laticifers at the node. *Bar* = 2.0 mm

Fig. 4. Cryostat section of shoot apex of a mature embryo. Bright fluorescent cells are laticifers. *Bar* = 30 μm

Table 1. Immunocytochemical identification of laticifers using anti-latex antiserum prepared from *Asclepias syriaca*

Species	Result[a]	Laticifer type
Asclepias syriaca		Nonarticulated
Embryo	+	
Shoot	+	
Asclepias tuberosa		Nonarticulated
Embryo	+	
Shoot	−	
Stapelia bella		Nonarticulated
Shoot	+	
Euphorbia tirucalli		Nonarticulated
Shoot	−	
Euphorbia marginata		Nonarticulated
Embryo	−	
Musa paradisiaca		Articulated
Petiole	−	
Cichorium intybus		Articulated
Shoot	−	

[a] + means laticifers were fluorescent and controls were negative. − means no fluorescence was observed in the section

of the interference of other cells and their components (such as cell storage components), the unique laticifer morphology is not readily visible in unstained material, even using phase-contrast or interference-contrast microscopy. The sensitivity and specificity of immunocytochemical techniques allow laticifer identification at the light microscopic level in zygotic tissues of a variety of laticifer-bearing plants (see Table 1; and Wilson et al. 1984).

2.2 Methods of Preparing Anti-Latex Antibodies

The same techniques that allow us to identify laticifers in zygotic material can be applied directly to laticifer cell identification in cultured tissues. Techniques described below permit the detection and observation of laticifer cells or of cells with laticiferlike metabolism in callus cells, in suspension cultures, and within embryoids.

The following immunization techniques are specific for preparing anti-latex antiserum, but may be generally applied to production of antibodies to any immunogen. The secret to immunocytochemical identification of cells or cell components is, first of all, to identify a cell or cell-specific antigen. If such an antigen cannot be purified then antibodies can still be prepared against a heterogeneous set of antigens and, in the right circumstances, the antibodies of interest can be purified. The single most important feature of the antigen is "foreignness" with respect to the animal in which the antibodies are prepared. The larger, the more complex, and the less inert chemically, the better the antigen will be at eliciting an immune response [see Tizard (1984) for basic theory of the immune response]. Such problems also are given further consideration by Jeffree et al. (1982), who point out that even though a single homogeneous polypeptide or carbohydrate is used as an antigen, it is likely that a heterogeneous group of antibodies will be synthesized during immunization. In fact, to prepare anti-latex antibodies, a latex-specific antigen need not be purified. Latex can be used as the antigen and the nonlatex specific antibodies, against cellular components common to all the cells in the plant, can be removed as described below.

New Zealand White rabbits are the animal of choice in which to prepare antisera because they are relatively inexpensive to buy and care for. Reagents to purify and test the sera are readily available. Finally, the animals grow large enough to allow one to take large blood samples with minimal trauma. Because latex is known to contain toxins, such as cardiac glycosides (see Selber et al. 1984), toxicity studies are run on white mice first to determine a nontoxic dosage for injection. Normally three rabbits are injected for each compound of interest so save time in immunization, that is, to insure a "hit" where at least one animal produces the antibodies needed. The latex used for injection is collected from *Asclepias syriaca*, the common milkweed, growing wild. Latex is "bled" into test tubes, covered, and stored on ice to reduce oxidation. The greater part of the rubber fraction of the latex is removed by ultracentrifugation for 1 h at 60000 × g in the cold, and the latex serum is used for injection (Wilson et al. 1976). Ultrastructural studies (Wilson, unpublished) indicate that this treatment does not remove all the rubber particles with their proein coats. Thus, the latex antigen represents what has been identified by ultrastructural studies (Wilson and Mahlberg 1980) as the vacuolar sap of the large central vacuole of the mature laticifer. This vacuole will contain cell constituents that originate in the cytoplasm. Therefore, the anti-latex antiserum should, theoretically, ultimately be able to mark the entire laticifer cell.

Immunization protocols are the same for most antigens and have been reviewed and discussed by several authors (Knox 1982; Knox and Clarke 1978; Mayer and Walker 1980; Jeffree et al. 1982). For immunization against latex, the protocol is to mix 2.22 mg (dry weight) latex serum in phosphate buffered saline (PBS) (pH 7.2) with an equal volume of Freund's complete adjuvant. Initially 0.5 ml of this mixture is injected into each rear foot pad and each thigh of 2–3 kg rabbits. A second and third injection of a 1:1 mixture of latex serum (8.88 mg dry weight in PBS) and Freund's incomplete adjuvant is administered 1 and 2 weeks after the first injection in the hind quarters (0.5 ml each site). A booster is administered as above and given 6 weeks after the third injection. Success also can be achieved by giving the initial injection subcutaneously in each of four locations along the length of either side of the spine followed by boosters given in the thighs.

To derive serum, rabbits usually are bled by heart puncture, a procedure that requires courage and several demonstrations at the side of an experienced immunologist. It is simpler to insert an 18 gauge sterile disposable syringe needle into the large vein on the lower ear. To accomplish this, the rabbit is immobilized by hand or in a restraining unit and the veins at the ear tip swabbed with xylene to dilate the vessels. After gently shaving the area over the lower vein, the needle is slipped into the vessel and blood collected in a test tube. The flow of blood will slow after a few minutes and can be restarted by turning the needle slightly. When the blood flow ceases completely or enough blood is collected the needle is removed. It is important to remove all xylene from the ear with ethyl alcohol and to check the rabbit within several hours after bleeding for complications. Large New Zealand White rabbits will yield safely about 50 ml blood each session. Care must be taken, however, not to bleed more than two or three times within a 6 week period to prevent anemia. In any case, all research animals should be checked regularly by a veterinarian. Once antibody production is detected, the investigator will appreciate healthy animals, especially since rabbits can yield antisera, with periodic boosters, for 2–3 years. After collection, blood is allowed to clot in the test tube about 1 h at room temperature. Tubes are then "rimmed" with a wooden stick, covered, and placed in the refrigerator overnight. The clot is removed carefully from the tube and the remaining cells and debris are centrifuged from the serum. Serum may be frozen, usually in 2 ml aliquots, without further treatment for future use.

One should test for the presence of antibodies before and after purification of IgG immunoglobulins. The simplest procedure is to place 5–10 µl of the serum in a central well of a circle of wells on an Ouchterlony plate. Various dilutions with PBS (for instance, full strength. 1:1, 1:25, 1:50, and 1:100) of the immunogen (in this case, latex) are distributed in the outer wells and the plate is incubated overnight in a moist box at room temperature. The presence of precipitation bands indicates successful immunization (see Mayer and Walker 1980).

Immunoglobulins are precipitated from thawed serum with saturated ammonium sulfate (1:1, which means in 50% ammonium sulfate) in the refrigerator overnight. The precipitate is centrifuged out at 6000 rpm for 5 min in a clinical centrifuge and washed two times with 50% ammonium sulfate. The precipitate is redissolved in 2 ml (or in the amount of serum started with) of PBS, and desalted on a Sephadex G-25 column. The protein fraction is concentrated in an Amicon cell (Amicon, Lexington, Massachusetts, USA) with a YM 30 filter to 2–5 ml. To put the fraction into a Tris-HCL buffer (0.5 M, pH 8.6) the fraction is mixed with that buffer and reconcentrated two times. To purify IgG from other immunoglobulins the fraction is passed over a DEAE Bio-Gel A column (Bio Rad Labs, Richmond, California, USA) and 280 nm fractions are collected. For anti-latex antiserum purification, an Affigel Blue DEAE column, which should deliver a purer IgG fraction, was found to bind the antibodies and, therefore, could not be used. However, that column may be quite suitable, and possibly more desirable, for antisera that do not bind. For a 2 ml starting sample about 19 ml are collected from the DEAE Bio-Gel column, concentrated in an Amicon cell to 4 ml in Tris-HCl buffer, and diluted for use in staining procedures if no further purification is needed. The mg ml^{-1} present in the sample is figured by reading A(280) for the sample off the final column and using the Mab sor (E280) (which is 1.38 for 1.0 mg ml^{-1} IgG) to calculate the milligram total in the sample. For instance, if $A(280) = 1.001$ for 19 ml of sample, then the milligram total $= (1.001/1.38) \times 19 = 13.78$ mg total.

The final purification step removes antibodies not specific to laticifer cells by absorption of the IgG fraction with cells derived from liquid cultures of *Asclepias syriaca*. Such cells are presumed not to contain laticifer cells. They are harvested in log phase from cell suspension cultures grown in a Murashige-Skoog medium (Murashige and Skoog 1962) supplemented with 1.0 mg l^{-1} kinetin, 1.0 mg l^{-1} 2,4-D, 20 mg l^{-1} adenine, 100 mg l^{-1} myo-inositol, and 400 mg l^{-1} casein hydrolysate. The culture system is described more fully in Sec. 3. Cells are lyophylized and then rehydrated (0.01 g cell in 1.0 ml distilled water) and added to 1.0 ml of the IgG fraction. The cell-antiserum mixture is incubated at room temperature for 30 min and then centrifuged to remove the cells. This procedure is repeated three times to yield a laticifer cell-specific antiserum.

2.3 Section Preparation

Cryotechniques are the techniques most likely to preserve antigen-antibody reactive sites and to preserve the antigen of interest in situ, although fixation may be necessary in some cases. The basic procedures and theory for preparation of sections of plant (and animal) tissues for immunocytochemical procedures at the light microscopic level have been described and discussed in several books and reviews (Knox and Clarke 1978; Sternberger 1979; Jeffree et al. 1982; Wilson et al. 1984). However, these techniques require modification to suit the special problems that arise in culture-grown material. Tissue culture samples are more delicate than most zygotic tissue and must be dried on slides longer, as well as washed more thoroughly than zygotic tissue sections. To prepare slides, samples of *Asclepias syriaca* and *A. tuberosa* culture material are quick frozen on copper stubs in the cryoprotectant, Tissuetec. 1.0% gelatin also has been used successfully as a protectant in which to freeze plant material. The stubs are placed in the cryostat (for instance, an IEC Model Minitome) and frozen at temperature settings between −12 °C and −8 °C. These settings are much higher than those used normally for sectioning animal tissue on the same instrument (about −22 °C). Sections are cut with a metal knife that is precooled or stored continuously in the freezer. The cryostat is set to take sections of between 20 and 22 μm, although the actual section thickness taken is about 10 μm. Sections are lifted from the knife by moving a prepared slide (at room temperature) up to the knife edge holding the section. Each section is collected individually and several sections can be collected on a single slide. Frozen sections are mounted on slides subbed with a gelatin (1.0%) and chromium potassium sulfate (2.5%) solution (modified from Berlyn and Miksche 1976). We have experimented with changes in the percent of chromium potassium sulfate as well as with the gelatin to optimize section adhesion. However, changes in these parameters have not solved the general unacceptable loss of sections during washing. In fact, the critical parameter in section adhesion to the slide is the amount of time the sections are allowed to dry after being placed on the slide. With animal tissue enough protein is present in the sections to stick them easily to the glass. Plant tissues generally contain high amounts of water and carbohydrates and less protein than animal tissue and this difference is even greater in callus tissue. Thus, sections are allowed to dry down on the slides for as long as 5 min before they are flooded with distilled water. After this drying period sections are not allowed to become dried out throughout the rest of processing.

Either a direct or an indirect method of immunocytochemical localization may be used to detect laticifer cells. Both methods require the use of a detectable marker. For light microscopy, either rhodamine or fluorescein commonly are used as fluorescent markers (Sternberger 1979; Mayer and Walker 1980). For electron microscopy, electron-dense markers, such as colloidal gold (Roman et al. 1974; Sternberger 1979; DeMay 1981; DeWaele et al. 1981; Geuze et al. 1981; Roth 1982, 1983) or ferritin (Sternberger 1979; Mayer and Walker 1980; Himmelhoch and Zuckermann 1982) are used. By the direct method these markers are attached directly to the antiserum of interest. The antiserum is then allowed to attach to the antigen on the section to localize its position. The indirect immunofluorescent method was first applied by Coons and Kaplan (1950) using a fluorescent secondary antibody. By this indirect method, secondary antibodies are made by immunizing an animal against the blood serum from the specific animal in which the primary antibody of interest is derived. This secondary antibody then may be attached to any one of the above markers. To localize a cell or cell component, first the primary antibody is attached to the antigen in the section and then the primary antibody is localized by the secondary antibody attached to a marker. The indirect method is more sensitive because the primary antibody can bind several secondary antibodies with their markers. The direct method allows detection of only the marker attached to the primary antibody (Jeffree et al. 1982).

Once the plant sections are collected on slides, for fluorescent staining, sections are flooded with an experimentally determined dilution of the antiserum with phosphate buffered saline (pH 7.2) and incubated 30 min at 37 °C in a moist chamber. When working with succulent plant material or laticifer-bearing material, the sections are prewashed in PBS three times, 3 min each rinse, in coplin jars to removed cell and latex contamination from sections. The moist chamber consists, in our case, of a pair of flat metal slide racks. One rack holds the slides and the other is used for a cover. The pair of racks are covered with foil for the incubation period. Slides are rinsed seven times in PBS in coplin jars at room temperature for 3 min each rinse, and then flooded with fluorescein-conjugated IgG fraction goat anti-rabbit IgG. Again, the dilution factor must be determined experimentally to optimize low background and bright staining. The slides are incubated again for 30 min at 37 °C in a moist chamber, rinsed seven times, drained, mounted in glycerin:PBS (1:2), and cover-slipped. That seven washes are needed for low background also is determined experimentally. For instance, only five rinses resulted in unacceptably high background, which obscures staining. Control slides are prepared in the same way, but with IgG fraction prepared from serum from uninjected rabbits. The best control is to use serum from the rabbit used for antibody production collected before injection. We observe and photograph our slides with a Leitz Dialux 20 fluorescent microscope (using the B cube) equipped with a Wild Photoautomat MPS45 camera system using Tri-X film, for black and white, pushed to 1000 ASA and push-processed with Acufine developer, or using color slide or print film rated at 1000 ASA.

In the methods described above one important consideration is that of the permeability of cells to the necessary reagents. We have had no trouble localizing laticifer cells in either zygotic or tissue culture-derived cells. Sections that are frozen and then thawed, as in our methods, apparently become permeabilized to markers and other reagents. The technique, probably because of quick freezing, also stabilizes the otherwise osmotically sensitive protoplast of the laticifer so that minimal plasmolysis and surging is observed (Fig. 4). This would suggest this technique would be useful for localizing components of other osmotically sensitive cells, such as phloem sieve elements. Where cryostat-prepared material either cannot be used or does not permeabilize tissue, methods common to animal tissue preparation can be applied. For instance, for tissue culture materials, cell clumps from suspension cultures can be attached to slides coated with polylysine and permeabilized by plunging the slides briefly into cold ethanol (Wick et al. 1981). Slides then can be treated as described above.

2.4 Laticifer Identification in Zygotic Material

Before immunocytochemistry the only methods to identify laticifers and monitor their differentiation were fluorescein and Hg-fluorescamine stains developed specifically for laticifers (Bruni et al. 1977; Dall'olio et al. 1978; Bruni and Tosi 1980). A study by Biesboer(1984) used these stains and showed a direct correlation between numbers of Hg-fluorescamine-stained cells in *Asclepias syriaca* suspension cultures and the levels of a specific laticifer secondary metabolite, β-amyrin. However, these stains were tested on a variety of laticifer-bearing plants, including members of the Asclepiadaceae and Euphorbiaceae, and were found, in both my laboratory and that of Dr. Biesboer (personal communication) to be nonspecific in some cases and therefore unreliable. In fact, these stains were purported to stain cells rich in cysteine (Dall'olio et al. 1978) and they apparently, in fact, do not bind specifically to unique laticifer products.

The immunocytochemical procedures described above overcome the nonspecificity of the above conventional stains and have been used successfully to identify laticifers in several laticifer-bearing plants (Wilson et al. 1984). Material included *Asclepias syriaca* and *A. tuberosa* shoot apices and mature embryos, *Stapelia bella* (Asclepiadaceae) stem pieces, *Euphorbia marginata* embryos, *E. tiru-*

Fig. 5. Cryostat section through cotyledon of zygotic embryo of *Asclepias tuberosa*. Fluorescent cells are laticifers. *Bar* = 50 μm

Fig. 6. Paraffin section through 18-day-old callus from a stem explant of *A. tuberosa*. *Arrows* indicate islands of relatively rapidly dividing cells in the callus, some of which become embryoids. *Bar* = 50 μm

Fig. 7. Cryostat section of callus of a 50-day-old *A. syriaca* viewed by Nomarski optics. *Arrows* indicate protoplasts of callus cells. *Bar* = 20 μm

Fig. 8. The same section as shown in Fig. 7, but by epifluorescent microscopy. *Arrows* correspond with those in Fig. 7. *Bar* = 20 μm

Fig. 9. Paraffin section through *Asclepias* callus more than 30 days old. Several embryoids are embedded among callus cells. *Bar* = 0.5 mm

Fig. 10. Cryostat section of *A. syriaca* embryoid derived from an 80-day-old callus culture. Xylem cells could be seen in this section. *Arrows* indicate fluorescent laticifers. *Bar* = 20 μm

calli stem pieces, *Musa paradisiaca* var.*sapientum* (Musaceae) petiole, and *Cichorium intybus* (Asteraceae) petiole. The latter two plants possess articulated laticifers, while the former possess nonarticulated branched laticifers. The results are summarized in Table 1. Figure 5 shows the typical appearance of fluorescent laticifers in the embryos of *A. tuberosa*.

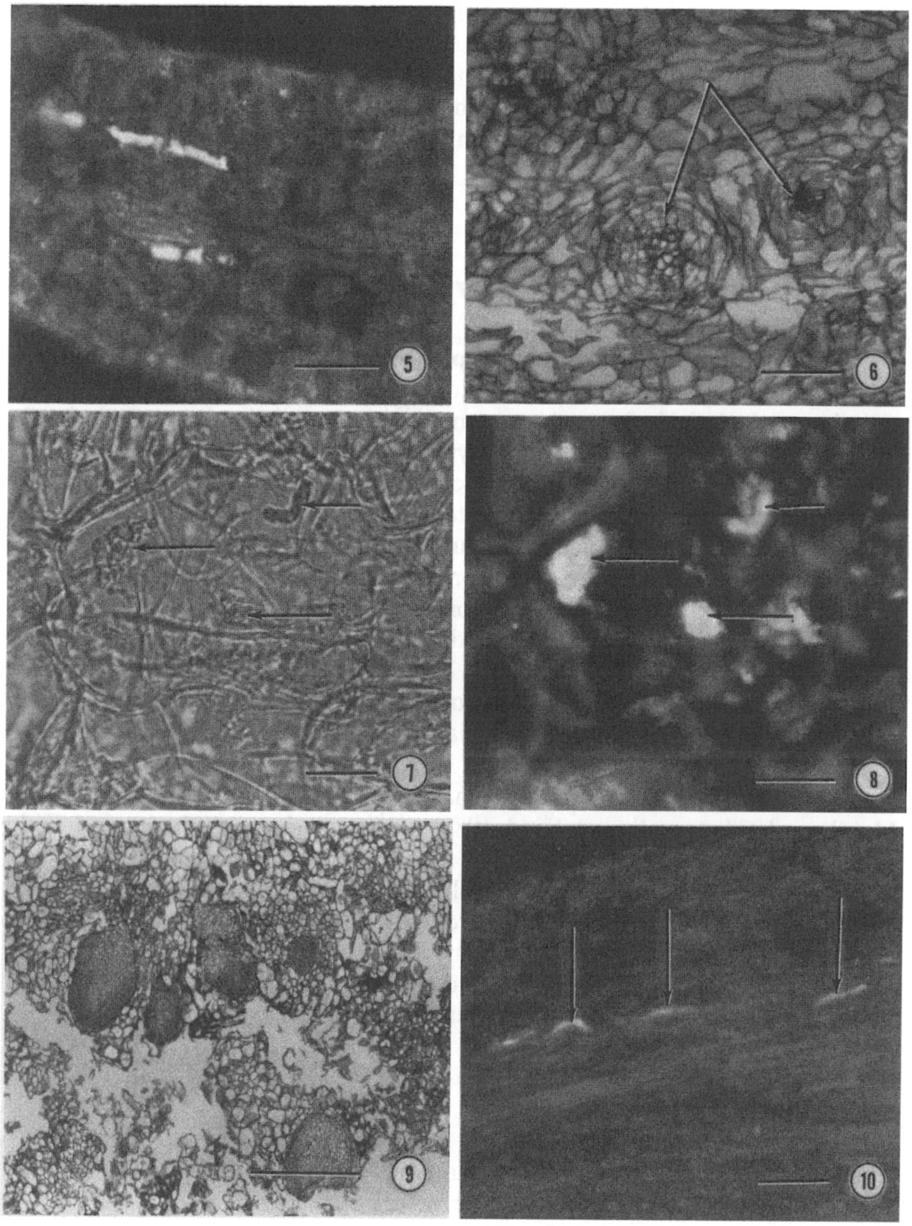

One of the more interesting aspects of these results is that they can provide some developmental information about the laticifer, at least for *A. tuberosa*. For instance, it is significant that nonarticulated laticifers in the shoot of the plant do not fluoresce using immunocytochemical methods, while laticifers in the embryos do fluoresce. Apparently, developmental events after germination alter the cell contents sufficiently to preclude, in some way, antigen-antibody reactions required for this method. It is notable that laticifer contents would differ so significantly at different plant developmental stages in one plant species, but not in another (as in *A. syriaca*). It is perhaps easier to understand why laticifers in *Euphorbia*, *Musa*, and *Cichorium* do not fluoresce since, in all cases, these plants are not closely related to *A. syriaca*, against whose latex the primary antibody was derived. Also, in *Musa*and *Cichorium*, the laticifers are articulated rather than nonarticulated and thus differ significantly in their mode of origin and development (Esau 1965).

3 The Tissue Culture System for the Asclepiadaceae

The tissue culture systems for the Asclepiadaceae provide excellent cultured material in which to study cell differentiation as well as embryogenesis. The differentiation of the laticifer is a unique event occuring in all members of this family studied to date. For the laticifer, a cell type that produces a large number of valuable commercial products, immunocytochemistry opens the possibility of specifically identifying, selecting, and, ultimately, of proliferating cells in culture that will produce valuable secondary products. Although plant tissue cultures are often inactive in secondary metabolite production, they possess the genes responsible for the biosynthesis of secondary compounds in unproductive cultures. Identification of cells capable of differentiating into latex-producing cells is the first step in controlling their developmental course and biosynthetic activity. As for embryogenesis, embryoids, as well as roots and shoots, have been derived from callus of several members of the Asclepiadaceae including *Tylophora indica* (Rao et al. 1970; Rao and Narayanaswamy 1972), *Pergularia minor*, and *Asclepias curassavica* (Prabhudesai and Narayanaswamy 1974), *A. syriaca* (Wilson and Mahlberg 1977; Groett and Kidd 1981), and *A. tuberosa* (Groett and Kidd 1981).

Asclepias tissue cultures are extremely easy to grow and, since embryogenesis occurs readily, it is clear that cultures provide excellent material in which to study the process of embryogenesis. Culture media for both *A. syriaca* and *A. tuberosa* have been developed that support rapid callus and suspension cell growth and that promote prolific asexual embryogenesis and are described below.

3.1 Culture Techniques

The techniques described in this section are specifically for *Asclepias tuberosa* culture establishment. Changes for *A. syriaca* are noted, but these techniques can be applied to most dicotyledonous plants with only minor changes in media. Stem explants used to derive callus cultures are obtained from sterile seedlings grown from excised embryos of *A. tuberosa* obtained from seeds, purchased from Park Seed Company, or collected from wild plants. Excised embryos are grown on a basic inorganic salt medium with no hormone supplements (Mahlberg 1968; Wilson and Mahlberg 1977) (0.9% agar, 2% sucrose, pH 5.8). Seeds are sterilized for 15 min in 3% calcium hypochlorite, the embryos excised under sterile conditions, sterilized for 1 min in 1% calcium hypochlorite and inoculated into 125 ml flasks or 25 × 100 mm test tubes covered with polypropylene. Experiments carried out to determine the optimal seedling age from which to start cultures for greatest callus growth, indicated that 25–30-day-old seedlings yield the most rapid callus growth. A 3–5 mm stem section from the top of the second internode up from the cotyledonary node is used routinely as the primary explant. To derive callus, stem explants are placed on a Murashige-Skoog medium (Murashige and Skoog 1962), supplemented with 0.1 g l^{-1} inositol, 4.0 g l^{-1} casein hydrolysate, and 5.0 mg l^{-1} adenine (MS medium). Plant hormone supplements are 0.5 mg l^{-1} 2,4-dichlorophenoxy acetic acid (2,4-D) and 0.5 mg l^{-1} benzyladenine (BA). For all *A. syriaca* cultures 1.0 mg l^{-1} kinetin is substituted for BA. Two percent sucrose and 0.9% agar are added to all solidified MS media and the pH is adjusted to 5.8. Cultures are maintained in 125 ml polypropylene-covered culture flasks or in test tubes under constant light at 22 °–24 °C. Cell suspension cultures are established in liquid MS (lacking agar) by transferring a small piece of 30-day-old callus derived on solid MS medium. Cultures are maintained in 125 ml polypropylene-covered flasks on a rotary or reciprocal shaker in constant light at 22 °C. Transfers of suspension cells must be made at least every 2 weeks and are accomplished by pipetting 5–10 ml of the old culture to 20 ml of fresh medium.

Embryoids appear spontaneously in *Asclepias tuberosa* callus and suspension cell cultures after 30 days of growth. At least 50 days are required for *A. syriaca* cultures, and embryoid development is less abundant. For both species more embryoids can be derived in liquid cultures by transferring cells and clumps of cells to MS medium supplemented with 0.5 mg l^{-1} BA (1.0 mg l^{-1} kinetin for *A. syriaca*) and no 2,4-D.

3.2 Embryoid Differentiation and Laticifer Identification

Callus and embryoid development and morphology have been carefully studied in cultures grown on solid MS medium (manuscript in preparation). Paraffin-embedded sections of callus, sampled at different stages, were stained with safranin and fast green, with a tannic acid-ferric chloride stain to enhance wall staining (Johansen 1940) for light microscopy. The progressive stages of embryoid development that begin in phloem parenchyma in explants cultured 11 days, ultimately lead to the production of rapidly dividing islands of cells within a disorganized mass of callus cells. These islands (Fig. 6) apparently give rise to embryoids, buried in the callus, that can be identified at several embryonic stages within a single callus by day 30 (Fig. 9).

An assessment of the morphology of culture development and embryogenesis, made in these and earlier studies (Wilson and Mahlberg 1977), determined that embryoids of *Asclepias tuberosa* and *A. syriaca* do not differentiate laticifer initials that can be identified by ordinary light microscopy under any culture conditions used in our laboratory to induce embryogenesis. Laticifers were not observed in any callus sections or in embryoids viewed at any stage. Clearly, laticifers may be present in such material, but do not display an obvious normal morphology (branched and coenocytic) and thus cannot be recognized. Such observations mean either that laticifers are not ever present or are somewhat variable in their differentiation in cultured material because of different microenvironmental conditions, or that they are present, but unrecognizable. Immunocytochemical techniques that do not depend on morphology for cell identification are ideal methods to use to differentiate these possibilities.

Although laticifers cannot be recognized in paraffin-embedded material, the correlation between β-amyrin and fluorescence mentioned earlier at least provides some evidence that laticifers may indeed be present. In fact, with the same rather unspecific Hg-fluorescamine, Bruni et al. (1981) have been successful in localizing what they believe to be laticiferlike cells within cultures of *Euphorbia marginata*.

Using the immunocytochemical techniques described above, we have been successful in detecting fluorescent cells that do not display the typical laticifer morphology (Figs. 7 and 8) within disorganized callus of *A. syriaca* (Dunbar, unpublished). That cells within callus are marked is a clear indication that there exist, within the callus, cell populations that have a laticiferlike metabolism. Similarly, we have identified laticifers present in late heart stage and torpedo stage embryoids *A. syriaca* derived in callus cultures that are 50 days old or older (Fig. 10). We have been unable to detect fluorescent markers in younger heart stage embryos from the same cultures. We have not been able to mark laticifers in embryoids from *A. tuberosa* cultures to date and the data remains preliminary. We have preliminary evidence that laticifer cells are not marked (although they may be present) in superficial shoots from the latter species or from germinated embryoids. The latter result would be expected if the results in cultured material parallel those in zygotic material where laticifers in embryos are marked, but shoot laticifers are not marked. In any case, for *A. syriaca* this recent evidence allows us to utilize anti-latex antiserum to tag and monitor latex-containing cells in suspension cultures and to investigate the differentiation of laticifers in callus and embryoids.

The availability of laticifer-specific, but heterogeneous immunocytochemical probes, alludes to the intriguing possibility of using more specific laticifer components to raise antibodies. To date this has not been possible because few laticifer-specific components have been identified that could easily be detected by histochemical means. β-Amyrin and asclepians are believed to be laticifer-specific compounds, but this has never been proved. Biesboer and Mahlberg (1979b) have isolated β-amyrin from whole latex as a fine white crystal, 98% pure. Using β-amyrin prepared by their procedures and running it against anti-latex antiserum on an Ouchterlony plate we have determined that β-amyrin antibodies are present among other anti-latex IgGs that we prepared from whole latex. Thus, because we know we have removed nonlatex-specific materials from the anti-latex antiserum, the β-amyrin probably is a laticifer cell-specific component. Therefore, β-amyrin antibodies could be used as a cell specific probe in cultures. Along the same lines, asclepains (A3 and B5) with MWs of 23 000 and 21 000, respectively, have been isolated and purified from whole latex of *A. syriaca* by Brockbank and Lynn (1979). Antibodies to this compound would provide a cell-specific probe that would be, like anti-β-amyrin, a more homogeneous marker. Such homogeneous anti-latex markers would allow one to study the production of such resinous compounds with respect to cell maturation.

4 Future Uses of Immunological-Cytochemical Techniques in Tissue Culture

Armed with a specific immunological marker, one is able to pick the proverbial needle, a laticiferlike or embryogenic cell, out of the haystack, a large, phenotypically-mixed population of cells. Up to this time only morphological criteria have been used to infer the state of differentiation of mixed-cell populations. However, it is obvious that any two mophologically identical cells might be physiologically distinct. A marker that will bind selectively to cells in a specific developmental state will elucidate the functional status, with respect to that marker and its role in further development. Once cells of a unique physiological status are identified it is possible to begin to answer such fundamental questions concerning in vitro morphogenesis as how the marked cell is activated or whether it acts as a sink for metabolites drawing growth factors and nutrients from its neighbors and embarking on its own developmental course, or even whether the marked cell releases substances that influence neighboring cells that cause them to divide or change. More specifically, there are a number of important questions, alluded to earlier, that may be easily and successfully submitted to immunocytochemical techniques. The single most interesting question to be answered by these techniques is what is the origin of embryogenic cell types in callus. Other questions, that develop from this question, revolve around the, as yet, unclear role of auxins and cytokinins in embryogenesis. With the invention of monoclonal antibodies against various forms of these hormones and the techniques to immobilize them at their proper cell sites, immunocytochemistry will become an increasingly important addition to the arsenal of investigatory techniques. This section is intended to suggest possible future targets of immunocytochemical investigations.

4.1 Embryogenesis and Plant Hormones

A number of questions remain unanswered regarding embryoid differentiation and proliferation in plant tissue cultures. It is unclear whether embryogenic cells originate in groups within the callus or as single cells that perhaps may recruit surrounding cells to embryogenesis. It is not known in any culture system whether embryoids arise in cultured tissues themselves or whether they originate from special cells in the explant that proliferate in culture.

Several investigators (Street 1978; Tisserat et al. 1979; Kato and Takeuchi 1966) believe that embryogenesis proceeds from cells that are predetermined in the original explant. These preembryonic cells, which may clone themselves in culture, await the synthesis of an inducer substance or removal of an inhibitory substance prior to resumption of mitotic activity and development of an embryoid. Sharp et al. (1980) agree and hypothesize that a distinct population of cells becomes embryogenic in one of the following ways: (1) by proliferation of a unique phenotypic population from the original explant; (2) by lengthening of the mitotic cell cycle time of a particular cell population by interference with one or more cell cycle control points; or (3) by determination and arrest of a distinct population of cells in the cell cycle with a blockage of G_1, G_2, or G_0 with subsequent release to the embryogenic developmental sequence by placement in the induction medium.

In both callus and suspension cultures of *Asclepias syriaca* and *A. tuberosa* it is unknown whether cells that participate in embryogeny are determined in the explant and simply proliferated or are determined at some point in time after proliferation of dedifferentiated totipotent cells derived from the explant. The results of Wernike et al. (1982) suggest that in *Sorghum* callus, the auxin, 2,4-D, prevents organ differentiation by suppressing already present primordia. We are able, in *A. tuberosa* suspension cell-cultures, to repress embryogenesis simply by raising 2,4-D levels in the maintenance medium from 0.5 mg l^{-1} to 4.0 mg l^{-1}. Subculturing 2–5 ml aliquots from inhibited suspension cell cultures (which include carry-over 2,4-D) to a non-2,4-D-supplemented medium induces cultures to become highly embryogenic within 2 weeks (Wilson, unpublished). Immunocytochemical approaches, using the presently available antibodies to forms of plant hormones (for instance, from Idetek, Inc., San Bruno, California, USA) should give insight into the role of auxins and cytokinins in embryoid induction in tissue culture. For example, it may prove or disprove the hypothesis that auxin causes the proliferation of a specific embryogenic cell type from the original explant. A marker specific for IAA will reveal the presence of cells producing this substance in the explant, if they exist, and allow us to determine if derivatives of this cell type proliferate in culture.

4.2 Embryogenesis and Embryogenic Protein

Related to the above questions is the question of whether embryoids arise from single cells or embryogenic clusters within callus or among suspension cells. Single cell origin of embryoids within cell aggregates has been reported, for instance, in suspension cultures of several plants (Halperin and Wetherell 1964; Halperin 1967; Halperin and Jensen 1967; McWilliam et al. 1974; Dunstan et al. 1982; Conger et al. 1983; Kononowicz et al. 1984). However, origin of embryoids from whole complexes of cells, not from single committed cells, has also been reported (Reuther 1977; Haccius 1978; Wernicke et al. 1982; Dos Santos et al. 1983). The problem is to identify some factor in embryogenic cultures related directly to a particular cell's participation in embryogenesis. Along these lines we have identified, isolated, and purified a unique protein that appears in *A. syriaca* suspension cultures 10 h after cells have been transferred to an auxin-deficient medium (Biesboer, Wilson and Petersen, manuscript in preparation). This "embryogenic" protein (E protein) is similar to that derived and studied by Sung and Okimoto (1981, 1983) in cultures of carrot. E protein from *A. syriaca* cultures is a glycoprotein of MW of about 55000. Such proteins are prime candidates against which to prepare antibodies. If markers can be derived from such nonhormonal factors, they may contribute to the elucidation of embryogenesis by identifying the cells that become involved in the process at very early, and morphologically undetectable, stages.

4.3 Cell Sorting

Recently it has been demonstrated that plant cells that can be immunocytochemically labeled can be sorted out from unlabeled cells in a flow cytometer (normally used for sorting blood cells). Galbraith et al. (1984) demonstrated that sorted protoplasts of suspension cultures of *Nicotiana tabacum* could be recovered and successfully cultured, in spite of the presence of fluorescent marker. It will soon be routine to sort out plants cells labeled for various compounds and the source of choice will be from tissue culture since cultured cells are easily macerated, sized (by mesh filtration), and remain viable. With the availability of antibodies against plant hormones it will soon be straightforward to sort out cells in culture that are enriched for IAA production. Theoretically, it should be possible to increase the rate of embryogenesis within cultures derived from such cells. If E protein is involved in inducing or recruiting cells to become embryogenic, it is not unlikely that E protein-producing enriched cell cultures will also be highly embryogenic. Finally, the ability to sort out such cells opens exciting possibilities which may be directly applied to crop plant culture systems where culture material is the most recalcitrant to growth and embryogenesis.

Acknowledgements. The author wishes to thank Dr. David Biesboer, Dr. Bruce Petersen, and Mr. Kerry Dunbar for help in developing the manuscript. Research represented here was supported by a grant from NSF (PCM 81-04092) to KJ Wilson and a grant from the Whitehall Foundation to KJ Wilson, BH Petersen, and DD Biesboer.

References

Berlyn GP, Mikshe JP (1976) Botanical microtechnique and cytochemistry, 1st edn. The Iowa State University Press, Ames, Iowa

Biesboer DD (1984) The detection of cells with a laticifer-like metabolism in *Asclepias-syriaca* L. suspension cultu res. Plant Cell Rep 2:137–139

Biesboer DD, Mahlberg PG (1979a) Sterol synthesis and identification in cultures of *Euphorbia tirucalli* L. In: Sala F, Parisi B, Cella R, Cifferri O (eds) Plant cell cultures: results and perspectives. Elsevier/North-Holland Biomedical, Amsterdam

Biesboer DD, Mahlberg PG (1979b) The effect of medium modification and selected precursors on sterol production by short-term callus cultures of *Euphorbia tirucalli*. J Nat Prod (Lloydia) 42:648–657
Brockbank WJ, Lynn K (1979) Purification and preliminary characterization of two asclepians from the latex of *Asclepias syriaca* L. (milkweed). Biochim Biophys Acta 578:13–22
Bruni A, Tosi B (1980) A method for localizing embryonal laticifers by combined conventional and fluorescent microscopy. Protoplasma 102:343–347
Bruni A, Dall'Olio G, Mares D (1977) Use of fluorescent labeling methods in morphological and histochemical studies of latex in situ. Caryologia 30:486–487
Bruni A, Vannini GL, Dall'Olio G (1981) Occurence of laticifers in tissue cultures derived from *Euphorbia marginata*: a study by fluorescence microscopy. Z Pflanzenphysiol 103:373–377
Buehrer T, Benson L (1945) Rubber content of native plants of the southwestern desert. Tech Bull 108, Univ Arizona Agr Exp Station
Chaleff RS (1983) Isolation of agronomically useful mutants from plant cell cultures. Science 219:676–682
Chauveaud G (1891) Recherches embryogeniques sur l'appareil laticifere des Euphorbiacees, Urticacees, Apocynacees, et Asclepiadacees. Ann Sci Nat Bot Biol Veg 14:1–161
Conger BV, Hanning GE, Gray DJ, McDanial JK (1983) Direct embryogenesis from mesophyll cells of orchardgrass. Science 221:850–851
Coons AH, Kaplan MH (1950) Localization of antigens in tissue cells II. Improvement in a method for the detection of antigen by means of fluorescent antibody. J Exp Med 91:1–11
Dall'Olio G, Tosi B, Bruni A (1978) A new rapid fluorescent labeling method for the detection of the latex tissues in Euphorbiaceae plants. Planta Med 34:183–187
DeMay J (1981) Colloidal gold probes in immunocytochemistry. In: Immunocytochemistry. Wright, Boston, pp 82–112
DeWaele J, DeMey J, Moerens M, VanCamp B (1981) The immuno-gold staining method: an immunocytochemical procedure for leukocyte characterization by monoclonal antibodies. In: Knapp W (ed) Leukemia markers. Academic Press, London, pp 173–176
Dos Santos A, Cutter EG, Davey MR (1983) Origin and development of somatic embryos in *Medicago sativa* L. (alfalfa). Protoplasma 117:107–115
Dunstan D, Short K, Merrick K, Collin HA (1982) Origin and early growth of celery embryoids. New Phytol 91:121–128
Esau K (1965) Plant anatomy, 2nd edn. Wiley, New York
Galbraith DW, Afonso DL, Harkins KR (1984) Flow sorting and culture of protoplasts: Conditions for high-frequency recovery, growth and morphogenesis from sorted protoplasts of suspension cultures of nicotiana. Plant Cell Rep 3:151–155
Geuze HJ, Slot J, VanderLey P, Scheffer C, Griffith J (1981) Use of colloidal gold particles in double labeling immuno-electron microscopy of ultrathin frozen tissue sections. J Cell Biol 89:653–665
Groett S, Kidd G (1981) Somatic embryogenesis and regeneration from milkweed cell cultures. Biomass 1:93–97
Haccius B (1978) Question of unicellular origin of non-zygotic embryos in callus cultures. Phytomorphology 28:74–81
Halperin W (1967) Population density effects on embryogenesis in carrot cell cultures. Exp Cell Res 48:170–173
Halperin W, Jensen W (1967) Ultrastructural changes during growth and embryogenesis in carrot cell cultures. J Ultrastruct Res 18:428–443
Halperin W, Wetherell D (1964) Adventive embryony in tissue culture of wild carrot, *Daucus carota*. Am J Bot 51:274–283
Heinrich G (1970) Electronenmikroskopische Untersuchung der Milchröhren von *Ficus elastica*. Protoplasma 70:317–323
Himmelhoch S, Zuckerman BM (1982) *Xiphinema index* and *Caenorhabditis elegans*: preparation and molecular labeling of ultrathin frozen sections. Exp Parasitol 54:250–259

Jeffree CE, Yeoman MM, Kilpatric DC (1982) Immunofluorescence in cells. Int Rev Cytol 80:231–265
Johansen D (1940) Plant microtechnique. McGraw-Hill, New York
Kato H, Takeuchi M (1966) Embryogenesis from the epidermal cells of carrot hypocotyl. Sci Pap Coll Gen Educ Univ Tokyo (Biol Part) 16:245–253
Knox RB (1982) Methods for locating and identifying antigens in plant tissues. In: Bullock GR, Petrusz P (eds) Techniques in immunochemistry, vol 1. Academic Press, London, pp 205–238
Knox RB, Clarke AE (1978) Localization of proteins and glycoproteins by binding to labeled antibodies and lectins. In: Hall JL (ed) Electronmicroscopy and cytochemistry of plant cells. Elsevier/North Holland Biomedical, Amsterdam, pp 149–185
Kononowicz H, Kononowicz AK, Janick JJ (1984) Asexual embryogenesis via callus of *Theobroma cacao*. Z Pflanzenphysiol 113:347–358
Mahlberg PG (1968) Growth response of the laticifer and the juvenile shoot apices of *Euphorbia marginata*. Phytomorphology 17:429–437
Marty F (1968) Infrastructure des laticiferes differencies d'*Euphorbia characias*. Compt Rend Hebd Seances Acad Sci 267:299–302
Marty F (1970) Role du systeme membranaire vacuolaire dans la differenciation des laticiferes d'*Euphorbia characias* L. Compt Rend Hebd Seances Acad Sci 271:2301–2304
Marty F (1971) Vesicles autophagiques des laticiferes differencies d'*Euphorbia characias* L. Compt Rend Hebd Seances Acad Sci 272:399–402
Mayer RJ, Walker JH (1980) Immunochemical methods in the biological sciences: enzymes and proteins. Academic Press, New York
McWilliam AA, Smith A, Street H (1974) The origin and development of embryoids in suspension cultures of carrot (*Dautus carota*). Ann Bot (Lond) 38:243–250
Murashige T, Skoog F (1962) A revised medium for rapid growth and bioassays with tobacco tissue culture. Physiol Plant 15:473–497
Nielsen PE, Nishimura H, Otvos H, Calvin M (1977) Plant crops as a source of fuel and hydrocarbon-like materials. Science 198:942–944
Nishimura H, Philip R, Calvin M (1977) Lipids of *Hevea brasiliensis* and *Euphorbia coerulescens*. Phytochemistry (Oxf) 16:1048–1049
Paul E, Blakers A, Watson R (1943) The rubber hydrocarbon of *Asclepias syriaca* L. Can J Res 21:219–223
Prabhudesai V, Narayanaswamy S (1974) Organogenesis in tissue cultures of certain Asclepiads. Z Pflanzenphysiol 71:181–185
Rao P, Narayanaswamy S (1972) Morphogenetic investigations in callus cultures of *Tylophora indica*. Physiol Plant 27:271–276
Rao P, Narayanaswamy S, Benjamin B (1970) Differentiation *ex ovulo* of embryos and plantlets in stem tissue cultures of *Tylophora indica*. Physiol Plant 23:140–144
Reuther G (1977) Embryoide Differenzierungsmeister im Kallus der Gattungen *Iris* und *Asparagus*. Ber Dtsch Bot Ges 90:417–437
Roman O, Stolinski C, Hughes-Jones N (1974) An antiglobulin reagent labelled with colloidal gold for use in electron microscopy. Immunochemistry 11:521–522
Roth J (1982) The protein A-gold (pAg) technique – a qualitative and quantitative approach for antigen localization on thin sections. In: Bullock GR, Petrusz P (eds) Techniques in immunocytochemistry, vol 1. Academic Press, London, pp 107–134
Roth J (1983) The colloidal gold marker system for light and electron microscopic cytochemistry. In: Bullock GR, Petrusz P (eds) Techniques in immunocytochemistry, vol 2. Academic Press, London, pp 217–284
Schulze C, Schnepf E, Mothes K (1967) Über die Lokalisation der Kautschukpartikel in verschiedenen Typen von Milchrohren. Flora, Abt A 158:458–460
Selber JN, Lee SM, Benson JM (1984) Chemical characteristics and ecological significance of cardenolides in *Asclepias syriaca* (milkweed) species. In: Ness WD, Fuller G, Tsai LS (eds) Isopentenoids in plants, biochemistry and function. M Dekker, NY, pp 563–589
Sharp WR, Sondahl MR, Caldos LS, Maraffer SB (1980) The physiology of in vitro asexual embryogenesis. In: Janick J (ed) Horticultural reviews, vol 2. AVI, Westport, CT, pp 268–230

Shukla OP, Murti DR (1971) The biochemistry of plant latex. I Sci Ind Res (India) 30:640
Snowcroft WR, Larkin PJ (1982) Somaclonal variation: a new option for plant improvement. In: Vasil IK, Snowcroft WR, Frey KJ (eds) Plant improvement and somatic cell genetics. Academic Press, NY, pp 159–178
Sternberger L (1979) Immunocytochemistry, 2nd edn. Wiley, New York
Street HE (1978) Differentiation in cell and tissue cultures – regulation at the molecular level. In: Schutte HR, Gross D (eds) Regulation of developmental processes in plants. Fischer, Jena, pp 192–218
Sung ZR, Okimoto R (1981) Embryonic proteins in somatic embryos of carrot. Proc Natl Acad Sci USA 78:3683–3687
Sung ZR, Okimoto R (1983) Coordinate gene expression during somatic embryogenesis in carrots. Proc Natl Acad Sci USA 89:2661–2665
Tisserat B, Esan EB, Murashige T (1979) Somatic embryogenesis in angiosperms. In: Janick J (ed) Horticultural reviews, vol 1, Westport, CT, pp 1–78
Tizard IR (1984) Immunology, an introduction. Saunders, Philadelphia
Wernicke W, Potrykus I, Thomas E (1982) Morphogenesis from cultured leaf tissue of *Sorghum bicolor* – the morphogenetic pathways. Protoplasma 111:53–62
Wick SN, Robert W, Osborn M (1981) Immunofluorescence microscopy of organized microtubule arrays in structurally stabilized meristematic plant cells. J Cell Biol 89:685–690
Wilson K, Ellis B (1982) Preliminary light and electron microscope studies of laticifers in *Stapelia bella* (Ascelepiadaceae). AIBS Meet – Am Bot Soc, Penn State Univ (abstract)
Wilson K, Frantz V (1981) The ultrastructure of nonarticulated branched laticifers in *Asclepias tuberosa* L. (butterfly weed). Invited paper, Symp Secretory Structures. Am Bot Soc Meet, Bloomington, IN
Wilson K, Mahlberg P (1977) Investigations of laticifer differentiation in tissue cultures derived from *Asclepias syriaca* L. Ann Bot (Lond) 41:1049–1054
Wilson K, Mahlberg P (1980) Ultrastructure of developing and mature nonarticulated laticifers in the milkweed *Asclepias syriaca* L. (Asclepiadaceae). Am J Bot 67:1160–1170
Wilson KJ, Nessler CL, Mahlberg PG (1976) Pectinase in *Asclepias* latex and its possible role in laticifer growth and development. Am J Bot 63:1140–1144
Wilson K, Petersen B, Biesboer D (1984) Immunocytochemical identification of laticifers. Protoplasma 122:86–90

Measurement of Oat Globulin by Radioimmunoassay

D. S. LUTHE

1 Introduction

Storage proteins, the proteins that accumulate during seed development, have been extensively studied because of their nutritional importance. For this reason, they are of interest to plant molecular biologists who would like to improve the amino acid composition of these proteins using genetic engineering technology. They are also of interest to developmental biologists because their accumulation during seed development and no other part of the plant life cycle provides such an excellent example of selective gene expression. In the effort to learn more about the structure, accumulation, and biosynthesis of storage proteins, the use of specific antibodies has been invaluable. For example, immunoaffinity columns have been used to isolate radiolabeled storage protein precursors from subcellular fractions, resulting in information about the site of synthesis and deposition into protein bodies (Chrispeels et al. 1982a, b). Antibodies have also been used to elucidate precursor-product relationships and posttranslational processing events for many different plant species. However, one of the most frequent uses of antibodies is to immunoprecipitate storage proteins from in vitro translation products. This technique is particularly useful when hybrid-select translation is used to identify specific messenger RNA molecules (McGrogan et al. 1979; Meinke et al. 1981). The most commonly used method of quantifying specific storage proteins during seed development has been immunoelectrophoresis (Weeke 1973; Sun et al. 1978; Crouch and Sussex 1981). However, the radioimmunoassay (RIA) (Felsted et al. 1982; Triplett and Quatrano 1982; Colyer and Luthe 1984) and a variation of the RIA (Domoney et al. 1980) using the ELSA technique have been used to quantify several other plant proteins.

1.1 Theory of the RIA

The RIA was developed by Rosalyn S. Yallow and Solomin A. Berson in 1959 (see Hawker 1973; for an extensive list of references) and they originally used it to detect insulin in human plasma. Typically, the RIA has been widely used in the biomedical sciences to measure substances such as hormones, vitamins, drugs, metabolites, etc. in biological specimens. The RIA can be used to detect any substance which is of sufficiently low concentration to make detection by biological, chemical, or instrumental means difficult. It is especially useful for detecting small amounts of nonradioactive material, even when it is present in a mixture of extraneous material (Friefelder 1982). It is a very sensitive assay and can quantitate 10^{-11} to 10^{-9} g of a substance per sample (Cooper 1977).

The major components of the RIA are the *antibody* and *antigen*. An antibody is a serum protein that is produced in response to the injection of a foreign substance, generally a protein, into a higher animal. These serum proteins are called immunoglobulins and they will specifically bind with the foreign substance, which is called the antigen. These immunoglobulins are primarily of the G class, thus they are called immunoglobulin G or IgG. If the antigen and antibody are incubated under the appropriate conditions, the antibody will specifically bind to the antigen and an antibody/antigen complex will form.

The RIA is a competitive assay, in which antigens that are identical (excepts that one is radiolabeled, and the other is not) compete for binding to a limited

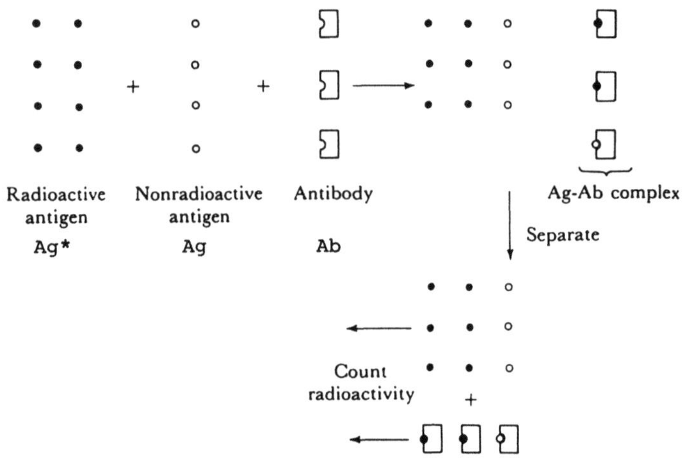

Fig. 1. Principles of the radioimmunoassay. In the radioimmunoassay (RIA) radioactive and nonradioactive antigen molecules compete for binding with a limited amount of antibody. After the binding takes place free and bound antigen are separated and the amount of antigen bound to the antibody is quantitated. The addition of nonradioactive antigen to the RIA will reduce the binding of radioactive antigen. If this is done in a stepwise fashion, a standard curve can be generated in which the decrease in binding of radioactive antigen is proportional to the amount of nonradioactive antigen added. (From Physical Biochemistry by David Freifelder 1982)

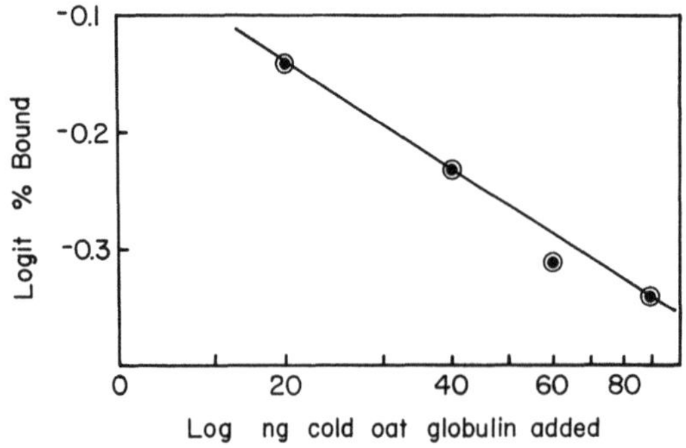

Fig. 2. A typical standard curve for the oat globulin radioimmunoassay. As the amount of nonradioactive (cold) oat globulin in the assay increased the percent of radioactive antigen bound to the antibody decreased

or fixed amount of antibody (see Fig. 1). As the amount of unlabeled antigen added to the reaction increases, there is a reduction in the amount of radiolabeled antigen bound to the antibody. If the bound antigen can be separated from the free or unbound antigen, then the amount of radioactivity bound to the antibody can be quantitated. By keeping the amount of radiolabeled antigen and antibody constant, and by adding increasing amounts of unlabeled antigen to a set of reaction tubes, a standard curve can be prepared (Fig. 2). One can then determine the amount of antigen in an unknown sample by comparing the reduction in the binding of radiolabeled antigen to antibody to values on the standard curve.

1.2 Oat Globulin

Seed storage proteins have traditionally been classified into four solubility fractions: albumins, water soluble; globulins, salt-soluble; prolamins, alcohol-soluble; and glutelins, acid or alkali-soluble (Osborne 1910). Globulin is the most abundant storage protein fraction in oat (*Avena sativa* L.) Oats are atypical among the cereals because the major portion of their storage protein is globulin and not prolamin or glutelin, as is the case for wheat, barley, corn and rye (Larkins 1981). There have been varying reports of the proportion of the oat seed protein that is globulin. Values ranging from 12% (Michael et al. 1961; Volker 1975; Wieser et al. 1980) to 80% (Brohult and Sandegren 1954) have been reported. Typically, it has been reported that globulin contributes about one-half of the total seed protein (Peterson 1976; Peterson and Smith 1976). The variation in these estimates may result from the use of the Osborne (1910) fractionation method. The use of this method often results in incomplete extractions and cross-contamination of fractions (Peterson and Brinegar 1984). For example, the glutelin fraction (the last fraction extracted when the Osborne method is used) generally contains incompletely extracted albumin, globulin, and prolamin. Consequently, the contribution of glutelin to the total seed protein has been overestimated (Peterson and Brinegar 1984; Robert et al. 1983). Electrophoretic analysis of oat glutelin extracts indicated that there are globulin polypeptides present in that fraction (Peterson and Brinegar 1984; Robert et al. 1983). This has led several researchers (Peterson and Brinegar 1984; Robert et al. 1983) to estimate that oat globulin may comprise as much as 75% of the total seed protein. To avoid the problem of cross-contamination of the Osborne fractions, the RIA, a technique that can be used to quantitate substances in complex mixtures, was used to determine the amount of globulin in developing and mature oat seeds.

2 Requirements for the Radioimmunoassay

Several components are needed to develop a radioimmunoassay. These are pure antigen, radiolabeled antigen, and antibody. The sections below will describe how each component was prepared for the RIA of oat globulin.

2.1 Preparation of Pure Antigen

"The success or failure of immunochemical methods is largely dependent on the initial step, preparation of the antigen" (Cooper 1977). The antigen will be used for three purposes: (1) it will be injected into the animal to elicit antibody production; (2) it will be radiolabeled; and (3) it will be used to prepare the standard curve. The immune systems of animals are capable of detecting small amounts of contaminating antigens (Cooper 1977); therefore, to insure that there are no immunological cross-reactions, and that the RIA is specific, it is important that the antigen be as pure as possible. Most standard protein purification methods are adequate for antigen preparation. The purity of the antigen should be checked by polyacrylamide gel electrophoresis and no contaminating bands should be evident (Cooper 1977). If the protein is comprised of subunits, each subunit is likely to have its own set of antigenic determinants and, therefore, a set of antibodies will be produced (Cooper 1977). In some cases, the specificity of the antibody preparation can be improved by injecting purified subunits (Cooper 1977). For the detection of in vitro translation products, Walburg and Larkins (1983) used the purified acidic subunit of oat globulin for antibody production. In most cases, this is not necessary; for example, antibody raised against whole globulin, and not purified subunits, was used for both the detection of in vitro translation products (Rossi and Luthe 1983; Brinegar and Peterson 1982; Matlashshewski et al. 1982) and the RIA (Colyer and Luthe 1984).

Because most storage proteins are relatively abundant in the seed, and because it is easy to obtain large quantities of most seeds, purification of seed storage proteins for use as antigens is not difficult. A modification of the classical Osborne (1910) procedure was used to extract oat globulin (Rossi and Luthe 1983). Albumins were first extracted by stirring the defatted flour in 10 vol water for 2 h at room temperature. After extraction, the mixture was centrifuged at 27000 × g for 30 min at room temperature. The supernatant containing the albumins was dis-

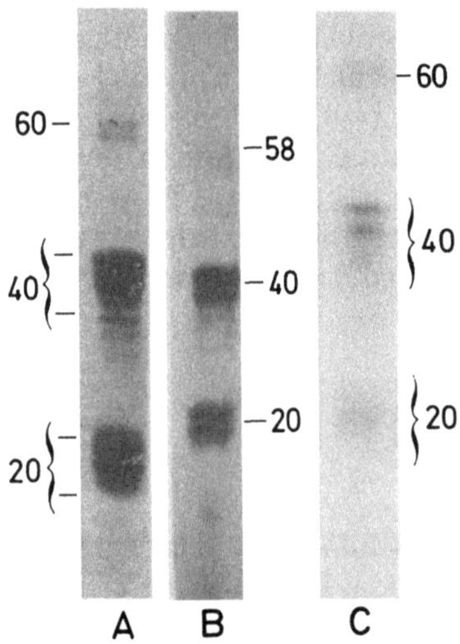

Fig. 3. SDS polyacrylamide gel electrophoresis of oat globulin. *Lane A* Blue-stained gel of authentic oat globulin. *Lane B* Fluorograph of oat globulin labeled in vitro with [^{14}C]-formaldehyde. *Lane C* Fluorograph of oat globulin labeled in vivo with [^{35}S]-sulfate and immunoprecipitated with anti-oat globulin IgG. *Numbers* refer to the apparent molecular weight of the polypeptides in kilodaltons (kD). *Note*: The enlargement of the photographs depicting the gels are not identical

carded. The pellet was extracted with globulin extraction buffer (1 M NaCl, 50 mM Tris-HCl (pH 8.5), and 1% (v/v) β-mercaptoethanol) (Peterson 1978). After stirring for 2 h at room temperature, the suspension was centrifuged at 27000 × g for 30 min at room temperature. The supernatant, globulin, was dialyzed overnight (4 °C) against several changes of distilled water. During dialysis the salt-soluble globulin precipitated. The precipitate was lyophilized and the amount of protein in the extract was determined by the Lowry method (Lowry et al. 1951). Analysis of oat globulin by sodium dodecyl sulfate gel electrophoresis (SDS-PAGE) is shown in Fig. 3 A. Under reducing conditions two major subunits of approximately 40 and 20 kD were present on the gel. The small band with molecular weight of approximately 60 kD is believed to be an unprocessed precursor of the 20 and 40 kD polypeptides (Luthe and Peterson 1977). Oat globulin extracted in this manner or by similar methods (Rossi and Luthe 1983; Matlashsheweski et al. 1982) has been successfully used for the production of anti-oat globulin antibodies (anti-oat globulin IgG).

2.2 Preparation of Radiolabeled Antigen

There are two things to keep in mind when preparing radiolabeled antigen for the RIA: (1) the sensitivity of the RIA is proportional to the specific activity of the antigen; and (2) the addition of the radiolabel should not damage the molecule, which would alter its antigenicity. It is important to have a high specific activity antigen, because to achieve maximum sensitivity in the RIA, the concentration of the labeled antigen must be lower than the minimum concentration of the antigen to be measured (Hawker 1973). The isotopes most frequently used to label proteins for the RIA are [^{131}I], [^{125}I], [^{14}C], and [^{3}H]. All of these isotopes emit β-particles, except for [^{125}I], which is a strong γ-ray emitter. The list in Table 1 shows various isotopes that can be used for radiolabeling proteins and it lists the number of atoms of each isotope that must be incorporated to produce an arbitrary count rate (Cooper 1977). [^{131}I] and [^{125}I], therefore, will produce the highest specific activity antigens. The disadvantage of [^{131}I] is that it has a short half-life of only 8.04 days; consequently, [^{125}I] is commonly used to label antigens, because it has a longer half-life (60 days) and it can be detected by a gamma counter. There are two principal methods for incorporating [^{125}I] (or [^{131}I] into purified antigens (Cooper 1977). With both methods one must be careful to avoid the loss of immunoreactivity (Cooper 1977). The covalent attachment of a large atom, such as iodine to a protein can alter its conformation; therefore, the addition of more than one I atom per protein molecule should be avoided (Cooper 1977). Kits are now commercially available for radioiodination.

When the RIA for oat globulin was designed, we (Colyer and Luthe 1984) decided to sacrifice sensitivity and use [^{14}C]-formaldehyde to label oat globulin. This was possible because oat globulin comprises such a large part of the total seed protein; that is, we were measuring a substance that was present in the seed in milligram quantities, so sensitivity was not as crucial as of the protein was present in much smaller quantities. Also, [^{14}C] was chosen because it is a safer isotope to work with than [^{125}I] or [^{131}I].

Table 1. Number of various isotopic atoms needed to produce a given counting rate[a]

Isotope	Number of isotopic atoms incorporated per macromolecule
^{131}I	1
^{32}P	1.8
^{125}I	7.5
^{35}S	10.9
^{3}H	557.0
^{14}C	261 672.0

[a] The values in Table 1 were calculated through the use of the equation $\lambda = \frac{0.693}{T^{1/2}}$, where λ is the rate of isotopic disintegration and $T^{1/2}$ is the half-life of the isotope in seconds. Disintegration rates may be converted to curies using the conversion factor $1\ Ci = 3.7 \times 10^{10}$ dps. (From Cooper 1977)

Oat globulin was radiolabeled with [^{14}C]-formaldehyde using a reductive alkylation method previously developed (Means and Freeney 1968; Rice and Means 1971). The method was designed as a gentle way of labeling a protein with [^{14}C] without altering its physiochemical structure (Rice and Means 1971). Presumably, a protein could also be labeled with [^{3}H] using the same method. To label the antigen 1 mg of oat globulin (1 mg ml^{-1} in 0.2 M sodium borate buffer, pH 9.0) was alkylated with 50 μCi of [^{14}C]-formaldehyde (specific activity 10 mCi $mmol^{-1}$). Unlabeled formaldehyde (100 μl of 0.04 M) was added to the ampule containing the [^{14}C]-formaldehyde and this mixture was added to the oat globulin solution while it was kept on ice. After 30 s four 2 μl aliquots of sodium borohydride (5 mg ml^{-1}) were sequentially added. To insure complete reduction of the formaldehyde an additional 10 μl of sodium borohydride was added after 1 min of incubation. Following the labeling, the unincorporated label was removed from the oat globulin by dialyzing the sample against borate-buffered saline (BBS) (5 mM boric acid, 2.5 mM Na borate, pH 8.4, 147.5 mM NaCl) overnight at 4 °C with several buffer changes. Following dialysis the globulin was divided into aliquots and stored at −20 °C. The specific activity of the labeled globulin was determined by removing an aliquot and determining the protein content (Lowry et al. 1951). The amount of radioactivity in another aliquot was determined by scintillation spectroscopy. When 50 μCi of [^{14}C]-formaldehyde was used to label 1 mg of oat globulin, the specific activity was 2.0 and 1.3×10^6 dpm mg^{-1} for two separate labeling experiments. When 250 μCi of [^{14}C]-formaldehyde were used to label 10 mg of oat globulin, the specific activity was 5.9×10^5 dpm mg^{-1}. To test that the alkylation did not structurally damage the oat globulin it was analyzed by SDS-PAGE followed by fluorography. Figure 3 B indicates that alkylated oat globulin had the same electrophoretic mobility as unlabeled oat globulin.

2.3 Antibody Production

For detailed instructions for antibody production an immunological methods book should be consulted. The two books listed as references (Cooper 1977; Garvey et al. 1977) were especially helpful in providing both theoretical and practical information. In view of animal care regulations, it is also a good idea to consult a veterinarian for information about care and handling of the animals used for antibody production. The methods discussed below are those that were used to produce anti-oat globulin antibody.

White New Zealand rabbits are most often used for antibody production; however, these rabbits were not available locally and a California-New Zealand cross was used (Rossi and Luthe 1983). To have sufficient serum it is generally best to use at least several rabbits. Before injections of the antigen are made, samples of blood from each rabbit should be taken for control or preimmune serum. Blood may be collected by either ear vein bleeding or cardiac puncture (Garvey et al. 1977).

When an animal is first injected with an antigen, there is a lag period of several days before antibody production begins. After the lag antibody production increases exponentially, reaches a steady state and then decreases. This response is called the primary response. If another injection is made after the production, antibody in the primary response has declined, a secondary response, or increase in antibody production occurs (Cooper 1977). This response occurs after a shorter lag time, results in higher levels of antibodies, and persists for a longer time (Cooper 1977). It is during this phase that the antisera should be collected. Repeated inoculations result in loss of antibody specificity (Cooper 1977).

To prepare the antigen for the first injection 0.1 to 1 mg of the protein should be mixed with an equal volume of Freund's complete adjuvant (Cooper 1977; Garvey et al. 1977). The adjuvant increases the antigenicity and prolongs the immune response (Cooper 1977). For anti-oat globulin preparation the antigen-adjuvant mixture was injected intramuscularly in several places on the back thigh of the rabbit. Three subsequent injections were made at 1 mo intervals following the first injections. For these last three injections, the antigen was mixed with Freund's incomplete adjuvant (Cooper 1977, Garvey et al. 1977). Two weeks after the final injection the rabbits were sacrificed by exsanquination and the blood was collected for serum preparation.

During the injection regime, blood samples should be taken and the titer of the serum should be determined. From this information one can determine the relative magnitude of antibody production during the primary and secondary immune responses. The relative amount of anti-oat globulin IgG produced by the rabbits during the immune response period was determined by checking the titer of the serum at 1 week intervals following the injection of the antigen. Small volumes of blood (about 5 ml) were removed from the rabbit each week. The blood was allowed to clot and the serum was collected by centrifugation (details of serum preparation will be described below). To test the titer an oat globulin solution [0.5 mg ml^{-1} dissolved in 0.085% NaCl solution, which is equivalent to an antigen dilution of 1/2000 (according to Garvey et al. 1977)] was serially diluted. One hundred μl of each antigen dilution was mixed with one drop of serum and the mixture was incubated at 37 °C for 1 h. The tubes were then checked for the presence of an immunoprecipitate. In the preparation of anti-oat globulin serum the greatest dilution at which immunoprecipitation occurred was the 1/16 dilution or 1/32,000 dilution of the antigen. This was defined as the titer, that is, the lowest concentration resulting in immunoprecipitation. The titer may also be determined by the Ouchterlony double diffusion method (Garvey 1977). An alternative way of defining the titer is the dilution of serum required to bind 50% of the total labeled antigen (Hawker 1973).

2.4 Serum and IgG Preparation

Because we found that there was little difference in the titer between the two rabbits used, their blood was pooled. Tubes containing the blood were ringed with a sterile wooden stick and the blood was allowed to clot at room temperature for 1–2 h. Following the clotting the blood was kept in the refrigerator at 4 °C overnight. The next day the antiserum was collected by centrifuging the blood at 5000 × g for about 15 min and the serum was carefully decanted from the tube. From two rabbits about 60 ml of antisera were obtained.

Serum can be stored frozen indefinitely at −20 °C or −70 °C; it is also possible to store the serum at 4 °C if anti-microbial agents are added to the sample (Cooper 1977). The antisera can be used directly for the RIA (Hawker 1973), but because the antibodies would be used for the immunoprecipitation of in vitro translation products, in addition to the RIA, the serum proteases were removed from the IgG fraction by chromatography or DEAE-Affigel Blue as described (Biorad Laboratories Technical Bulletin No. 1062, Richmond, California, USA; Williamson 1981). For this method the antiserum (approximately 5 ml) was dialyzed overnight at 4 °C against starting buffer (20 mM Tris-HCl, pH 8.0, and 28 mM NaCl) to remove excess salt from the serum. Prior to use, the DEAE-Affigel Blue column (25 ml) was washed with a least five bed volumes of 100 mM acetic acid, pH 3.0, 1.4 M NaCl and 40% isopropanol. Following this treatment the column was equilibrated with at least 10 bed volumes of starting buffer. Then 1 ml of dialyzed serum was applied to the column and the protein was eluted from the column with starting buffer until all 280 nm absorbing material was eluted from the column. Before another sample was applied to the column, it was regenerated with 5 bed volumes of a solution containing 8 M urea and 1.4 M NaCl followed by starting buffer. Once the column was regenerated, another 1 ml aliquot was applied to the column. This was repeated until 5 ml of serum was run through the column with reequilibration between each 1 ml sample. The material eluted with starting buffer was pooled and precipitated by making the solution 50% of saturation with ammonium sulfate. The ammonium sulfate precipitate was collected by centrifugation at 14000 × g for 10 min, dissolved in 0.1 M Na bicarbonate and 0.5 M NaCl, and dialyzed against the same solution at 4 °C overnight. The protein content of the sample was determined using the method of Murphy and Kies (1964) and was 306 mg ml^{-1}. This fraction contained IgG and transferin. Alternative methods for immunoglobulin isolation and concentration are ammonium sulfate fractionation, column chromatography (Cooper 1977), or immunochromatography (Shapiro et al. 1974).

It is important to check the specificity of the antiserum or purified IgG fraction before it is used for the RIA or any other immunological technique. The specificity of the anti-oat globulin IgG fraction purified by the DEAE-Affigel Blue was tested by two methods. One was by Ouchterlony double diffusion (Garvey et al. 1977; Cooper 1977) in which the IgG fraction was placed in the center well of an agarose gel and solutions of the four storage protein fractions, albumins globulins, prolamins, and glutelins were placed in the outer wells. Lines of identity, where antigen and antibody met and immunoprecipitated, were present only between the center wells and the wells containing oat globulin. The second way of checking the specifity of the fraction was to use it to immunoprecipitate an extract from oat seeds labeled in vivo with [^{35}S]-sulfate (Rossi and Luthe 1983). The immunoprecipitate was then analyzed by SDS-PAGE and fluorography (Rossi and Luthe 1984). The pattern of the immunoprecipitated globulin was identical to authentic globulin, with the exception that a slightly greater amount of the 60 kD polypeptides were immunoprecipitated (Fig. 3 C). However, the 60 kD polypeptides are known to be precursors of the 40 and 20 kD polypeptides, and hence immunologically similar to the 20 and 40 kD polypeptides (Brinegar and Peterson 1982; Walburg and Larkins 1983; Adeli and Altosaar 1983).

2.5 Preparation of Oat Seed Extracts

The major objective for developing the RIA was to quantitate oat globulin in developing and mature seeds (Colyer and Luthe 1984). Plant material for these experiments was collected from field-grown oats. The individual panicles were tagged at anthesis, harvested at 3-day-intervals from 3 to 24 days-post-anthesis (DPA), and at maturity (30 DPA). The seeds were dehulled, frozen in liquid N_2, and lyophilized. Seed extracts for the RIA were prepared in the following manner using a Teckmar Tissumizer: five seeds were extracted at room temperature in 1 ml of BBS (two times); 1 ml globulin extraction buffer without β-mercaptoethanol (four times); and 0.1 N NaOH (one time). After each extraction the homogenate was centrifuged at 13 000 × g in a microcentrifuge and the pellet was rehomogenized in the appropriate buffer. All of the supernatants were pooled with the exception of the NaOH extract. To insure that all the globulin was extracted, the protein content of the pooled and NaOH supernatants were determined (Lowry et al. 1951). In all cases the amount of protein in the NaOH fraction was negligible, indicating that 93% to 97% of the total seed protein was extracted with the BBS and globulin extraction buffer. The pooled supernatants were frozen until use in the RIA, at which time they were diluted 100- to 1000-fold in BBS.

3 Designing the RIA

When determining the amounts of the components to be used in the RIA it is important to keep several things in mind: (1) the amount of radiolabeled antigen used should be sufficient to result in a statistically valid count rate in a relatively short period of time; (2) the antibody should be present in a limiting amount; and (3) a method of separating free from bound antigen should be available.

3.1 Selecting the Amount of Radiolabeled Antigen

The amount of [^{14}C]-labeled globulin used in the RIA was arbitrarily chosen to give a count rate of 2000 to 3000 cpm. Assuming that the amount of antibody added to the RIA is sufficient to bind about 40% of the [^{14}C]-labeled antigen, then the total yield of counts in 10 min would be between 10000 to 15000 counts. This count rate would result in less than 2% error at the 95% probability level. Conservation of the radiolabeled antigen was also a consideration when the amount was chosen.

3.2 Titration of the Antibody

The principle of the RIA requires that the antibody be present in limiting amounts (Freifelder 1982). The antibody concentration should be such that 10% to 80% of the antigen is bound. Ideally, for maximal sensitivity the lowest antibody concentration that will bind a measurable fraction of the antigen should be used. It is generally best if the amount of antibody added to the RIA binds 40% to 50% of the antigen. To determine the amount of anti-oat globulin IgG needed to bind 40% of the [^{14}C]-oat globulins, serial dilutions of the antibody preparation were made and added to a constant amount of [^{14}C]-globulin in the RIA reaction. We found that a 1/16 dilution of the anti-oat globulin IgG fraction (purified from DEAE-Affigel Blue) was sufficient to bind 40% of the antigen. The final concentration of the anti-oat globulin IgG fraction was 9.2 μg per assay.

3.3 Precipitation of the Antigen/Antibody Complexes

The conditions for the RIA are generally not optimal for immunoprecipitation of the antigen/antibody complexes. For immunoprecipitation to occur the ratio of antigen and antibody must be such that a network of antibody and antigen molecules forms and thus precipitates. The ratio of antibody and antigen where this occurs is called the equivalence point (Cooper 1977). If either antigen or antibody is present in excess, no immunoprecipitation will occur. However, the presence of excess antigen in the RIA does not preclude the binding of antigen to antibody; it is that the conditions are not optimal for immunoprecipitation. Consequently, it is necessary to have some method for facilitating precipitation so that bound antigen can be separated from unbound. Several methods are available for separating free from bound antigen. One method is based on the removal of free antigen, such as the use of dextran-coated, activated charcoal (Freifelder 1982). Another frequently used method is the use of a second antibody to precipitate the bound antigen. For example, if the antisera used in the RIA was prepared in rabbits, then the second antibody would be raised against the rabbit IgG in another animal. The second antibody might be goat-anti-rabbit IgG. This second antibody is added to the RIA at the equivalence point with the rabbit IgG, and then the antibody/antigen complexes will immunoprecipitate (Freifelder 1982).

An alternative procedure which is similar to the second antibody method is the use of protein A of *Staphylococcus aureus* cells. The cell wall of most strains of *Staphylococcus aureus* contains protein A, which binds to the constant portion of the IgG molecules of all mammals (Kronvall et al. 1970); therefore, it can be substituted for the second antibody in immunoprecipitation reactions. We used a commercial preparation of heat-killed *Staphylococcus aureus* cells (IgG-Sorb, The Enzyme Center, Boston, Massachusetts, USA, to precipitate the antibody/antigen complexes; however, techniques for preparing these cells are available (Kessler 1975). The IgG-binding capacity known for each batch of IgG-Sorb is provided by the manufacturer; therefore, sufficient amounts can be added to quantitatively precipitate the antigen/antibody complexes (Colyer and Luthe 1984).

3.4 Conducting the RIA

This section describes the mechanics of conducting the RIA for oat globulin. A diagram of the procedure is shown in Fig. 4. Each RIA tube contained 9.2 μg of the anti-oat globulin IgG fraction, standard oat globulin, or an aliquot of the seed extract. The final volume of the reaction was made to 200 μl with BBS. Reactions were conducted in 1.5 ml Eppendorf centrifuge tubes. Each sample was run in triplicate. Control tubes to determine the amount of nonspecific binding of radiolabeled antigen to IgG-Sorb were included in each assay. These tubes contained all the components of the RIA except antibody and were processed like all other RIA tubes. In addition, a standard curve was run with each assay. The tubes were incubated at 37 °C for 1 h, followed by incubation at 4 °C for 24 to 48 h. The antigen/antibody complexes were precipitated by adding the appropriate amount of IgG-Sorb at the end of the incubation period and incubating for 30 min at 4 °C. Following this step the samples were centrifuged at 13 000 × g in a microcentrifuge for 1 min at room temperature. To remove unbound antigen, the pellets were carefully washed (at room temperature) four times with 0.5 ml of a buffer (NET-2) containing 0.15 M NaCl, 50 mM Tris-HCl, pH 7.5, and 0.05% (v/v) Nonidet P-40. The tubes were vortexed briefly to break up the pellet prior to the addition of the washing buffer.

After the washing the precipitates were prepared for scintillation counting by dissolving them in a buffer containing 62.5 mM Tris-HCl, pH 6.8, 2% (w/v) SDS, and (v/v) β-mercaptoethanol. This is identical to the sample buffer used for solubilizing for SDS-PAGE and it was adequate for solubilizing the RIA samples. Others (Triplett and Quatrano 1982) have used Protosol (New England Nuclear), a formulation designed to digest samples prior to scintillation counting. We did not use it because its use required an overnight incubation of the samples. Following solubilization, the samples were added to 3 ml of Aquasol (New England Nuclear) in mini-vials. The amount of radioactivity was determined in a Tracor

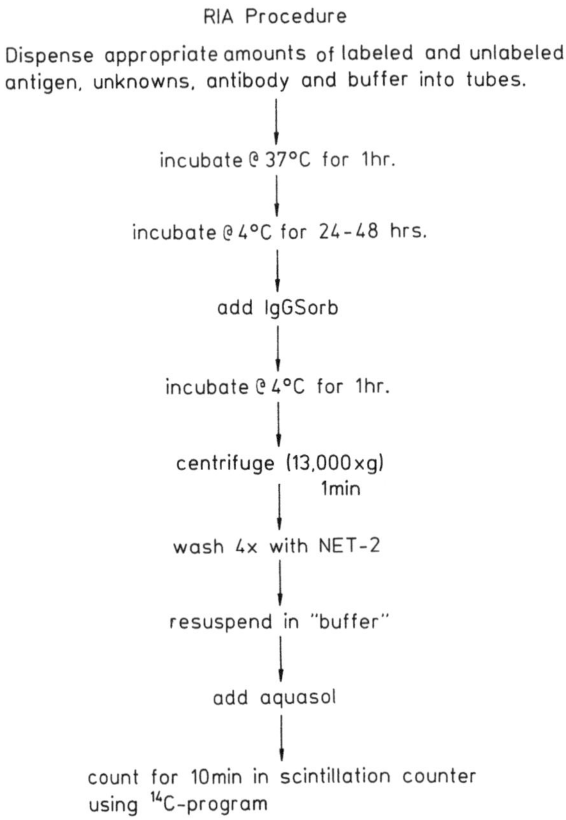

Fig. 4. The procedure used for the oat globulin radioimmunoassay

Mark III scintillation counter. At this time aliquots of [^{14}C]-labeled globulin, equivalent to the amount added to each RIA were added to Aquasol and counted. Samples were counted for 10 min.

During the formulation of the RIA, experiments were done to insure that oat globulin, which is maximally soluble in solutions containing in 1 M NaCl, did not precipitate during the assay. We calculated that the highest possible concentration of oat globulin in the RIA would be about 15 μg/l ml in BBS which has a NaCl concentration of 0.147 M. According to Peterson (1978) the maximum amount of oat globulin that can be dissolved in 0.1 M NaCl is 1 mg ml^{-1}. Since the concentration of oat globulin in the RIA is far less than this (15 μg ml^{-1}), it is unlikely that the globulin will precipitate during the reaction (Colyer and Luthe 1984).

3.5 Preparing the Standard Curve

Before unknown samples were analyzed in the RIA, several experiments were done to determine the sensitivity and the range of the assay. By adding increasing amounts of unlabeled oat globulin to the RIA tubes, we found that the standard curve was linear between 20 and 90 ng. This was a smaller range than that of 20 to 200 ng reported for wheat germ agglutinin (Triplett and Quatrano 1982). We later determined that if the data had been plotted "Logit % Bound vs Log cold antigen added" (see below), that the standard curve would have been linear for a wider range. However, since we knew the approximate protein content of each unknown sample, it was relatively easy to make the correct dilutions of the samples so that the unknown oat globulin value would fall within the range of the standard curve. A standard curve in this range was subsequently included with each assay.

3.6 Data Treatment

Typical data for an RIA are shown in Table 2. From these data the percentage of radiolabeled antigen bound to the antibody can be determined using the equations in Table 2. To obtain a linear standard curve the data was converted to the logit % bound, which can be calculated by the following equation (Hawker 1977):

$$\text{Logit \% bound} = \ln \frac{\text{\% bound}}{(100 - \text{\% bound})} .$$

If the logit % bound is plotted on the ordinate and the log of the unlabeled antigen concentration on the abscissa, a linear standard curve will be obtained (Hawker 1977). A typical standard curve for the oat globulin RIA is shown in Fig. 2.

We also noticed variability in the assay values and that is why the samples were run in triplicate. If values were obviously out of place, they were discarded

Table 2. Representative data from the RIA of oat globulin

Sample	CPM	Av CPM	% Bound[a]	Logit % bound[b]
A. ^{14}C-Oat globulin	1821	1861	–	–
	1878	–	–	–
	1886	–	–	–
B. Nonspecific binding (no antibody added to the assay)	255	242	–	–
	261	–	–	–
	211	–	–	–
C. Values for 15 DPA seeds	939	932	42.6	−0.298
	925	–	–	–

[a] % Bound was calculated in the following manner: $\text{\% bound} = \frac{C-B}{A-B}(100)$, where *A* is [^{14}C]-oat globulin per assay (CPM), *B* is the nonspecific binding (CPM), and *C* is the experimental values for 15 DPA seeds (CPM).
[b] Logit % bound was calculated using the following equation:
$\text{Logit \% bound} = \ln\left(\frac{\text{\% bound}}{100 - \text{\% bound}}\right)$ (see Hawker 1973).

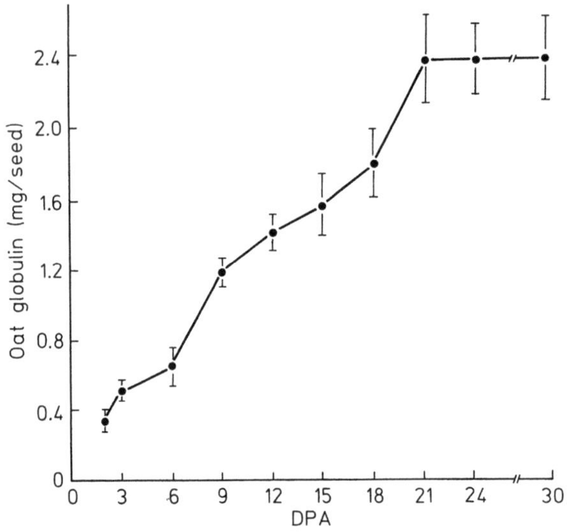

Fig. 5. Oat globulin content of developing oat seeds. The amount of globulin in seeds harvested at 3-day-intervals from 3 to 24 DPA (days-post-anthesis) was quantitated by RIA. Each value is the average of three to six determinations. Each determination was run in triplicate. (From Colyer and Luthe 1984)

and the assay was repeated. Some of the variability may have resulted from incomplete washing of the sample, or loss of a portion of the pellet during the assay. In any case, for each unknown sample several determinations (between three to six) were made and the data (Fig. 5) presented are an average of those values. Rodbard et al. (1976) has presented a method of dealing with random errors in the RIA.

4 Summary

The RIA proved to be a useful technique for measuring the amount of oat globulin in developing oat seeds. Using this technique we (Colyer and Luthe 1984) found that oat globulin increased from 0.52 ± 0.04 mg per seed at 3 DPA to 2.40 ± 0.20 mg per seed at maturity (Fig. 5). From micro-Kjeldahl analysis it was determined that a mature oat seed contained 3.15 mg per seed, therefore, globulin accounted for 75% of the total seed protein.

The use of the RIA enabled us to quantitate the amount of globulin in oat seeds at early stages of development, when it would have been difficult to get enough seeds to quantitate oat globulin using conventional extraction methods. No complicated extraction methods were needed to prepare the samples. The advantages of the technique are that small numbers of seeds are needed and that the problems with cross-contamination of Osborne fractions were eliminated.

If the appropriate assay conditions are determined, and if specific antibody is available, the RIA should be useful for quantitating any plant proteins that are soluble in aqueous solutions.

References

Adeli K, Altosaar I (1983) Role of endoplasmic reticulum in biosynthesis of oat globulin precursors. Plant Physiol (Bethesda) 73:949–955

Brinegar AC, Peterson DM (1982) Synthesis of oat globulin precursors. An analogy to legume 11S storage protein synthesis. Plant Physiol (Bethesda) 70:1767–1769

Brohult S, Sandegren (1954) Seed protein. In: Neurath H, Bailey K (eds) The proteins, vol 2, part A. Academic Press, New York

Chrispeels MJ, Higgins TJV, Spencer D (1982a) Assembly of storage protein oligomers in the endoplasmic reticulum and processing of the polypeptides in the protein bodies of developing pea cotyledons. J Cell Biol 93:306–313

Chrispeels MJ, Higgins TJV, Craig S, Spencer D (1982b) Role of the endoplasmic in the synthesis of reserve proteins and the kinetics of their transport to protein bodies in developing pea cotyledons. J Cell Biol 93:5–14

Colyer TC, Luthe DS (1984) Quantitation of oat globulin by radioimmunoassay. Plant Physiol (Bethesda) 74:455–456

Cooper T (1977) Immunological techniques. In: Tools of biochemistry. Wiley, New York

Crouch M, Sussex IM (1981) Development and storage protein synthesis in *Brassica napus* L. embryos in vivo and in vitro. Planta (Berl) 153:64–74

Domoney C, Davies DR, Casey R (1980) The initiation of legumin synthesis in immature embryos of *Pisum sativum* L. grown in vivo and in vitro. Planta (Berl) 149:454–460

Felsted RL, Pokrywka G, Chen C, Egorin M, Bachur NR (1982) Radioimmunoassay and immunochemistry of *Phaseolus vulgaris* phytohemagglutinin: verification of isolectin subunit structures. Arch Biochem Biophys 215:89–99

Freifelder D (1982) Immunological methods. In: Physical Biochemistry, 2nd edn. Freeman, San Francisco, CA

Garvey JS, Cremer NE, Sussdorf DH (1977) Methods in immunology. A laboratory text for instruction and research, 3rd edn. Benjamin, Reading MA

Hawker CD (1973) Radioimmunoassay and related methods. Anal Chem 45:878–888

Kessler SW (1975) Rapid isolation of antigens from cells with a staphylococcal protein A-antibody absorbent: parameters of the interaction of antibody/antigen complexes with protein A. J Immunol 115:1617–1624

Kronvall G, Seal US, Finstead J, Williams RJ Jr (1970) Phylogenetic insight into evolution of mammalian Fc fragment of γG globulin using staphylococcal protein A. J Immunol 104:140–147

Larkins BA (1981) Seed storage proteins: characterization and biosynthesis. In: Stumpf PK, Conn EE (eds) The biochemistry of plants, vol 6. Academic Press, New York

Lowry OH, Rosebrough NJ, Farr AL, Randall RJ (1951) Protein measurement with the folin phenol reagent. J Biol Chem 193:265–275

Luthe DS, Peterson DM (1977) Cell-free synthesis of globulin by developing oat (*Avena sativa* L.) seeds. Plant Physiol (Bethesda) 59:836–841

Matlashshewski GJ, Adeli K, Altosaar I, Shewry PR, Miflin BJ (1982) In vitro synthesis of oat globulin. FEBS Lett 145:208–212

McGrogan M, Spector DJ, Goldenberg CJ, Halbert D, Raskas HJ (1979) Purification of specific adenovirus 2 RNAs by preparative hybridization and selective thermal elution. Nucleic Acids Res 6:593–607

Means GE, Freeney RE (1968) Reductive alkylation of amino groups in proteins. Biochemistry 7:2192–2201

Meinke DW, Chen J, Beachy RN (1981) Expression of storage protein genes during soybean seed development. Planta (Berl) 153:130–139
Michael G, Blume E, Faust H (1961) Die Eiweißqualität von Körnern verschiedener Getreidearten in Abhängigkeit von Stickstoffversorgung und Entwicklungszustand. Z Pflanzenernaehr Bodenkd 92:106–116
Murphy JB, Kies MW (1964) Note on spectrophotometric determination of protein in dilute solutions. Biochim Biophys Acta 45:382–384
Osborne TB (1910) Die Pflanzenproteine. Ergeb Physiol Biol Chem Exp Pharmakol 10:47–215
Peterson DM (1976) Protein concentration, concentration of protein fractions and amino acid balance in oats. Crop Sci 16:663–666
Peterson DM (1978) Subunit structure and composition of oat seed globulin. Plant Physiol (Bethesda) 62:506–509
Peterson DM, Brinegar AC (1986) Oat storage proteins. In: Webster F (ed) Oat chemistry and production. Am Assoc Cereal Chemists, St Paul MN
Peterson DM, Smith D (1976) Changes in nitrogen and carbohydrate fractions in developing oat groats. Crop Sci 16:67–71
Rice RH, Means GE (1971) Radioactive labeling of proteins in vitro. J Biol Chem 246:831–832
Robard D, Lenox RH, Wray HL, Ramseth D (1976) Statistical characterization of the random errors in the radioimmunoassay dose-response variable. Clin Chem 22:350–358
Robert LS, Nozzolillo C, Cudjoe A, Altosaar I (1983) Total solubilization of groat proteins in high protein oat (*Avena sativa* L. cv Hinoat): evidence that glutelins are a minor component. Can Inst Food Sci Technol J 16:196–200
Rossi HA, Luthe DS (1983) Isolation and characterization of oat globulin messenger RNA. Plant Physiol (Bethesda) 72:578–582
Shapiro DJ, Taylor JM, McKnight SG, Palacios R, Gonzalez C, Kiely ML, Schimke RT (1974) Isolation of hen oviduct ovalbumins and rat liver albumin polysomes by indirect immunoprecipitation. J Biol Chem 249:3665–3671
Sun SM, Mutschler MA, Bliss FA, Hall TC (1978) Protein synthesis and accumulation in bean cotyledons during growth. Plant Physiol (Bethesda) 61:918–923
Triplett BA, Quatrano RS (1982) Timing, localization and control of wheat germ agglutinin synthesis in developing wheat embryos. Dev Biol 91:491–496
Volker T (1975) Untersuchungen über den Einfluß der Stickstoffdüngung auf die Zusammensetzung der Weizen- und Haferproteine. Arch Acker Pflanzenbau Bodenkd 19:267–276
Walburg G, Larkins BA (1983) Oat seed globulin. Subunit Characterization and demonstration of its synthesis as a precursor. Plant Physiol (Bethesda) 72:161–165
Weeke B (1973) Rocket immunoelectrophoresis. In: Axelson NH, Kroll J, Weeke B (eds) A manual of quantitative immunoelectrophoresis. Universitetsforlaget, Oslo, Norway
Wieser H, Seilmeier W, Belitz HD (1980) Vergleichende Untersuchungen über partielle Aminosäuresequenzen von Prolaminen und Glutelinen verschiedener Getreidearten. I. Proteinfraktionierung nach Osborne. Z Lebensm Unters Forsch 170:17–26
Williamson GG (1981) Autoimmune antibodies in systemic rheumatic disease: characteristics of antigenic nuclear particles. MS Thesis, Mississippi State University, Mississippi State, MS

Immunocytochemistry of Chloroplast Antigens

K. C. Vaughn

1 Introduction

Although immunocytochemical procedures have been available since the early 1940s, the first study in which immunofluorescence was used as a technique on plants was not reported until 1964 (Knox et al. 1980). It was not until 1972 that Billecocq et al. utilized a peroxidase-linked antibody to detect the sites on the thylakoid and envelope membranes where the galactocyldiglycerides are located. Since these pioneering works with immunocytochemical analysis of plant systems, a number of studies, both at the light and electron microscopic levels, have utilized immunocytochemical approaches to detect chloroplast antigens (Table 1). This increase in use of antibodies in plant science is probably related to the improvements made in electron microscopic immunocytochemistry, the availability of labeled secondary antibodies from commercial sources, and the more widespread use of immunology in all phases of plant science.

Table 1. Chloroplast antigens detected by immunocytochemistry

	Method[a]	Ref.
Protein antigens		
Ribulose bisphosphate carboxylase	FL	Hattersley et al. (1977)
Ribulose bisphosphate carboxylase	FL	Madhavan and Smith (1982)
Ribulose bisphosphate carboxylase	CG	Lacoste-Royale and Gibbs (1982)
Coupling factor complex	FE	Miller and Staehelin (1976)
P700 Chlorophyll *a* protein	PAP	Vaughn et al. (1983)
Polyphenol oxidase	PAP	Vaughn and Duke (1984)
Phosphoenolpyruvate carboxylase	FL	Madhavan and Smith (1984)
Phosphoenolpyruvate carboxylase	FE	Gadal et al. (1983)
Nitrate reductase	FL	Vaughn et al. (1984)
Datura lectin	FL	Jefree et al. (1982)
Superoxidase dismutase	FL	Salin and Lyons (1983)
Cytochrome b_6–f complex	FE	Allred and Staehelin (1985)
Photosystem II polypeptides	CG	Goodchild et al. (1985)
Plastocyanin	CG	Vaughn and Hauska (this report and in preparation)
Nonprotein antigens		
Galactolipids	EN	Billecocq et al. (1972)
Sulfolipids	EN	Billecocq (1975)
Abscisic acid	PAP	Sotta et al. (1985)

[a] Abbreviations: *FL* = fluorescence; *FE* = ferritin-labeling; *PAP* = peroxidase anti-peroxidase; *CG* = colloidal gold; *EN* = enzyme-linked antibody.

In this chapter, those studies that utilized immunocytochemical methods for chloroplast antigens will be detailed and methodology for these procedures will be outlined. A few procedures recently developed in our laboratory and by others will also be described.

2 Light Microscopy

Although the yield of pure chloroplasts from Percoll™ gradients has eliminated much of the controversy as to whether an enzyme is localized within the chloroplast, the tissue distribution of a particular enzyme has, in most cases, not been well established. Until recently, techniques for the isolation of such tissue as guard cell or paraveinal mesophyll chloroplasts have not been available so that determination of the presence or absence of a particular enzyme in the chloroplasts of these cell types was limited to histochemical approaches (e.g., Vaughn and Outlaw 1983). Light microscopy, using immunofluorescence, immunoenzyme, or immunogold procedures can satisfactorily and easily determine the distribution of an enzyme within a tissue slice. The procedures are quick, relatively easy, and highly reproducible. Generally, such procedures are attempted prior to electron microscopic cytochemistry procedures.

An example of the type of problem that has been solved using immunofluorescence is the distribution of ribulose bisphosphate carboxylase/oxygenase in C_4 plants by Hattersley et al. (1977). Attempts at an unequivocal localization of this enzyme previous to the Hattersley et al. work, using mesophyll-bundle sheath fractionation and subsequent enzyme assay, were hampered by the difficulty in obtaining complete fractionation of the two chloroplast types and the presence of high levels of polyphenol oxidase (which generates highly reactive quinones) in the mesophyll cell plastids (Vaughn and Duke 1981). The lack of ribulose bisphosphate carboxylase activity in the mesophyll cells was believed to be due to enzyme inactivation by quinones and the presence of activity in bundle sheath cells due to a lack of inactivation. Immunofluorescence revealed, however, that ribulose bisphosphate carboxylase was truly limited to the bundle sheath cells (Hattersley et al. 1977). Recently, this procedure has been used to demonstrate the presence or distribution of this enzyme in guard cells (Madhavan and Smith 1982) and in hybrids and intermediates of C_3 and C_4 plants (Keefe and Mets 1983).

2.1 Procedures for Indirect Immunofluorescence

1. Fix and permeabilize small pieces of tissue (1 mm^2) by immersion in 70% (v/v) ethanol for 1 h.
2. Wash the tissue briefly in phosphate buffered saline [PBS, 0.10 M sodium phosphate (pH 7.5) with 0.20 M sodium chloride].

3. Incubate sections in normal serum of the animal that will be used in the secondary antibody (step No. 6) in PBS for 1 h. This will block nonspecific sites. Crude bovine serum albumin (1%), nonfat dry milk (3%), or gelatin (1–3%) may also be used in this step. Some blocking agents work better for one tissue than another.
4. Incubate sections in primary antibody diluted in the same mixture used in step No. 3. Dilutions and times of incubations are determined empirically. A 2 h incubation in a 1:30–200 dilution of primary serum is standard for many polyclonal sera from rabbits.
5. Wash in PBS with the blocking reagent used in step No. 3.
6. Incubate in a 1:20–1:100 dilution of secondary antibody (e.g., goat anti-rabbit) labeled with fluorescein, rhodamine, or Texas Red™. Sources of secondary serum vary greatly in titer and several dilutions should be tried for best results.
7. Wash in several changes of PBS, 10–15 min each.
8. Mount sections on slides, add a drop 50% glycerol in PBS and cover with a cover slip.
9. Examine by epifluorescence microscopy. Most microscopes include detailed listings of the properties of the filter sets that are appropriate for a given fluorescent molecule. Texas Red™ is so new that at the time of this writing, filter sets are not available specifically for it. Filter sets meant for rhodamine will work with Texas Red™, however.

Photograph the results immediately as the fluorescence decays with time. Phase contrast or Nomarski differential interference microscopy of the same tissue piece is helpful for sections with such low autofluorescence that cell walls are not detectable.

There are many modifications of this basic procedure. Tissue may be fixed briefly in paraformaldehyde, glutaraldehyde, or a mixture of the two fixatives. Washing in glycine or lysine-containing buffers or treatment with dilute (0.01–0.1%) sodium borohydride after the aldehyde fixation is necessary to remove residual aldehydes so that antibodies are not bound nonspecifically to the tissue or to suppress the fluorescence of the tissue caused by these aldehydes. Toluidine blue, which complexes the highly fluorescent polyphenols common in most plant cells will suppress the fluorescence due to phenols (Knox et al. 1980). Other modifications of this procedure include obtaining thin pieces of tissue either by cryostat sectioning or using unfrozen tissue in the Sorvall TC-2 tissue chopper or the Oxford Vibratome. The TC-2 tissue chopper has the advantages that it is rapid and antigens that are particularly unstable (e.g., nitrate reductase or phytochrome) may be quickly reacted so that the antigenicity of the protein is not lost in the relatively longer processes of freeze-sectioning or sectioning with the Vibratome.

If penetration of the antibodies into the tissues is a problem, low concentration of Triton X-100 (0.01–0.1%) with the fixative and incubation solutions improve the permeability of the tissue to the antibodies (Vaughn, unpublished). Examination of these permeabilized tissues via transmission electron microscopy reveals some cellular perturbation, but the organelles are still clearly recognizable, indicating that these procedures are not so drastic as to completely obscure cellular detail. Colloidal gold and peroxidase anti-peroxidase procedures, which may be used at both the light and electron microscopic levels (Raikhel et al. 1984), will be discussed under the electron microscopic section (see below).

Another application of antibodies to chloroplast antigens at the light microscopic level involves the ability of antibodies to agglutinate chloroplasts into a mass if the antigen occurs on the surface of the membrane being investigated. For example, the position of at least some of the antigenic determinants of the light-harvesting complex were shown to be on the stroma (exposed) surface of the thylakoids by its ability to agglutinate the thylakoids into a mass (Andersson et al. 1982). Such procedures can be used for the right-side out and inside-out vesicles prepared during grana-stroma fractionation or whole chloroplasts (for detection of antigens from the outer envelope). These agglutination procedures are simple to perform and greatly augment the analysis of results made by other procedures.

3 Electron Microscopy

Many of the questions asked by plant scientists concerning the localization of proteins in the chloroplast are beyond the resolution capability of the light microscope. Some research areas, such as the distribution of the photosystem complexes along the thylakoid membrane are major research concerns of laboratories around the world and immunocytochemical procedures have been invaluable in working out the details of this system. Miller and Staehelin (1976) showed, in a classic work, that the chloroplast coupling factor (CF_1) complex was present only along the stroma lamellae and the exposed ends of the grana stacks. Later, localizations of the photosystem I apoprotein (Vaughn et al. 1983), the cytochrome b_6/f complex (Allred and Staehelin 1985), and the polypeptides involved in O_2 evolution (Goodchild et al. 1985) were localized to complete our picture of the organization of the various complexes.

The choices for electron microscopy cytochemistry have been greatly expanded in the last few years due to improvements in post-embedding immunolabeling procedures. Two procedures that are useful for electron microscopy localization of antigens, the peroxidase anti-peroxidase (PAP) und colloidal gold (CG) procedures may be used at the light microscopic level as well. Kits are now available containing all the components for PAP localization, including pretitered secondary antisera and peroxidase anti-peroxidase complex. With such a kit, even novist microscopists can produce fine light and electron microscopic immunolocalization of many antigens.

3.1 The Peroxidase Anti-Peroxidase (PAP) Method (Post-Embedding)

The PAP method has the advantage over other immunocytochemical procedures that a single antigen-antibody reaction in the tissue is amplified over the simple labeled-antibody procedures that preceded it. By reacting the primary antibody with an unlabeled secondary antibody, which, in turn, is linked to a number of soluble antibody-peroxidase complexes (made to the same IgG species as the primary antiserum), a great number of peroxidase molecules are bound for each antigen detected by the primary IgG. Thus, it is the method of choice for detecting antigens that are present in relatively small quantities or for antigens that are easily denatured during the fixation and embedding procedures for microscopy. For example, Coleman and Pratt (1974) were able to immunolocalize the very labile protein phytochrome using the peroxidase anti-peroxidase procedure on tissue sections that had been fixed with paraformaldehyde and embedded in either paraffin or polyethylene glycol.

The chloroplast is bounded by a double envelope membrane and antibodies are not normally able to cross such a barrier unless the membranes had been permeabilized prior to antibody reactions. Thus, post-embedding procedures, such as PAP or collodial gold are particularly valuable for chloroplast antigens when using intact chloroplasts or tissue pieces. Pre-embedding labeling of chloroplast antigens have been used, however. Miller and Staehelin (1976) and Vaughn and Duke (1984) have used pre-embedding labeling of isolated thylakoids to localize and determine the distribution of antigens that are known to be present on the thylakoids. A pre-embedding procedure that has been utilized only at the light microscopic level for chloroplast antigens is the use of ethyl diaminopropylcarbodiimide (EDC) and saponin, a fixation procedure that also permeabilizes the membranes to antibodies (Salin and Lyon 1983). A procedure that combines pre-embedding and post-embedding procedure described by Raikhel et al. (1984), has general applicability to both light and electron microscopy and makes use of cryosectioned tissue pieces that are then processed through the colloidal gold procedures. This procedure has yet to be utilized for chloroplast antigens, however. Because these pre-embedding techniques utilize immunolabeling procedures that are essentially the same as those used in the post-embedding procedures described below, readers should consult these papers for details. The PAP method and collodial gold procedures described below work successfully with a large range of chloroplast antigens.

3.2 Peroxidase Anti-Peroxidase Procedure

1. Fix small tissue pieces in 1–3% glutaraldehyde and 2% paraformaldehyde in 0.05 M cacodylate buffer (pH 7.2) for 1–2 h at 0 °–4 °C.
2. Wash in 0.10 M cacodylate buffer (pH 7.2) at 0 °–4 °C, rinse in cold distilled H_2O.
3. Dehydrate in acidified 2,2′-dimethoxypropane (Postek and Tucker 1976), acetone or ethanol, according to standard dehydration procedures. Cold temperatures may be maintained until dehydration in 70% acetone or ethanol is reached. Dimethoxypropane dehydration is carried out at room temperature.
4. Embed in Spurr's, Ladd's ultralow viscosity or Epon resins according to the manufacturer's directions.

5. Cut thin (gold-silver reflectance) sections and mount the sections on gold or nickel grids.
6. Etch the sections by floating the grids, specimen side down, on 3% (v/v) H_2O_2 for 5 min.
7. After a thorough washing in distilled H_2O, the grids are dried by touching to filter paper and then incubated on 0.10% (v/v) lyophilized goat serum in 0.10 M Tris(pH 7.6)–0.50 M NaCl (TBS) for 5 min to coat the grids and to block nonspecific sticking.
8. The grids are blotted with filter paper, but not washed, and transferred to a drop of diluted antisera. The concentrations are determined empirically but, because of the sensitivity of this procedure compared to others, much more dilute (1:500–1:20000) sera may be used. Dilutions are made in TBS with or without 0.1% normal goat serum. We normally incubate the specimens from 18–48 h at 4 °C and then an additional 2 h at room temperature. A petri dish sealed with Parafilm™ is a convenient method of storage during this long incubation. Wash with TBS from a small wash bottle after the incubation.
9. Float grids on 0.1% (w/v) goat serum for 5 min. Blot, but do not wash.
10. Incubate on a 1:10 dilution of goat anti-rabbit IgG in TBS for 5 min. Wash in TBS.
11. Float grids on 0.1% (w/v) goat serum for 5 min. Blot, but do not wash.
12. Incubate on peroxidase-antiperoxidase (PAP) complex (0.066 mg ml^{-1}) diluted in TBS for 5 min. Wash with TBS, but do not blot.
13. Suspend grids in a solution of diaminobenzidene (DAB) and 0.0025% H_2O_2 in TBS. Place the solution on a stir plate and stir at approximately 100 rpm. Incubate for no more than 7 min.
14. Wash each grid thoroughly in H_2O either by transferring from drop to drop or from a spray bottle. Repeat this several times as DAB is highly osmiophilic and will obscure results.
15. Place in 2% (w/v) OsO_4 in 0.10 M cacodylate (pH 7.2) for 30 min. Wash with water from a spray bottle, blot dry, and observe.

An example of PAP staining of the P700 chlorophyll *a* protein using a 1:800 primary antisera dilution is shown in Fig. 1. All of the thylakoids react strongly as a consequence of the generation of the highly osmiophilic oxidized DAB precipitate that is the result of the reaction of the peroxidase using DAB as the donor. Although the other membranes are lightly contrasted due to the last step osmication or inherent electron opacity of the tissue, the strong reaction over the thylakoids and lack of reaction in other portions of the cell give a striking positive localization of this protein.

Although these reactions in rather dilute (1:800) antisera solutions give the impression of an apparent equal distribution of the P700 chlorophyll *a* protein throughout the membrane, further dilutions of the sera (up to 1:8000) allow for a difference in antigen concentration across the thylakoid membrane to be determined. For example, serial sections to those shown in Fig. 1 A reacted in sera diluted 1:3000 show much more positive staining in the stroma lamellae and grana end membranes than in the stacked regions of the grana lamellae (Fig. 1 B). Even in these results from very dilute antisera incubations, however, substantial reac-

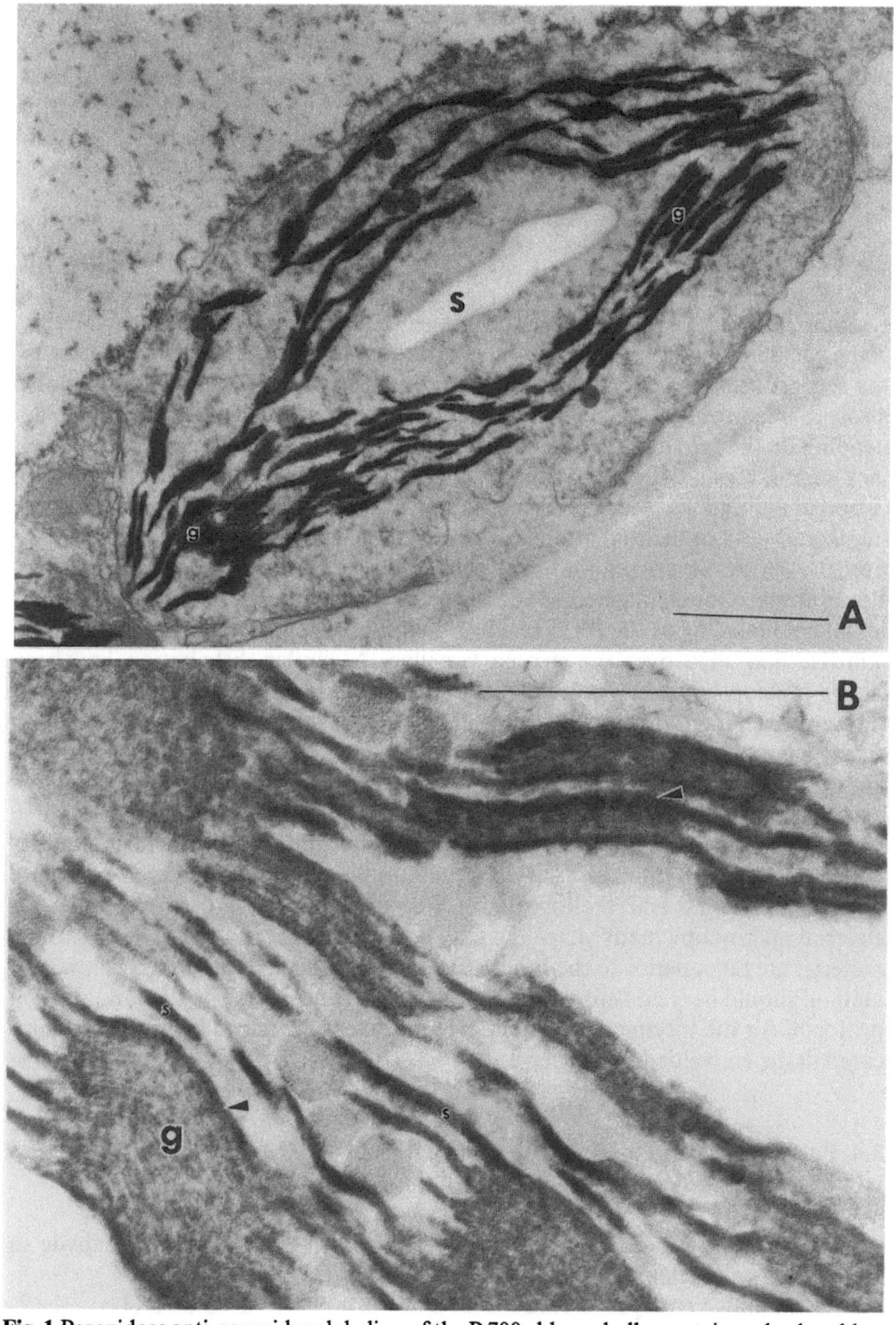

Fig. 1. Peroxidase anti-peroxidase labeling of the P 700 chlorophyll *a* protein on barley chloroplasts. **A** Relatively even staining in both the grana (*g*) and stroma lamellae are obtained at primary antibody dilutors of 1:800. *S* = starch. **B** Immunolabeling is stronger in the stroma lamellae (*s*) and the grana end membranes (*arrows*) than in the central portions of the grana stock at a 1:3000 dilution of the primary sera. *Bar* = 1.0 μM in **A** and 0.5 μM **B**

tivity is found within the grana stack. Relative quantitation can be obtained by densitometry traces of the thylakoid membranes using either negatives (in the transmittance mode) or prints (in the reflectance mode). It is clear from these studies, however, that a series of dilutions should be tested in the determination of the relative quantitative distribution of a given antigen using the PAP technique.

3.3 Colloidal Gold

Despite the sensitivity of the PAP technique, the tissue sections are etched free of some plastic to that results cannot be quantitated strictly on a μm^2 basis, because the section itself has depth. The amplification due to the PAP technique is also a source of a problem; the oxidized DAB reaction product is amorphous and, depending on the extent of the peroxidase reaction, can obscure precise suborganellar location. Colloidal gold has the advantage that it is particulate, making its location unambiguous. Because the tissue sections are not etched with peroxide, the number of gold particles that one counts over a particular structure can be quantitated with the assurance that only the surface-exposed antigens have reacted. For example, Goodchild et al. (1985) were able to determine that 84% of the collodial gold labeling of the PS II reaction center polypeptides were over the grana lamellae and 16% over the stroma lamellae. A similar distribution had been shown previously, based upon PS II activity (Armond and Arntzen 1977) and cytochemical staining for the PS II partial reaction (Vaughn et al. 1983). Although nearly all antigens that we have investigated can be detected by colloid gold staining of unosmicated tissue embedded in Lowicryl K4-M or L.R. White resin (as detailed below), a few antigens are abundant enough and resistant enough to treatment so that they may be fixed using standard glutaraldehyde – osmium fixation, embedded in Epon and still have immunological activity after *m*-periodate treatment (Doman and Trelease 1985). Because grids of material prepared for electron microscopy many years ago may be used and the structural details of the material are far superior to the briefly fixed material such as described below, this method should be tried before other procedures are utilized. The immunolabeling protocols for the specimens are identical to the steps described below for unosmicated tissue embedded in Lowicryl.

3.4 Post-Embedding Labeling with Colloidal Gold on Lowicryl-Embedded Tissue

1. Fix small tissue pieces in 1% glutaraldehyde – 1% paraformaldehyde in 0.05 M cacodylate buffer (pH 7.2) for 2 h at 4 °C.
2. Wash in two changes of cold buffer and wash briefly in cold H_2O.
3. Dehydrate in acidified 2,2′-dimethoxypropane (Postek and Tucker 1976) for 30 min.
4. Transfer to 100% acetone (two 10 min changes). The tissue must be colorless or nearly so at this point or the Lowicryl will not polymerize. L.R. White

resin will polymerize around tissue that has some opacity and should be used on such tissue.

5. Embed in Lowicryl by gradual increase in Lowicryl solvent 30-min incubations at 33% (v/v) 66% (v/v) and two 2-h changes of 100% Lowicryl (made from 2 g monomer A, 13 g monomer B, and 75 mg of the initiator).
6. Transfer specimens to BEEM or gelatin capsules and seal the capsules to exclude O_2.
7. Polymerize at either 10 cm from a germicidal lamp or black light (for a quick, 45-min polymerization) or at a greater distance (30 cm) for a slower (24–48 h) polymerization. Although the quick polymerization preserves the antigenicity the best, the plastic inside the tissue pieces is sometimes not evenly polymerized. Very small pieces of tissue or pellets of isolated organelles work better using the short polymerization procedure than do larger tissue pieces. Polymerization may take place at any temperature, although the quick polymerization appears to work best at room temperature. At lower temperatures, pockets of gummy, unpolymerized plastic are formed.
8. Cut sections with gold-silver reflectance using glass or diamond knifes and collect on gold or nickel grids. The water level in the boat should be adjusted lower than for sectioning epoxy resins so as not to wet the block face.
9. Float the grids, section-side down, on 2% (w/v) bovine serum albumin (BSA) or nonfat dry milk in PBS for 1 h at room temperature to block nonspecific binding.
10. Transfer to antiserum diluted 1:10–1:20 in PBS with either BSA or nonfat dry milk for 1 h. Although others have reported high background staining using this length for incubation, we have had no problem by using the relatively long (1 h) blocking step (No. 9) in BSA.
11. Transfer grids through six drops of PBS-BSA.
12. Incubate on a protein A or anti-primary IgG solution labeled with colloidal gold (5–20 nm particles). Although the concentration has to be determined empirically, a good working solution is a 1:10–1:20 dilution of either of the commercially available solutions in PBS-BSA. The color of diluted solution is light pink, like pale rosé wine. Directions for preparation of the colloidal gold-Protein A or collodial gold-IgG complexes may be found in Raihkel et al. (1984).
13. Transfer grids through six drops of PBS-BSA.
14. Wash in distilled H_2O from a wash bottle or transfer through 6–10 drops of H_2O.
15. Post-stain in 1% (w/v) uranyl acetate and Reynold's lead citrate.

Material prepared for electron immunocytochemistry using the procedures described above looks very different than the tissue treated with the PAP procedure. Although the plastid membranes stand out in negative relief after glutaraldehyde fixation and Lowicryl embedding, grana and stroma lamellae and the boundaries of the plastid are clearly distinct (Fig. 2). Immunolabeling of the section with antiplastocyanin from rabbit followed by protein A-colloidal gold reveals specific labeling in both the grana and stroma lamellae. Tissue slices reacted with preimmune sera are practically unstained, except for an occasional gold par-

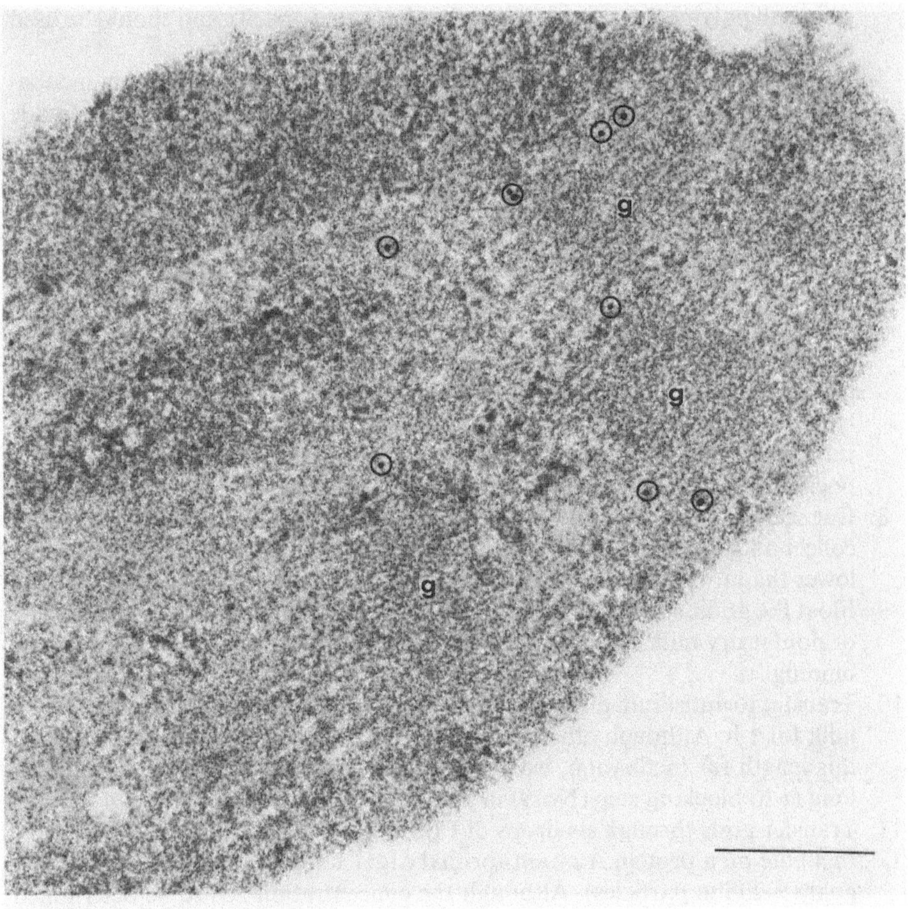

Fig. 2. Colloidal gold labeling of thin sections of Lowicryl-embedded tissue reacted either with anti-plastocyanin (1:10) as the primary antigen. Note the presence of colloidal gold labeling (*circled*) on the thylakoids of both grana (*g*) and stroma lamellae. *Bar* = 0.5 μM

ticle along the wall or near a hole in the section. Although one can obtain an impression of the distribution of plastocyanin from such electron micrographs, the results do not have the same visual impact as the PAP procedure (Fig. 1). Quantitation of the immunogold particles may be determined in a μm^2 basis for stromal proteins (Lacoste-Royale and Gibbs 1985) and on a linear μm basis for thylakoid proteins (Allred and Staehelin 1985). The Lowicryl-embedded chloroplast makes membranes that stand out in negative relief which makes it difficult to discern obliquely-sectioned stroma and grana lamellae. If the distribution of grana and stroma thylakoids appears abnormal compared to literature values, calculation of the relative areas of these two types of thylakoid can be determined on material prepared by standard microscopic procedures and the values obtained compared with those of tissues embedded in Lowicryl.

4 Conclusions

Chloroplast antigens may be detected by several light and electron microscopic immunocytochemical procedures. The particular technique one chooses is often determined by the question of the level of resolution required for the procedure. Although simple immunofluorescence may answer questions of the tissue distribution (Hattersley et al. 1977), localizations at the electron microscopic level are required to discriminate whether an antigen is on the grana and stroma thylakoids (Vaughn et al. 1983; Allred and Staehelin 1985; Goodchild et al. 1985). As procedures from animal immunocytochemistry become more adopted in plant immunocytochemistry, a greater understanding of the distribution of antigens in the chloroplast and a greater resolution of structural details concomitant with the ultrastructural localization should be forthcoming. Exciting developments, such as the localization of the water-soluble, mobile plant hormone abscisic acid using the PAP procedure (Sotta et al. 1985) should give us an even more detailed picture of the workings of the chloroplast.

Acknowledgements. Thanks are extended to my colleagues, E. Vierling, R.S. Alberte, S.O.Duke, E.A. Funkhouser, D. Allred, and A. Staehelin, who provided either antiserum, counsel, or both during the course of my own immunocytochemical studies.

References

Allred DR, Staehelin LA (1985) Lateral distribution of the cytochrome b_b/f and CF_0/CF_1 complexes of thylakoid membranes. Plant Physiol (Bethesda) 78:199–202

Andersson B, Anderson JM, Ryrie IF (1982) Transbilayer organization of the chlorophyll-proteins of spinach thylakoids. Eur J Biochem 123:465–472

Armond PA, Arntzen CJ (1977) Spatial relationship of photosystem I, photosystem II, and the light-harvesting complex on chloroplast membranes. J Cell Biol 73:400–418

Billecocq A (1975) Structure des membranes, biologiques: localization du sulfoquinovosyl-diglyceride dans les diverses membranes des chloroplastes au moyen des anticorp specifiques. Ann Immunol 126C:337–352

Billecocq A, Douce R, Faure M (1972) Structure des membranes biologiques. Localization des galactosyldiglycerides dans les chloroplastes au moyen des anticorp specifiques. C R Acad Sci Paris 275D:1135–1137

Coleman RA, Pratt LH (1974) Subcellular localization of red-absorbing form of phytochrome by immunocytochemistry. Planta (Berl) 121:119–131

Doman DC, Trelease RN (1985) Protein A-gold immunocytochemistry of isocitrate lyase in cotton seeds. Protoplasma 124:157–167

Gadal P, Perrot-Rechenmann P, Vidal T (1983) Immunocytochemical visualization of phosphoenolpyruvate carboxylase in higher plants. Physiol Veg 21:1055–1062

Goodchild DJ, Andersson B, Anderson JM (1985) Immunocytochemical localization of polypeptides associated with the oxygen evolving system of photosyntheses. Eur J Cell Biol 36:294–298

Hattersley PW, Watson L, Osmond CB (1977) In situ immunofluorescent labelling of ribulose-1,5 bisphosphate carboxylase in leaves of C_3 and C_4 plants. Aust J Plant Physiol 4:523–539

Jefree CE, Yeoman MM, Kilpatrick DC (1982) Immunofluorescence studies on plant cells. Int Rev Cytol 80:231–265

Keefe D, Mets L (1983) Bundle sheath cell development in *Flaveria palmeri*. Plant Physiol (Bethesda) 72s:43
Knox RB, Vithange HIMV, Howlett BJ (1980) Botanical immunocytochemistry: A review with special reference to pollen antigens and allergens. Histochem J 12:247–272
Lacoste-Royal G, Gibbs SP (1985) *Ochromonas* mitochondria contain a specific chloroplast protein. Proc Natl Acad Sci USA 82:1456–1459
Madhavan S, Smith BN (1982) Localization of ribulose bisphosphate carboxylase in the guard cells by an indirect, immunofluorescence technique. Plant Physiol (Bethesda) 69:273–277
Madhavan S, Smith BN (1984) Phosphoenolpyrmate carboxylase in guard cells of several species as determined by an indirect, immunofluorescent technique. Protoplasma 122:157–161
Miller KR, Staehelin LA (1976) Analysis of the thylakoid outer surface: coupling factor is limited to unstacked membrane regions. J Cell Biol 68:30–47
Postek MT, Tucker SC (1976) A new short chemical dehydration method for light microscopy preparations of plant material. Can J Bot 54:872–875
Raikhel NV, Mishkind M, Palevitz BA (1984) Immunocytochemistry in plants with colloidal gold conjugates. Protoplasma 121:25–33
Salin MT, Lyon DS (1983) Iron containing superoxide dismutases in eukaryotes: localization in chloroplasts from water lily, *Nuphar luteum*. In: Cohen G, Greenwald RA (eds) Oxy radicals and their scavenger systems, vol 1. Molecular aspects. Elsevier, Amsterdam, p 344
Sotta B, Sossountzov L, Maldiney R, Sabbagh I, Tachon P, Miginiac E (1985) Abscisic acid localization by light microscopic immunohistochemistry in *Chenopodium polyserum* L. Effects of water stress. J Histochem Cytochem 33:201–208
Vaughn KC, Duke SO (1981) Tissue localization of polyphenol oxidase in *Sorghum*. Protoplasma 108:319–327
Vaughn KC, Duke SO (1984) Tentoxin stops the processing of polyphenol oxidase into an active protein. Physiol Plant 60:257–261
Vaughn KC, Outlaw WH (1983) Cytochemical and cytofluorometric evidence for guard cell photosystems. Plant Physiol (Bethesda) 71:420–424
Vaughn KC, Vierling E, Duke SO, Alberte RS (1983) Immunocytochemical and cytochemical localization of photosystems I and II. Plant Physiol (Bethesda) 73:203–207
Vaughn KC, Duke SO, Funkhouser EA (1984) Immunochemical characterization and localization of nitrate reductase in norflurazon-treated soybean cotyledons. Physiol Plant 62:481–484

Subject Index

MIX
Papier aus verantwortungsvollen Quellen
Paper from responsible sources
FSC® C105338

If you have any concerns about our products,
you can contact us on
ProductSafety@springernature.com

In case Publisher is established outside the EU,
the EU authorized representative is:
Springer Nature Customer Service Center GmbH
Europaplatz 3, 69115 Heidelberg, Germany

Printed by Libri Plureos GmbH
in Hamburg, Germany